Experimental Particle Physics
실험 입자 물리학

Experimental Particle Physics
실험 입자 물리학

박환배 자연과학대학 물리학과 교수
오용석 자연과학대학 물리학과 교수
조연정 자연과학대학 물리학과 교수
양기원 교육혁신본부 교양교육센터 강의초빙교수

Experimental Particle Physics 실험 입자 물리학

초판 1쇄 2023년 2월 27일 | **초판 2쇄** 2024년 1월 31일
지은이 박환배, 오용석, 조연정, 양기원| **펴낸이** 홍원화 | **펴낸 곳** 경북대학교출판부
출판등록 1973년 10월 10일 ㉑97호 | **주소** 대구광역시 북구 대학로 80
전화 053-950-3830 | **팩스** 053-953-4692 | **E-mail** press@knu.ac.kr
Homepage http://knupress.com
ISBN 978-89-7180-610-4 93420

값은 표지 뒷면에 있습니다.
파본은 바꾸어 드립니다.

머리말

입자물리학은 자연계의 기본 원리를 탐구하는 학문으로 우주의 기본 물질과 그 들 사이의 상호작용을 이해하고자 하는 학문이다. 현재의 입자물리학은 유럽연합의 CERN에 위치한 거대강입자충돌기(Large Hadron Collider, LHC)를 비롯한 거대 실험장치와 함께, 중성미자를 검출하기 위해 지하에 건설한 실험장치로부터 저 멀리 우주에 있는 국제 우주정거장에 설치되어 우주로부터 날아오는 입자를 검출하는 장치에 이르기까지 아주 작은 미시세계와 아주 큰 스케일인 우주의 구조를 함께 연구한다.

입자물리학의 근본적인 이론 체계는 표준 모형(Standard Model)이라 불리는 이론이며 최근 이를 넘어선 새로운 물리현상을 찾고자 하는 노력이 진행되고 있다. 이를 위해서는 우선 표준 모형을 잘 이해해야 한다.

이 책은 그동안 경북대학교 물리학과 대학원에서 학생들에게 제공되었던 실험 입자물리학 과목의 내용을 정리한 것이다. 이 책에서는 기본입자에 관한 기본적인 설명에서 출발하여 그들 사이의 상호작용에 관한 이론, 불변의 원리와 보존 법칙 등, 현대의 표준 모형을 공부하기 위한 기초를 다루며 또한 입자물리학의 현상론에 해당하는 쿼크와 렙톤 그리고 그들 사이의 충돌 현상과 양자색소역학(QCD)를 소개한 다음 약한 상호작용을 도입하여 현재 입자물리학의 근본 이론인 표준 모형을 다룬다.

이 책에서는 복잡한 수식을 사용하기 보다는 가능한 직관적인 설명과 실험 데이터로 학생들의 이해를 돕고자 하였다. 물론 이 책만으로 표준 모형을 다 이해할 수는 없으며, 이 책이 더욱 심화된 내용의 공부와 연구를 진행하기 위한 길잡이 역할을 할 수 있기를 바란다. 마지막으로 이 책이 나올 수 있도록 도움을 주신 물리학과와 경북대 출판사에게 감사를 드립니다.

2023년 2월

경북대학교 복현 캠퍼스에서

Contents

1. Physical Constants

Table 1.1: Revised 2021 by D. Robinson (LBNL). Reviewed by P. Mohr (NIST). Mainly from "CODATA Recommended Values of the Fundamental Physical Constants: 2018," E. Tiesinga, D.B. Newell, P.J. Mohr, and B.N. Taylor, NIST SP961 (May 2019) [1]. The electron charge magnitude e, and the Planck, Boltzmann, and Avogadro constants h, k, and N_A, now join c as having defined values; the free-space permittivity and permeability constants ϵ_0 and μ_0 are no longer exact. These changes affect practically everything else in the Table. Figures in parentheses after the values are the 1-standard-deviation uncertainties in the last digits; the fractional uncertainties in parts per 10^9 (ppb) are in the last column. The full 2018 CODATA Committee on Data for Science and Technology set of constants are found at `https://physics.nist.gov/constants`. The last set of constants (beginning with the Fermi coupling constant) comes from the Particle Data Group. See also "The International System of Units (SI)," 9th ed. (2019) of the International Bureau of Weights and Measures (BIPM), `https://www.bipm.org/utils/common/pdf/si-brochure/SI-Brochure-9-EN.pdf`.

Quantity	Symbol, equation	Value	Uncertainty (ppb)
speed of light in vacuum	c	299 792 458 m s^{-1}	exact
Planck constant	h	$6.626\ 070\ 15 \times 10^{-34}$ J s (or J/Hz) §	exact
Planck constant, reduced	$\hbar \equiv h/2\pi$	$1.054\ 571\ 817\ldots \times 10^{-34}$ J s	exact*
		$= 6.582\ 119\ 569\ldots \times 10^{-22}$ MeV s	exact*
electron charge magnitude	e	$1.602\ 176\ 634 \times 10^{-19}$ C	exact
conversion constant	$\hbar c$	197.326 980 4... MeV fm	exact*
conversion constant	$(\hbar c)^2$	0.389 379 372 1... GeV^2 mbarn	exact*
electron mass	m_e	0.510 998 950 00(15) MeV/c^2 = $9.109\ 383\ 7015(28) \times 10^{-31}$ kg	0.30
proton mass	m_p	938.272 088 16(29) MeV/c^2 = $1.672\ 621\ 923\ 69(51) \times 10^{-27}$ kg	0.31
		= 1.007 276 466 621(53) u = 1836.152 673 43(11) m_e	0.053, 0.060
neutron mass	m_n	939.565 420 52(54) MeV/c^2 = 1.008 664 915 95(49) u	0.57, 0.48
deuteron mass	m_d	1875.612 942 57(57) MeV/c^2	0.30
unified atomic mass unit**	u = (mass ^{12}C atom)/12	931.494 102 42(28) MeV/c^2 = $1.660\ 539\ 066\ 60(50) \times 10^{-27}$ kg	0.30
permittivity of free space	$\epsilon_0 = 1/\mu_0 c^2$	$8.854\ 187\ 8128(13) \times 10^{-12}$ F m^{-1}	0.15
permeability of free space	$\mu_0/(4\pi \times 10^{-7})$	1.000 000 000 55(15) N A^{-2}	0.15
fine-structure constant	$\alpha = e^2/4\pi\epsilon_0\hbar c$	$7.297\ 352\ 5693(11) \times 10^{-3}$ = 1/137.035 999 084(21)† ‡‡	0.15
classical electron radius	$r_e = e^2/4\pi\epsilon_0 m_e c^2$	$2.817\ 940\ 3262(13) \times 10^{-15}$ m	0.45
(e^- Compton wavelength)/2π	$\lambda\!\!\!^{-}_e = \hbar/m_e c = r_e \alpha^{-1}$	$3.861\ 592\ 6796(12) \times 10^{-13}$ m	0.30
Bohr radius ($m_{\text{nucleus}} = \infty$)	$a_\infty = 4\pi\epsilon_0\hbar^2/m_e e^2 = r_e\alpha^{-2}$	$0.529\ 177\ 210\ 903(80) \times 10^{-10}$ m	0.15
wavelength of 1 eV/c particle	$hc/(1\ \text{eV})$	$1.239\ 841\ 984\ldots \times 10^{-6}$ m	exact*
Rydberg energy	$hcR_\infty = m_e e^4/2(4\pi\epsilon_0)^2\hbar^2 = m_e c^2\alpha^2/2$	13.605 693 122 994(26) eV	1.9×10^{-3}
Thomson cross section	$\sigma_T = 8\pi r_e^2/3$	0.665 245 873 21(60) barn	0.91
Bohr magneton	$\mu_B = e\hbar/2m_e$	$5.788\ 381\ 8060(17) \times 10^{-11}$ MeV T^{-1}	0.30
nuclear magneton	$\mu_N = e\hbar/2m_p$	$3.152\ 451\ 258\ 44(96) \times 10^{-14}$ MeV T^{-1}	0.31
electron cyclotron freq./field	$\omega^e_{\text{cycl}}/B = e/m_e$	$1.758\ 820\ 010\ 76(53) \times 10^{11}$ rad s^{-1} T^{-1}	0.30
proton cyclotron freq./field	$\omega^p_{\text{cycl}}/B = e/m_p$	$9.578\ 833\ 1560(29) \times 10^{7}$ rad s^{-1} T^{-1}	0.31
gravitational constant‡	G_N	$6.674\ 30(15) \times 10^{-11}$ m^3 kg^{-1} s^{-2}	2.2×10^4
		$= 6.708\ 83(15) \times 10^{-39}$ $\hbar c\ (\text{GeV}/c^2)^{-2}$	2.2×10^4
standard gravitational accel.	g_N	9.806 65 m s^{-2}	exact
Avogadro constant	N_A	$6.022\ 140\ 76 \times 10^{23}$ mol^{-1}	exact
Boltzmann constant	k	$1.380\ 649 \times 10^{-23}$ J K^{-1}	exact
		$= 8.617\ 333\ 262\ldots \times 10^{-5}$ eV K^{-1}	exact*
molar volume, ideal gas at STP	$N_A k$ (273.15 K)/(101 325 Pa)	$22.413\ 969\ 54\ldots \times 10^{-3}$ m^3 mol^{-1}	exact*
Wien displacement law constant	$b = \lambda_{\max} T$	$2.897\ 771\ 955\ldots \times 10^{-3}$ m K	exact*
Stefan-Boltzmann constant	$\sigma = \pi^2 k^4/60\hbar^3 c^2$	$5.670\ 374\ 419\ldots \times 10^{-8}$ W m^{-2} K^{-4}	exact*
Fermi coupling constant‡‡	$G_F/(\hbar c)^3$	$1.166\ 378\ 8(6) \times 10^{-5}$ GeV^{-2}	510
weak-mixing angle	$\sin^2\hat{\theta}(M_Z)$ ($\overline{\text{MS}}$)	0.231 21(4)††	1.7×10^5
$W^\pm$ boson mass	m_W	80.377(12) GeV/c^2¶	1.5×10^5
Z^0 boson mass	m_Z	91.1876(21) GeV/c^2	2.3×10^4
strong coupling constant	$\alpha_s(m_Z)$	0.1179(9)	7.6×10^6
$\pi = 3.141\ 592\ 653\ 589\ 793\ 238\ldots$		$e = 2.718\ 281\ 828\ 459\ 045\ 235\ldots$	$\gamma = 0.577\ 215\ 664\ 901\ 532\ 860\ldots$
1 in ≡ 0.0254 m	1 G ≡ 10^{-4} T	1 eV = $1.602\ 176\ 634 \times 10^{-19}$ J (exact)	kT at 300 K = $[38.681\ 727\ 0718\ldots]^{-1}$eV (exact*)
1 Å ≡ 0.1 nm	1 dyne ≡ 10^{-5} N	(1 kg)c^2 = $5.609\ 588\ 603\ldots \times 10^{35}$eV(exact*)	0 °C ≡ 273.15K
1 barn ≡ 10^{-28} m^2	1 erg ≡ 10^{-7} J	1 C = $2.997\ 924\ 58 \times 10^9$ esu	1 atmosphere ≡ 760 Torr ≡ 101 325Pa

§CODATA recommends that the unit be J/Hz to stress that in $h = E/\nu$ the frequency ν is in cycles/sec (Hz), not radians/sec.

*These are calculated from exact values and are exact to the number of places given (i.e. no rounding).

**The molar mass of ^{12}C is 11.999 999 9958(36) g.

†At $Q^2 = 0$. At $Q^2 \approx m_W^2$ the value is $\sim 1/128$.

‡Absolute laboratory measurements of G_N have been made only on scales of about 1 cm to 1 m.

‡‡See the discussion in Ch. 10, "Electroweak model and constraints on new physics."

††The corresponding $\sin^2\theta$ for the effective angle is 0.23153(4).

¶See the "Mass and width of the W boson" review.

https://pdg.lbl.gov/2022/reviews/contents_sports.html

- Four fundamental forces

gravitation	$G_N m_p^2 = 5.9 \times 10^{-39}$
electric force	$\alpha = \dfrac{e^2}{4\pi} = \dfrac{1}{137} = 0.007297$
weak force	$G_F m_W^2 = 0.0754$
strong force	$\alpha_s = \dfrac{g_s^2}{4\pi} = 0.118$

Chapter 1

Fundamental Particles

■ Why we need high energies?

It is for resolving smallest dimensions and producing massive particles.

① for probing small scale: $\lambda(\text{wavelength}) = \dfrac{h(\text{Planck's const.})}{p(\text{momentum})}$,

$$\text{where } h = 6.63\times10^{-34}\ \text{J}\cdot\text{s} = \frac{6.63\times10^{-34}}{1.6\times10^{-19}}\ \text{eV}\cdot\text{s}$$

$$= 4.14\times10^{-24}\ \text{MeV}\cdot\text{s}$$

(c.f. reduced wavelength: $\lambdabar = \dfrac{\lambda}{2\pi}$, reduced Planck constant: $\hbar = \dfrac{h}{2\pi}$)

$$p = \hbar k = \frac{h}{2\pi}\frac{2\pi}{\lambda} \Rightarrow \lambda = \frac{h}{p}$$

$$\Rightarrow \therefore \text{ as } p\uparrow,\ \lambda\downarrow$$

ex (i) for Tevatron, $E = 10^{12}$ eV (1 TeV),

$$E = h\nu = \frac{hc}{\lambda} \Rightarrow \lambda = \frac{2\pi\hbar c}{E}$$

$$\therefore\ \lambda = \frac{2\pi\times197\ \text{MeV}\cdot\text{fm}}{10^{12}\ \text{eV}} = 1.2\times10^{-18}\ \text{m}$$

(※ a nucleus radius: $1\ \text{fm} = 10^{-15}\ \text{m}$)

(ii) for LHC, $E = 7\times10^{12}$ eV $\Rightarrow$ $\lambda = 1.7\times10^{-19}$ m

② for producing massive particles: $E = mc^2$

■ Units in HEP

length: 10^{-15} m = 1 fm

mass: $m = 1.78\times10^{-27}\ \text{kg} = \dfrac{1.78\times10^{-27}\times c^2}{c^2} = \dfrac{1.78\times10^{-27}\times(3\times10^8)^2\text{J}}{c^2}$

$$= \frac{1.6\times10^{-10}\ \text{J}}{c^2} = \frac{1.6\times10^{-10}}{1.6\times10^{-19}}\ \frac{\text{eV}}{c^2}$$

$$(\because\ 1\ \text{eV} : 1.6\times10^{-19}\ \text{J} = x : 1.6\times10^{-10}\ \text{J}$$

$$\Rightarrow x = \frac{1.6\times10^{-10}}{1.6\times10^{-19}} = 10^9\ \text{eV}\ )$$

$\therefore\ mc^2 = 10^9\ \text{eV} = 1\ \text{GeV}$ for $m = 1.78\times10^{-27}$ kg

ex for proton $m_p = 1.67\times10^{-27}$ kg

$$m_pc^2 = \frac{1.67\times10^{-27}\times(3\times10^8)^2}{1.6\times10^{-19}}\ \text{eV} = 939\times10^6\ \text{eV} = 939\ \text{MeV}$$

for electron $m_e = 9.11\times10^{-31}$ kg

$$m_ec^2 = 0.5\ \text{MeV}$$

Set $\hbar = c = 1$: "natural units"

$$[c] = [\text{velocity}] = \text{LT}^{-1}$$

$$[\hbar] = [\text{length}\times\text{momentum}] = \text{ML}^2\text{T}^{-1} \qquad \left(\leftarrow \lambda = \frac{h}{p}\right)$$

$$[E] = [\text{mass}\times\text{velocity}^2] = \text{ML}^2\text{T}^{-2} \qquad (\leftarrow E = mc^2)$$

$$\Rightarrow \begin{cases} \text{L} = [\hbar][c][E]^{-1} \rightarrow [E]^{-1} \\ \text{M} = [E][c]^{-2} \rightarrow [E] \\ \text{T} = [\hbar][E]^{-1} \rightarrow [E]^{-1} \end{cases}$$

So,

$$1\ \text{m} = \frac{1}{\hbar c} = \frac{1}{197\times10^6\times10^{-15}} = 5.07\times10^{-3}\times10^{-6}\times10^{15} = \frac{5.07\times10^{15}}{\text{GeV}}$$

$$\Rightarrow 0.2\ \text{fm}\times1\ \text{GeV} = (0.2\times10^{-15})\times(5.07\times10^{15}) = 1$$

	natural units	actual dimension
length	GeV^{-1}	$\dfrac{\hbar c}{\mathrm{GeV}}$
mass	GeV	$\dfrac{\mathrm{GeV}}{c^2}$
time	GeV^{-1}	$\dfrac{\hbar}{\mathrm{GeV}}$

• Conversion factor

$$\rightarrow \begin{cases} 1\ \mathrm{m} = 5.07\times 10^{15}\ \mathrm{GeV}^{-1} \\ 1\ \mathrm{kg} = 5.61\times 10^{26}\ \mathrm{GeV} \\ 1\ \mathrm{sec} = 1.52\times 10^{24}\ \mathrm{GeV}^{-1} \end{cases}$$

ex a proton radius of $4\ \mathrm{GeV}^{-1}$

$$\Rightarrow r = 4(\hbar c)\ \mathrm{GeV}^{-1}$$

$$= 4\times\frac{6.63\times 10^{-34}}{2\pi}\times 3\times 10^{8}\times\frac{1}{1.6\times 10^{-19}\times 10^{9}} = 8\times 10^{-16}\ \mathrm{m} = 0.8\ \mathrm{fm}$$

[※ $\hbar c$ units ⇒ actual dimension:

$$\frac{4}{\mathrm{GeV}}\times\hbar c = \frac{4}{\mathrm{GeV}}\times 197\ \mathrm{MeV}\cdot\mathrm{fm} = \frac{788\times 10^{-3}\ \mathrm{GeV}}{\mathrm{GeV}}\mathrm{fm} \approx 0.8\ \mathrm{fm}\ ]$$

"Energetic" means that "special relativity" is important, and we should use Einstein`s equation;

$$E^2 = m^2c^4 + \vec{p}^{\,2}c^2 \xleftrightarrow[\text{units}]{\text{natural}} E^2 = m^2 + \vec{p}^{\,2}$$

c.f. space-time coordinates: $\tilde{x} = (\vec{x}, it)$

4-momentum: $\tilde{p} = (\vec{p}, iE)$ in Minkowski convention

then, $\tilde{p}^2 = p^2 - E^2 = -m^2$ (i.e. $\tilde{p}^2$ is the relativistic invariant)

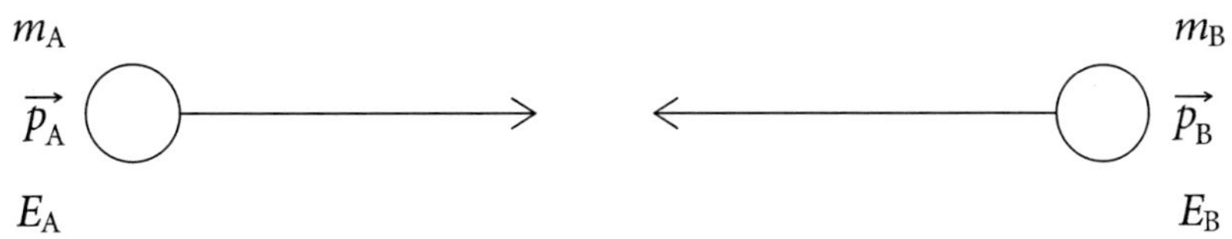

$$\tilde{p}^2 = (\tilde{p}_A + \tilde{p}_B)^2 = \tilde{p}_A^2 + \tilde{p}_B^2 + 2\tilde{p}_A\tilde{p}_B = -m_A^2 - m_B^2 + 2\vec{p}_A \cdot \vec{p}_B - 2E_A E_B$$

$$\Downarrow$$

$$-m^2 = \vec{p}_{cms}^{\,2} - E_{cms}^2 = -E_{cms}^2$$

($\because$ total 3-momentum is zero in the center-of-momentum system(cms))

① for fixed-target, m_B is at rest $\Rightarrow$ $\vec{p}_B = 0$, $E_B = m_B$

then $E_{cms}^2 = m_A^2 + m_B^2 + 2E_A m_B \Rightarrow E_{cms} \approx \sqrt{2E_A m_B}$

→ Therefore, cms energy rises as the square root of the incident energy.

② for colliding beams, $\vec{p}_A \cdot \vec{p}_B = p_A p_B \cos\pi = -p_A p_B$

$$E_{cms}^2 = m_A^2 + m_B^2 + 2p_A p_B + 2E_A E_B$$

$$= 2(E_A E_B + p_A p_B) + (m_A^2 + m_B^2) \approx 4E_A E_B$$

for $m_A, m_B \ll E_A, E_B \rightarrow E_A^2 = p_A^2 + m_A^2 \approx p_A^2$

$$\Rightarrow E_{cms} \approx 2E_A$$

→ Therefore, cms energy rises linearly with E_A.

■ Fundamental particles

- three generation

quarks			leptons		
1st	2nd	3rd	1st	2nd	3rd
u	c	t	e^-	μ^-	τ^-
d	s	b	ν_e	ν_μ	ν_τ

- leptons, quarks, and gauge bosons

particle		charge	mass	lifetime/width	principal decays
leptons	ν_e, ν_μ, ν_τ	0	0	stable	-
	e^-	-1	0.511 MeV	stable	-
	μ^-	-1	106 MeV	2.2×10^{-6} sec	$e^-\bar{\nu}_e\nu_\mu$

particle		charge	mass	lifetime/width	principal decays
	τ^-	-1	1780 MeV	2.9×10^{-13} sec	$\pi^-\pi^0\nu_\tau$, $\mu^-\bar{\nu}_\mu\nu_\tau$, $e^-\bar{\nu}_e\nu_\tau$
quarks	u	2/3	3 MeV	-	-
	c	2/3	1.3 GeV	-	-
	t	2/3	172 GeV	-	-
	d	-1/3	5 MeV	-	-
	s	-1/3	100 MeV	-	-
	b	-1/3	4.2 GeV	-	-
gauge bosons	photon	0	0	stable	-
	$W^\pm$	±1	80.4 GeV	2.1 GeV	$W^+\to\ell^+\nu_\ell$, $u\bar{d}$, $c\bar{s}$
	Z	0	91.2 GeV	2.5 GeV	$\ell^+\ell^-$, $\nu\bar{\nu}$, $q\bar{q}$
	gluon	0	0	-	-

- leptons and quarks are fermions ⇔ bosons
- leptons exist as free particles but quarks are in combinations.

↳ quark confinement

quark combination: meson $q\bar{q}$

baryon qqq; $|p>=|uud>$, $|n>=|udd>$

- each flavor of quark has three different colors (green, red, blue: three primary colors of light)
- total charge of all fermions;

$$\text{u-type quarks} = 3\times3\times\frac{2}{3}=6,$$

where the first is the number of u-type quarks, the second is the number of colors and the third is the charge for u-type quark.

$$\text{d-type quarks} = 3\times3\times\left(-\frac{1}{3}\right)=-3$$

$$\text{leptons} = 3\times(-1)=-3$$

⇒ The sum of the above all charges is $6-3-3=0$.

• Statistics determines how the wave function behaves under interchange of any pair of particles.
  - Fermions obey Fermi-Dirac statistics
  - Bosons obey Bose-Einstein statistics

Under exchange of identical fermions(bosons), Ψ is antisymmetric(symmetric).
Pauli exclusion principle; two or more identical fermions cannot exist in the same quantum state.

■ Four types of fundamental interaction

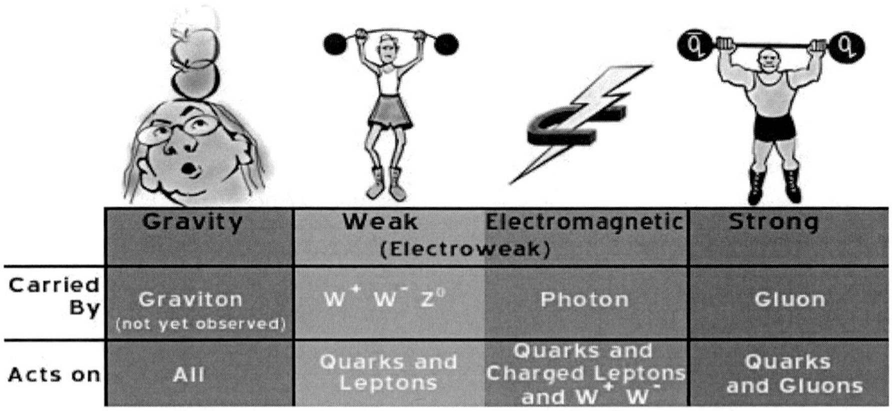

(https://particleadventure.org/inter_summary.html)

interactions	mediators	strength
Strong	a massless particle, gluon	1
Electromagnetic	a photon	10^{-2}
Weak	massive $W^{\pm}$ and Z bosons (about order 100 times the proton mass)	10^{-7}
Gravitational	graviton(spin 2 boson)	10^{-39}

※ c.f. Grand Unification Theory: $10^{16} \sim 10^{17}$ GeV

For a strong interaction, mean life time is $\tau \sim 10^{-23}$ s and can not be measured directly.

$\left(\because \text{ range of strong interaction } \sim 10^{-15}\text{ m} \Rightarrow \dfrac{10^{-15}\text{ m}}{3\times10^{8}\text{ m/s}} = \dfrac{1}{3}\times10^{-23}\text{ sec}\right)$

but $\Gamma\tau = \hbar \Rightarrow \tau = \dfrac{\hbar}{\Gamma}$; τ can be inferred from Γ measurement.

ex width of top quark ~ 2 GeV $\Rightarrow \tau \approx 0.33\times10^{-24}$ sec

$$\left(\because \frac{\hbar}{\Gamma} = \frac{6.58\times10^{-22}\text{ MeV}\cdot\text{s}}{2\text{ GeV}} = 3.3\times10^{-25}\text{ sec}\right)$$

In Standard Model(SM), neutrinos are assumed to be massless.

① but experiments (neutrino oscillations) show that they have finite masses.

② SM contains 17 arbitrary parameters: masses, mixing angles, coupling constants, etc.

③ origin of six quarks and six lepton flavors

⇒ To date, most of experimental results are in good agreement with SM predictions, but SM does not cover all physical phenomena.

⇒ require physics beyond the SM(New Physics).

	fermions			bosons			
	leptons (e, μ, τ)	neutral leptons (ν_e, ν_μ, ν_τ)	quarks	g	γ	$W^{\pm}$	Z^0
Strong			O	O			
EM	O		O		O		
Weak	O	O	O			O	O
Gravitational	O	O	O				

• In quantum mechanics(QM),

$$\Psi = Ae^{\frac{i}{\hbar}(px - Et)} \text{ or } \Psi = Ae^{i(kx - \omega t)}: \text{ travelling along } +x\text{-axis with } \vec{p},\ \omega = \frac{E}{\hbar},\ k = \frac{p}{\hbar}$$

↳ As t increases, phase advances in the direction of increasing x.

Consider

$$\Psi = Ae^{\frac{i}{\hbar}[(-p)(-x) - (-E)(-t)]}: \text{ energy } -E \text{ and momentum } -p \text{ travelling in the negative } x\text{-direction and backwards in the time}$$

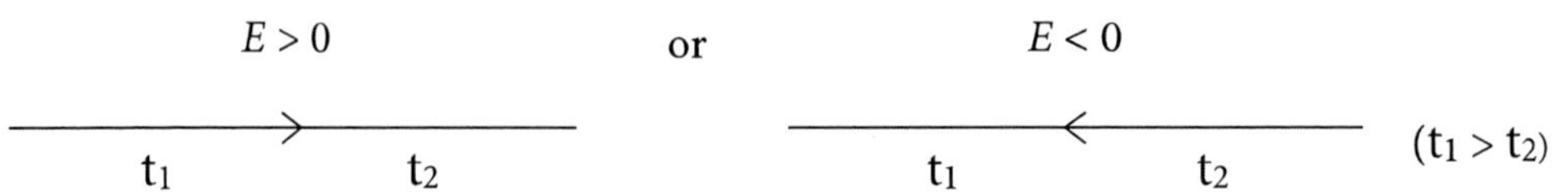

Hence, negative energy particle states are connected with the existence of positive energy antiparticles of exactly equal but opposite electric charge and magnetic moment, and otherwise equal. (ex electron ↔ positron)

■ Free particle wave equations

$$E^2 = \vec{p}^2 c^2 + m^2 c^4 \xrightarrow[E \to i\hbar\frac{\partial}{\partial t},\ \vec{p} \to \frac{\hbar}{i}\vec{\nabla}]{QM} -\hbar^2 \frac{\partial^2 \Psi}{\partial t^2} = -\hbar^2 c^2 \nabla^2 \Psi + m^2 c^4 \Psi$$

$$\Rightarrow \frac{1}{c^2}\frac{\partial^2 \Psi}{\partial t^2} = \nabla^2 \Psi - \frac{m^2 c^2}{\hbar^2}\Psi: \text{ Klein-Gordon wave equation}$$

※ Solving this equation, there are two solutions corresponding to positive and negative energy states. However, what is physically acceptable is the positive energy state.

$$\xrightarrow[\text{units}]{\text{natural}} \frac{\partial^2 \Psi}{\partial t^2} = (\nabla^2 - m^2)\Psi$$

Since no spin variable has been introduced, this equation is suitable for describing spinless bosons.

Note that the Klein-Gordon equation is the second order in the derivatives, while the Schrödinger equation is the first order in time and the second order in space.

※ Schrödinger equation: $E = K + U \Rightarrow i\frac{\partial\Psi}{\partial t} = -\frac{1}{2m}\nabla^2\Psi$ (for free particle)

⇒ This is unsatisfactory when we are dealing with high energy particles.

※ Dirac thought that the difficulty of the Klein-Gordon wave equation was due to the thought of only one wave function, and proposed the wave function of $\Psi = (\Psi_1, \Psi_2, \cdots\cdots \Psi_n)$.

⇒ Dirac sets out to formulate a wave equation that is symmetric in space and time, which was the first order in both derivatives.

$$\Rightarrow \frac{\partial\Psi}{\partial t} = \pm\vec{\sigma}\cdot\frac{\partial\Psi}{\partial\vec{r}} = \pm\left(\sigma_1\frac{\partial\Psi}{\partial x} + \sigma_2\frac{\partial\Psi}{\partial y} + \sigma_3\frac{\partial\Psi}{\partial z}\right) \text{ for massless particles}$$

※ In high energy physics, fermions have extreme relativistic velocities so that fermion masses can be neglected.

To satisfy the Klein-Gordon equation,

$$\Rightarrow \frac{\partial^2\Psi}{\partial t^2} = \sigma_1^2\frac{\partial^2\Psi}{\partial x^2} + \sigma_2^2\frac{\partial^2\Psi}{\partial y^2} + \sigma_3^2\frac{\partial^2\Psi}{\partial z^2} + \sigma_1\sigma_2\frac{\partial^2\Psi}{\partial x\partial y} + \sigma_2\sigma_1\frac{\partial^2\Psi}{\partial y\partial x}$$

$$+ \sigma_2\sigma_3\frac{\partial^2\Psi}{\partial y\partial z} + \sigma_3\sigma_2\frac{\partial^2\Psi}{\partial z\partial y}$$

$$+ \sigma_3\sigma_1\frac{\partial^2\Psi}{\partial z\partial x} + \sigma_1\sigma_3\frac{\partial^2\Psi}{\partial x\partial z}$$

$$\left[※\ \sigma_1\sigma_2\frac{\partial^2\Psi}{\partial x\partial y} + \sigma_2\sigma_1\frac{\partial^2\Psi}{\partial y\partial x} = (\sigma_1\sigma_2 + \sigma_2\sigma_1)\frac{\partial^2\Psi}{\partial x\partial y}\right]$$

for massless particle,

$$\frac{\partial^2\Psi}{\partial t^2} = (\nabla^2 - m^2)\Psi = \nabla^2\Psi$$

- Pauli matrices $\vec{\sigma}$: 2×2 complex matrix, Hermite & Unitary matrices

※ Hermitian $A_{ij} = A_{ji}^*$ $(A = A^+)$, Unitary $A^+ = A^{-1}$

If $A^+ = A$, its eigenvalues are real numbers.

$$\sigma_1 = \begin{pmatrix} 0 & 1 \\ 1 & 0 \end{pmatrix},\ \sigma_2 = \begin{pmatrix} 0 & -i \\ i & 0 \end{pmatrix},\ \sigma_3 = \begin{pmatrix} 1 & 0 \\ 0 & -1 \end{pmatrix}$$

$$\Rightarrow \begin{cases} \sigma_1^2 = \sigma_2^2 = \sigma_3^2 = 1 \Rightarrow ① \ \sigma_i^2 = 1 \\ \sigma_1\sigma_2 + \sigma_2\sigma_1 = 0,\ \sigma_2\sigma_3 + \sigma_3\sigma_2 = 0,\ \sigma_3\sigma_1 + \sigma_1\sigma_3 = 0 \\ m = 0 : \text{massless particle} \end{cases}$$

and

② $\{\sigma_i, \sigma_j\} = 2\delta_{ij}$ or $\{\sigma_i, \sigma_j\} = 0$: they are not commute.

$\Rightarrow$ $\sigma's$ can not be numbers.

③ $\sigma_i\sigma_j = \delta_{ij} + i\epsilon_{ijk}\sigma_k$

Pauli matrices are associated with the spin quantum number of the electron;

$$S_i = \frac{\hbar}{2}\sigma_i \Rightarrow \vec{S} = \frac{\hbar}{2}\vec{\sigma}$$

$$\frac{\partial \Psi}{\partial t} = \pm \vec{\sigma} \cdot \frac{\partial \Psi}{\partial \vec{r}} \xrightarrow{\times i\hbar} i\hbar\frac{\partial \Psi}{\partial t} = \pm i\hbar\vec{\sigma} \cdot \vec{\nabla}\Psi = \mp \vec{\sigma} \cdot \left(\frac{\hbar}{i}\vec{\nabla}\right)\Psi$$

$$\Rightarrow \begin{cases} E\chi = -\vec{\sigma} \cdot \vec{p}\chi, \\ E\phi = +\vec{\sigma} \cdot \vec{p}\phi, \end{cases}$$

where E and $\vec{p}$ are the energy and momentum operators, respectively.

χ and ϕ, spinors, are two component wave functions.

$\Rightarrow$ four solutions, corresponding to particle and antiparticle states with two spin substates of each.

> Now, let`s include the fermion mass,
> $E\Psi = (\vec{\alpha} \cdot \vec{p} + \beta m)\Psi$: **Dirac equation**
> Here, matrices $\vec{\alpha}$ and β are 4×4 matrices, operating on four-component (spinor) wave functions (particles, antiparticle and two spin substates for each).

- standard form of $\vec{\alpha}$ and β; $\vec{\alpha} = \begin{pmatrix} 0 & \vec{\sigma} \\ \vec{\sigma} & 0 \end{pmatrix},\ \beta = \begin{pmatrix} I & 0 \\ 0 & -I \end{pmatrix}$

then

$$i\hbar\frac{\partial}{\partial t}\begin{pmatrix}\Psi_1\\ \Psi_2\\ \Psi_3\\ \Psi_4\end{pmatrix}=(-i\hbar c\vec{\alpha}\cdot\vec{\nabla}+\beta mc^2)\begin{pmatrix}\Psi_1\\ \Psi_2\\ \Psi_3\\ \Psi_4\end{pmatrix}=\begin{pmatrix}mc^2 I & -i\hbar c\vec{\sigma}\cdot\vec{\nabla}\\ -i\hbar c\vec{\sigma}\cdot\vec{\nabla} & -mc^2 I\end{pmatrix}\begin{pmatrix}\Psi_1\\ \Psi_2\\ \Psi_3\\ \Psi_4\end{pmatrix}$$

$$\xrightarrow[\text{units}]{\text{natural}} i\frac{\partial}{\partial t}\begin{pmatrix}\Psi_1\\ \Psi_2\\ \Psi_3\\ \Psi_4\end{pmatrix}=\begin{pmatrix}mI & -i\vec{\sigma}\cdot\vec{\nabla}\\ -i\vec{\sigma}\cdot\vec{\nabla} & -mI\end{pmatrix}\begin{pmatrix}\Psi_1\\ \Psi_2\\ \Psi_3\\ \Psi_4\end{pmatrix}$$

- Let`s look at the Klein-Gordon and Dirac expressions in non-relativistic(NR) limit.

$$E\begin{pmatrix}\chi\\ \phi\end{pmatrix}=\begin{pmatrix}m & \vec{\sigma}\cdot\vec{p}\\ \vec{\sigma}\cdot\vec{p} & -m\end{pmatrix}\begin{pmatrix}\chi\\ \phi\end{pmatrix}$$

$$\Rightarrow\begin{cases}E\chi=m\chi+\vec{\sigma}\cdot\vec{p}\phi & ①\\ E\phi=\vec{\sigma}\cdot\vec{p}\chi-m\phi\Rightarrow\phi=\dfrac{\vec{\sigma}\cdot\vec{p}}{E+m}\chi\approx\dfrac{\vec{\sigma}\cdot\vec{p}}{2m}\chi & ②\end{cases}$$

$$(\because \text{ for NR limit, } E^2=p^2+m^2\Rightarrow E\approx m)$$

Substituting Eq. ② into ①,

$$E\chi\approx m\chi+\frac{(\vec{\sigma}\cdot\vec{p})^2}{2m}\chi\quad\Rightarrow\quad m+\frac{\vec{p}^{\,2}}{2m}$$

$$\therefore\ E\approx m+\frac{\vec{p}^{\,2}}{2m}$$

Meanwhile, for Klein-Gordon equation

$$E^2=\vec{p}^{\,2}+m^2\Rightarrow E=\sqrt{\vec{p}^{\,2}+m^2}$$

$$=m\left(1+\frac{\vec{p}^{\,2}}{m^2}\right)^{1/2}\approx m\left(1+\frac{\vec{p}^{\,2}}{2m^2}\right)=m+\frac{\vec{p}^{\,2}}{2m}$$

Therefore, Dirac and Klein-Gordon equations are equivalent in the NR limit.

■ Helicity States: helicity conservation

$$E\chi=-\vec{\sigma}\cdot\vec{p}\chi\Rightarrow|\vec{p}|\chi=-\vec{\sigma}\cdot\vec{p}\chi$$

$$(\because \text{ for a massless fermion of positive energy, } E=|\vec{p}|)$$

$$\Rightarrow\frac{\vec{\sigma}\cdot\vec{p}}{|\vec{p}|}\chi=-\chi$$

in which $\frac{\vec{\sigma} \cdot \vec{p}}{|\vec{p}|}$ is defined as the helicity (or handedness) which measures the sign of the component of the spin of a particle in the direction of motion.

$\Rightarrow$ helicity = +1; corresponds to a right-handed (RH) screw

helicity = −1; corresponds to a left-handed (LH)

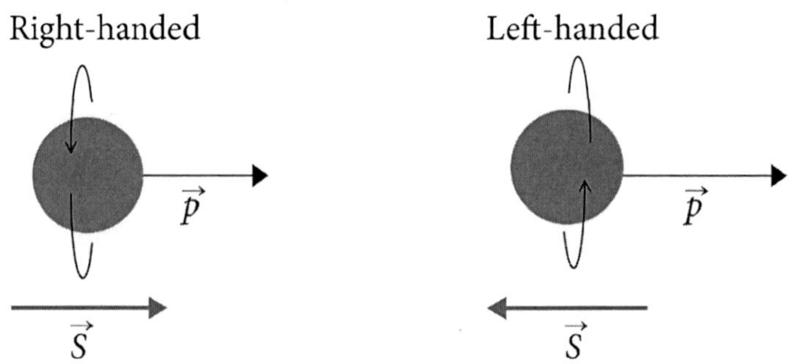

Thus,

$\Rightarrow E\chi = -\vec{\sigma} \cdot \vec{p}\chi$ corresponds to a LH particle or a RH antiparticle state

$(\because\ (-E)\chi = -\vec{\sigma} \cdot (+\vec{p})\chi \Rightarrow \text{helicity} = \frac{\vec{\sigma} \cdot (+\vec{p})}{|-\vec{p}|}\chi = \chi)$

On the other hand,

$$E\phi = +\vec{\sigma} \cdot \vec{p}\phi$$

corresponds to a RH particle or a LH antiparticle state

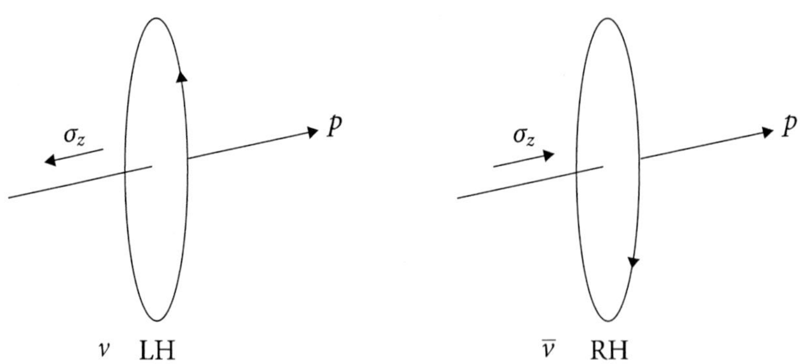

The helicity is conserved in the strong (gluon= $J^P = 1^-$), weak ($W = 1 = Z$) and electromagnetic ($\gamma = 1^-$) interactions, which are all mediated by vector or axial bosons.

⇒ together with conservation of angular momentum, this determines angular distributions in many interactions.

■ Lepton flavors

neutral lepton = neutrino = assumed to be massless in the SM

The neutrinos are unique in that they are completely longitudinally polarized.

⇒ only $j_z = -\frac{1}{2}\hbar$ is observed

⇒ neutrino is a LH particle, (antineutrino $\bar{\nu}$ is RH)

three flavors in the lepton: $L_e,\ L_\mu,\ L_\tau = \begin{cases} +1 & \text{for each lepton} \\ -1 & \text{for each antilepton} \end{cases}$

Lepton flavor (or family) numbers are conserved in the SM.

ex ① $\pi^+ \rightarrow \mu^+ + \nu_\mu \quad \sim 100\%$

L_μ: 0 -1 +1

② $\nu_e + n \rightarrow p + e^-$

L_e: +1 0 0 +1

③ $\mu^+ \rightarrow e^+ + \nu_e + \bar{\nu}_\mu \quad \sim 100\%$

L_e: 0 -1 +1 0

L_μ: -1 0 0 -1

④ $\mu^+ \rightarrow e^+ + \gamma \quad < 5.7 \times 10^{-13}$ at 90% CL

L_e: 0 -1 0

L_μ: -1 0 0

■ Quark flavors

So far we have looked at the lepton flavor. Now that let`s look at quark flavors.

• quark combinations: ① Meson= $q\bar{q} \rightarrow \begin{cases} \pi^0(u\bar{u}), \quad K^-(s\bar{u}) \\ \pi^+(u\bar{d}), \quad K^0(d\bar{s}), \quad \rho^-(\bar{u}d) \\ \pi^-(\bar{u}d) \end{cases}$

② Baryon= $qqq \rightarrow \begin{cases} p(uud) \\ \Lambda(uds), \quad \Omega^-(sss) \end{cases}$

The baryons have half-integer spin and mesons have integer spin because quarks have half-integer spin.

The flavor quantum number, S, C, B, T, are conserved in the strong interaction, but not conserved in weak interaction.

flavor	quantum number
u, d	-
s	$S = -1$
c	$C = +1$
b	$B = -1$
t	$T = +1$

ex $K^- + p \rightarrow \Lambda + \pi^0$

$(s\bar{u} + uud \rightarrow sud + u\bar{u})$: S is conserved.

-1 -1 $\Delta S = 0$

■ Particle physics in cosmology

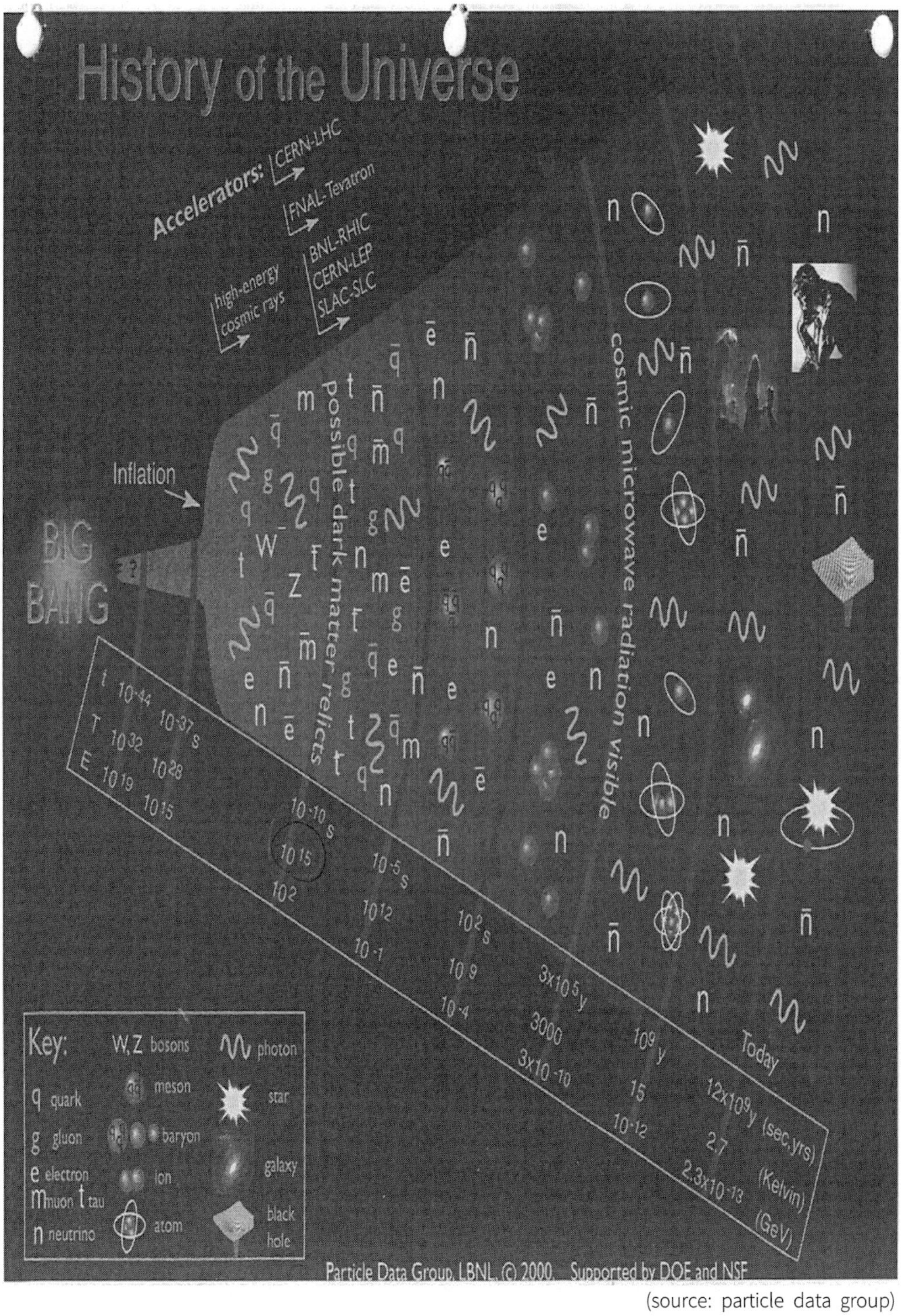

(source: particle data group)

The relationship equation for obtaining energy $E(\text{GeV})$ in the above figure is as follows:

$$k_B T = \text{constant} \times \frac{1\ \text{MeV}}{\sqrt{t}},$$

where constant≈ 1, k_B is the Boltzmann`s constant and t is the time of expansion in seconds.

$$k_B \approx 8.6 \times 10^{-5}\ \frac{\text{eV}}{\text{K}} \Rightarrow T \approx \frac{10^{10}}{\sqrt{t}}\ \text{K}$$

$\Rightarrow$ for $t = 10^{-12}$ sec,

① $k_B T \approx 10^6\ \text{MeV} = 10^3\text{GeV} = 1\ \text{TeV}$;

approximately the energy limit attainable with present-day accelerators

② towards higher energies $\Rightarrow$ look backward in time to the early moments of creation

③ for $t \approx 10^{18}$ sec (present time)

$T \approx 10\ K$, radiation is the cosmic microwave radiation

④ for $T = 10^{15}$ K, $t = 10^{-10}$ sec $\Rightarrow$ 100 GeV $\Rightarrow$ electroweak is broken

■ History

~1928. **Paul Dirac** prediction: antimatter (won Nobel prize in physics in 1933)

develop relativistic Q.M. for electron

※ one for electron with positive energy

one for electron with negative energy → corresponding antiparticle

i.e. same mass, opposite charge → positron

require negative energy solutions

1933. positron discovered in a cloud chamber (**A. Anderson**):

→ (shared Nobel prize with Victor Hess in 1936)

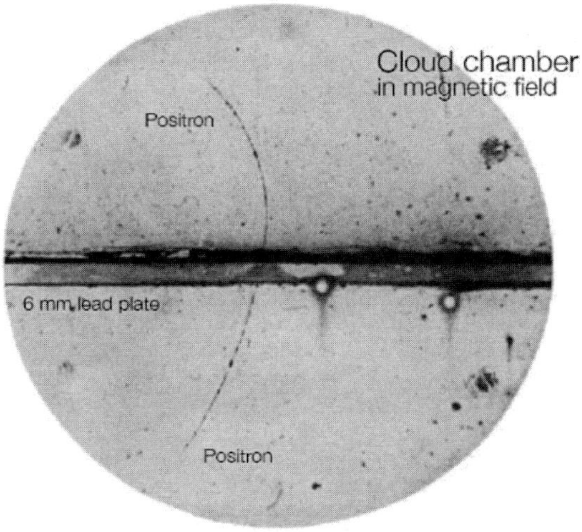

A 63 MeV positron passes through a 6 mm lead leaving the plate with energy 23 MeV. The ionization of the particle, and its behavior in passing through the foil the same as that of an electron.

~1935. **Yukawa** theory of strong interaction

(the range of this force

$$\text{result} \xrightarrow{\text{hypothesis}} \text{meson} \sim 200 \times m_e \rightarrow \pi \text{ meson}$$

prediction of neutral current: weak interaction

Glashow and Weinberg + Salam, they predict existence of

Z^0 $(m_Z \approx 100 \times m_p)$

unification of weak+E.M. interactions (electroweak theory)

1937. $206 \times m_e$, called μ(muon), which is actually turned out the lepton

μ is weakly interacting particle

※ μ was discovered in the 1930ies and was first believed to be Yukawa`s meson that mediates the strong force. But the long range of the muon caused contradictions with this hypothesis.

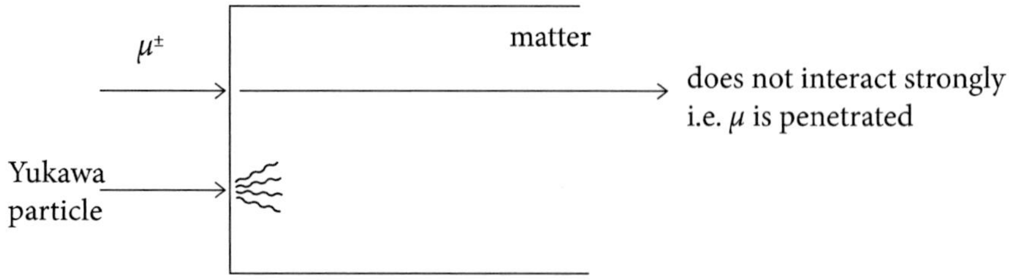

1947. π meson

$\pi^+ \rightarrow \mu^+\nu_\mu$ 100% (**Perkins**+colleagues, first saw π interactions)

※ see " **ex** in Lepton flavors" of the Lecture note

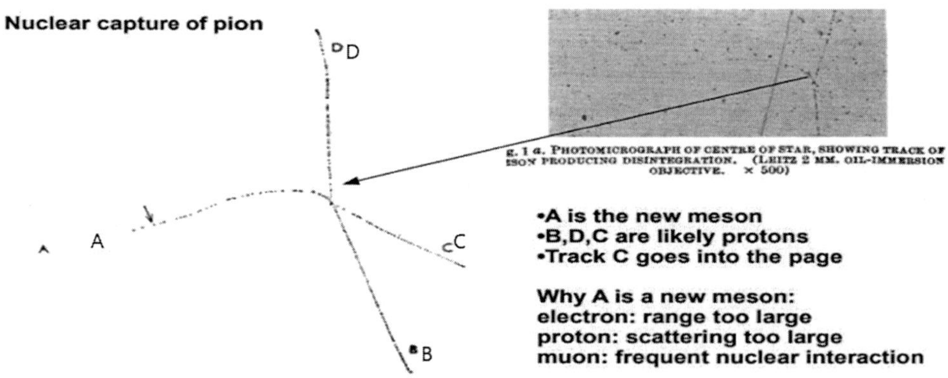

(1947. first evidence for strangeness by Rochester and **Butler**, Manchester group)

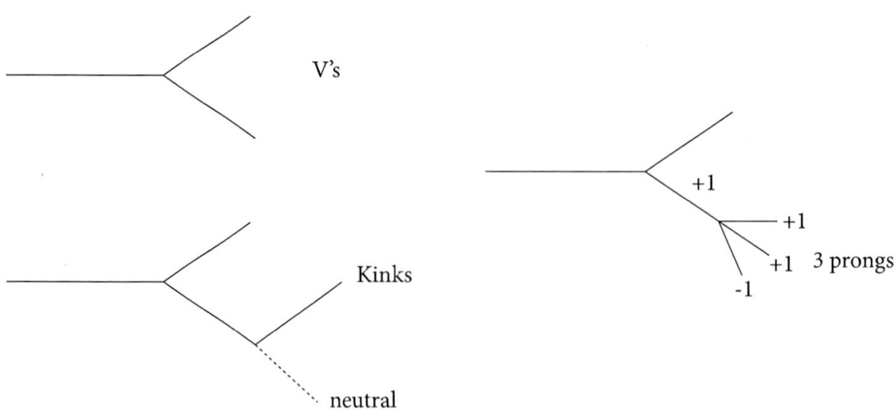

※ kaons(490 MeV): mass between π and p

hyperons(1200 MeV): heavier than p

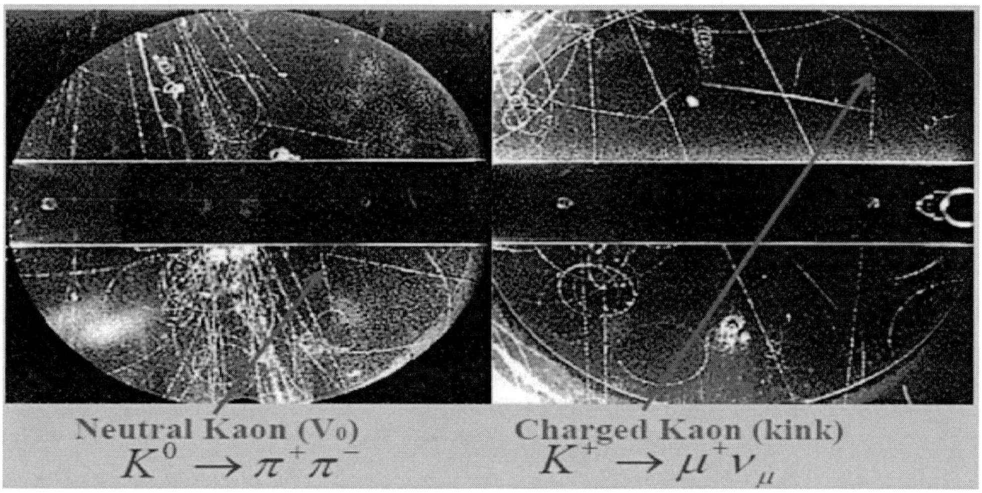

(Rochester and Butler, Manchester group)

1950. Neutral particle $\rightarrow \gamma\gamma$; $\pi^0 \rightarrow \gamma\gamma$

produced with same 0 as it`s change

1953. $K^- p \rightarrow \pi^- \Sigma^+$

strangeness $-1+0=0-1$: conserved $\Rightarrow$ strong interaction

strangeness is conserved in the strong interaction but can be violated in the weak interaction.

Gell-Mann (1953)

$\Lambda \rightarrow p\,\pi^-$

$(uds) \rightarrow (uud)(d\bar{u})$ ※ see "Quark flavors" of the Lecture note

strangeness $-1 \neq 0+0$: non-conserved $\Rightarrow$ is not the strong interaction

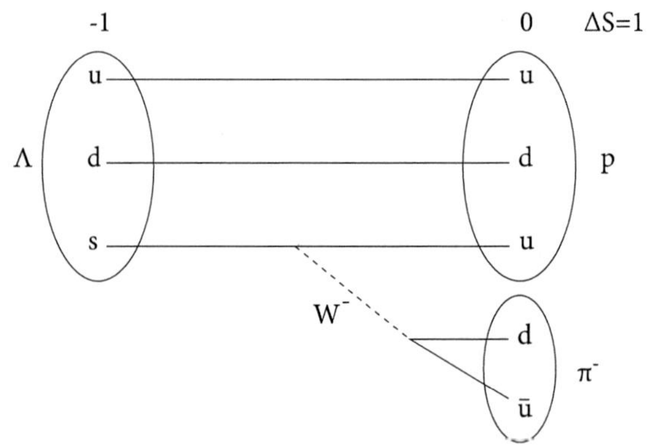

strangeness can be violated in the weak interaction.

1962. non-existence of $\mu^- \rightarrow e^- \gamma$

$\left(※ \begin{cases} \beta-\text{decay: } n \rightarrow pe^-\overline{\nu} \\ \pi^- \rightarrow \mu^-\overline{\nu} \end{cases} \right)$: these are different because of non-existence of $\mu^- \rightarrow e^- \gamma$

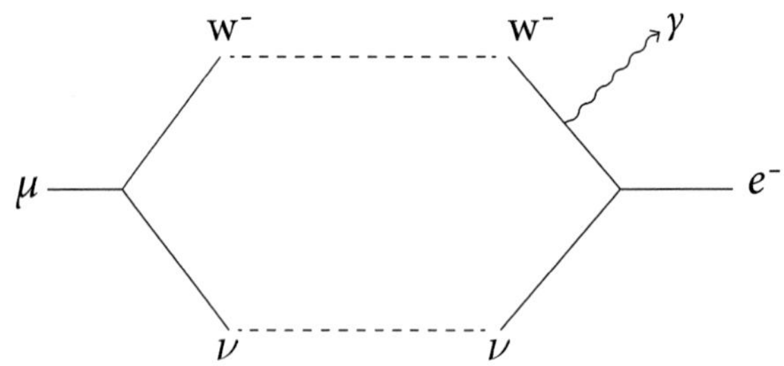

$(\mu^- \rightarrow e^- \overline{\nu}\nu)$

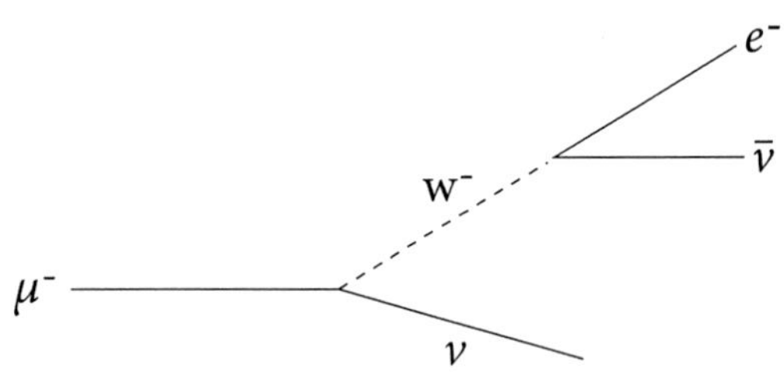

if $\nu_e = \nu_\mu$

$\dfrac{\mu \rightarrow e\gamma}{\mu \rightarrow e\nu\overline{\nu}} \approx 10^{-4}$ is expected but experimentally $< 10^{-10}$

$\Rightarrow \nu_\mu$ is different from ν_e

Chapter 2

Interactions and Fields

■ Yukawa theory of quantum exchange

Klein-Gorden equation:

$$\frac{1}{c^2}\frac{\partial^2\Psi}{\partial t^2} = \nabla^2\Psi - \frac{m^2c^2}{\hbar^2}\Psi$$

※ for $m=0$, the equation becomes the wave equation describing the propagation of an electromagnetic wave:

Substitute $\Psi = e^{i(kx-\omega t)}$ into the equation, we get $\frac{\omega^2}{c^2} = k^2$,

which is $k = \frac{2\pi}{\lambda} = \frac{2\pi\nu}{c} = \frac{\omega}{c}$

Ψ can be interpreted as potential or wave amplitude

$$\xrightarrow[\text{time}-\text{independent}]{\text{for static potential}} \nabla^2 U(r) = \frac{m^2c^2}{\hbar^2}U(r)$$

$$\xrightarrow[\text{in spherical coord.}]{\text{Laplacian}} \nabla^2 U(r) = \frac{1}{r^2}\frac{\partial}{\partial r}\left(r^2\frac{\partial U}{\partial r}\right) = \frac{m^2c^2}{\hbar^2}U(r)$$

$$\Rightarrow \frac{d^2U}{dr^2} + \frac{2}{r}\frac{dU}{dr} - \frac{U}{R^2} = 0,$$

where $R = \frac{\hbar}{mc}$,

$$[R] = \frac{\hbar c}{mc^2} = \frac{\text{energy}\cdot\text{length}}{\text{energy}} = \text{length}$$

nuclear potential $\Rightarrow U(r) = \frac{g_0}{4\pi r}e^{-\frac{r}{R}}$ ...... (1), where g_0 is the constant of integration.

Meanwhile, for electromagnetism

$$\nabla^2 U(r) = 0, \quad U(r) = \frac{Q}{4\pi r} \quad \cdots\cdots (2),$$ where Q is the charge at the origin.

From Eq. (1)

$$U(r) \xrightarrow{r \to 0} \frac{g_0}{4\pi r} = \frac{Q}{4\pi r};$$ g_0 can be interpreted as the "strong nuclear charge".

For nuclear forces, $R \approx 10^{-15}$ m

Yukawa hypothesis predicted a spinless quantum of mass which is

$$mc^2 = \frac{\hbar c}{R} = \frac{197 \text{ Mev} \cdot \text{fm}}{1 \text{ fm}} \approx 100 \text{ MeV}$$

$$\left(\because R = \frac{\hbar}{mc} \Rightarrow \frac{R}{c} = \frac{\hbar}{mc^2} \Rightarrow mc^2 = \frac{\hbar c}{R} \right)$$

- pion mass is 140 MeV (This is the reason, in the history summary we learned last time, we first thought of muons as pions.) and pion has spin 0

Actually, interquark potential contains two terms;

one corresponding to the form Eq. (1) with $m = 0$

+

second term proportional to r (responsible for quark confinement)

■ The boson propagator

associated amplitude for scattering of particle (scattering amplitude)

$$f(\vec{q}) = g \int U(\vec{r}) e^{i\vec{q} \cdot \vec{r}} dV:$$ $\vec{q}$ is the momentum transfer and

g is the intrinsic coupling strength of particle to potential

For central potential,

$$f(\vec{q}) = g\int U(r)e^{iqr\cos\theta}r^2\sin\theta drd\theta d\phi$$

$$= 2\pi g\int_0^\infty U(r)r^2dr\int_0^\pi e^{iqr\cos\theta}\sin\theta d\theta, \quad \left(\int_0^\pi e^{iqr\cos\theta}\sin\theta d\theta = \frac{2\sin qr}{qr}\right)$$

$$= 4\pi g\int U(r)\frac{\sin qr}{qr}r^2dr$$

(substitute Yukawa potential)

$$= 4\pi g\int \frac{g_0}{4\pi r}e^{-\frac{r}{R}}\frac{\sin qr}{qr}r^2dr = g_0 g\int \frac{\sin qr}{q}e^{-\frac{r}{R}}dr$$

$$= g_0 g\int \frac{\sin qr}{q}e^{-mr}dr \quad (\because \text{ introduce natural units } R = \frac{\hbar}{mc} \Rightarrow R = \frac{1}{m})$$

$$= g_0 g\int_0^\infty e^{-mr}\frac{e^{iqr} - e^{-iqr}}{2iq}dr$$

$$= \frac{g_0 g}{|\vec{q}|^2 + m^2}$$

$$\because \frac{1}{2iq}\left[\int_0^\infty e^{(iq-m)r}dr - \int_0^\infty e^{-(iq+m)r}dr\right]$$

$$= \frac{1}{2iq}\left[\frac{e^{(iq-m)r}}{iq-m}\Bigg|_0^\infty + \frac{e^{-(iq+m)r}}{iq+m}\Bigg|_0^\infty\right]$$

$$= \frac{1}{2iq}\left[-\frac{1}{iq-m} - \frac{1}{iq+m}\right] = \frac{-1}{2iq}\left[\frac{iq+m+iq-m}{(iq)^2 - m^2}\right]$$

$$= \frac{1}{|\vec{q}|^2 + m^2}$$

scattering amplitude (or matrix element)

$$\Rightarrow f(\vec{q}) = \frac{g_0 g}{|\vec{q}|^2 + m^2},$$

where g_0 is the coupling of the boson to the scattered particle,

g is the coupling of the boson to the scattering particle,

q is the 4-momentum transfer,

$\frac{1}{q^2 + m^2}$ is the "propagator"

In general, not only the moment but also the energy is transferred, so if you consider not only space but also time in the first equation,

$$\Rightarrow f(q^2) = \frac{g_0 g}{q^2 + m^2};$$

rate of a particular reaction mediated by the boson exchange is proportional to $|f|^2$ multiplied by a phase-space factor

The above equation is the basic formula describing the interaction of two particles via single-boson exchange.

■ Feynman diagram

is graphical representation of interactions between particles and fields.

ex e^+e^- scattering via photon exchange in the first-order

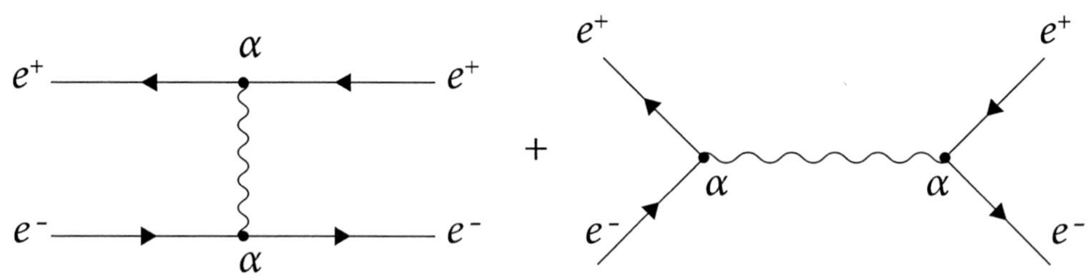

- charge, energy and momentum are conserved at the vertex
- $\sqrt{\alpha}$; coupling constant (in this case, α is the fine structure constant)
- joining vertices represent "virtual" particle

(virtual: satisfying the condition $\Delta E \Delta t \sim \hbar$)

→ virtual particles have energy and momentum such that the rest mass does not correspond to that of a real and free particle

⇒ Therefore, scattering amplitude is a function of the coupling constants at the vertices and the propagators associated with the virtual particles running between vertices.

■ Electromagnetic interactions

Coupling constant between charged particles and photon($\gamma = J^P = 1^-$, vector particle) is

$$\alpha(\text{dimensionless}) = \frac{e^2}{4\pi\hbar c} \approx \frac{1}{137};$$

fine structure constant which determines fine structure due to spin-orbit coupling

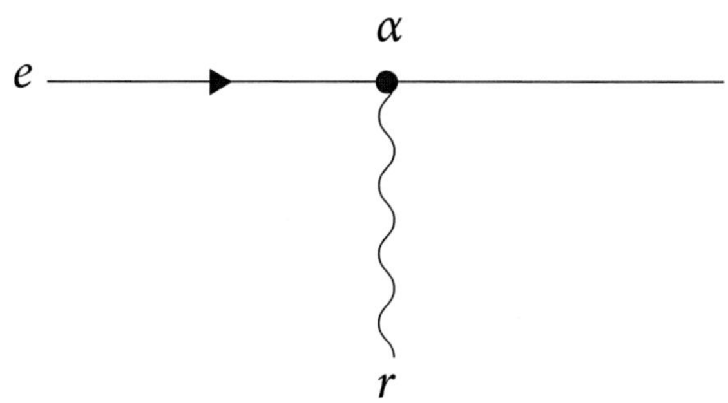

(absorption or emission of a photon by an electron)

- photon couples to the electron with amplitude $\sqrt{\alpha}$

$\Rightarrow$ cross-section is proportional to α (or e^2)

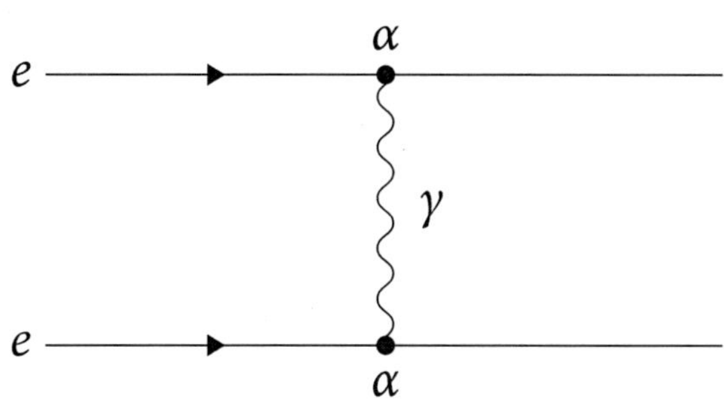

(Coulomb scattering between two electrons via exchange of a single virtual photon of momentum q)

- scattering amplitude $\propto \frac{1}{q^2}$ (with zero mass of it were a free particle)

$\Rightarrow$ differential cross-section $\left(\frac{d\sigma}{dq^2}\right)$ is proportional to $\frac{\alpha^2}{q^4}$

Rutherford scattering formula $\frac{d\sigma}{d\Omega} \propto \frac{e^4}{m^2 v^4} csc^4(\theta/2)$,

$$\alpha = \frac{e^2}{4\pi\hbar c} \Rightarrow \alpha^2 \propto e^4$$

[I] $e^+e^- \rightarrow e^+e^-$ scattering (**Bhabha scattering**)

corresponds to the first-order Feynman diagram for $e^+e^- \rightarrow e^+e^-$ scattering;

t-channel(scattering process), s-channel(annihilation process)

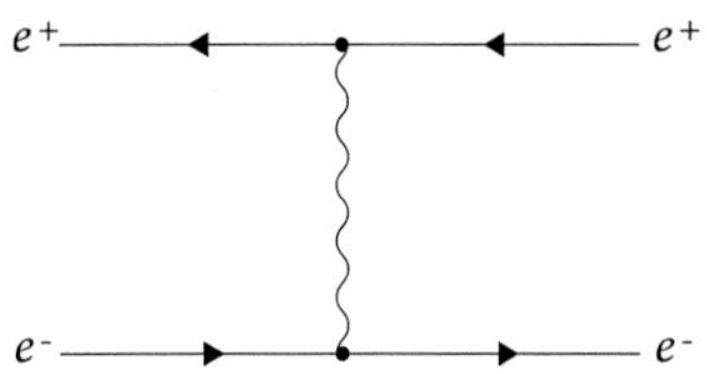

$(t-\text{channel})$

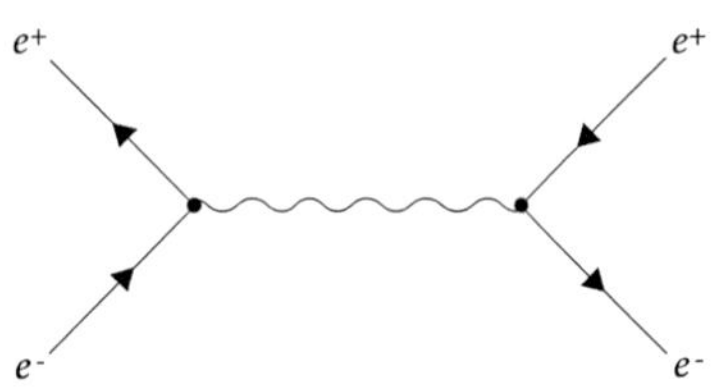

$(s-\text{channel})$

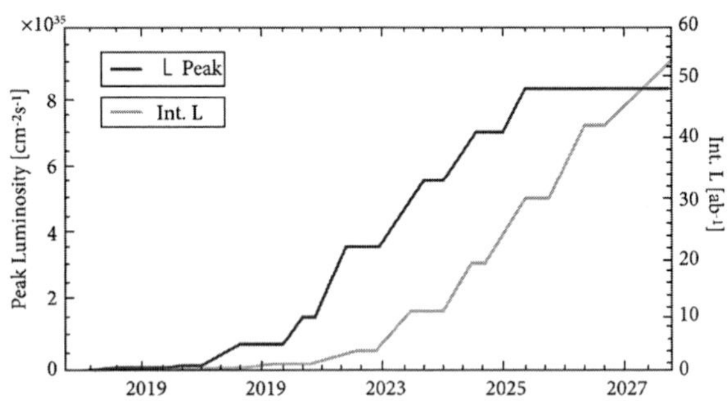

https://www-superkekb.kek.jp/

- The luminosity in electron-positron collider is measured by Bhabha scattering rate.

[II] **Bremsstrahlung**(braking radiation) **process**: emission of a real photon by an electron that has undergone deceleration in the electric field of atomic ucleus(nucleus charge Ze).

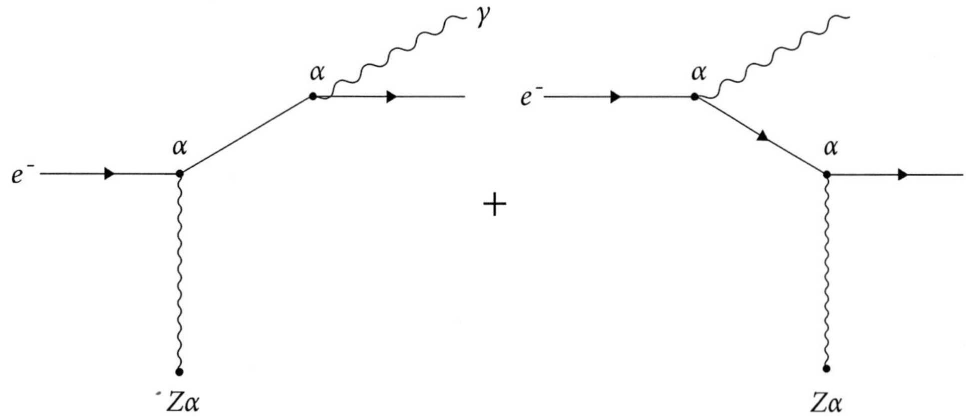

virtual electron because an electron can not emit real photon and conserve nergy and momentum without reducing its rest mass (**"off mass shell"**)

- The cross-section of the process: $\sim \alpha^3$

[III] e^+e^- pair production

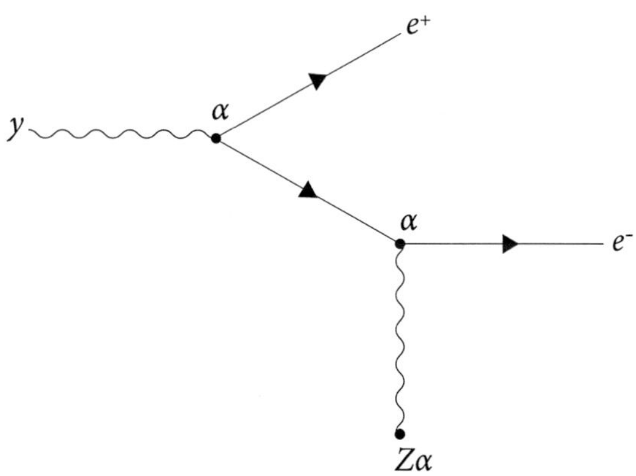

(e^+e^- pair production by a photon in the field of a nucleus)

These interactions were first observed in Patrick Blackett's counter-controlled cloud chamber (1948 Nobel Prize).

- The cross-section of the process: $\sim \alpha^3$

[IV] "self-energy" Feynman diagrams

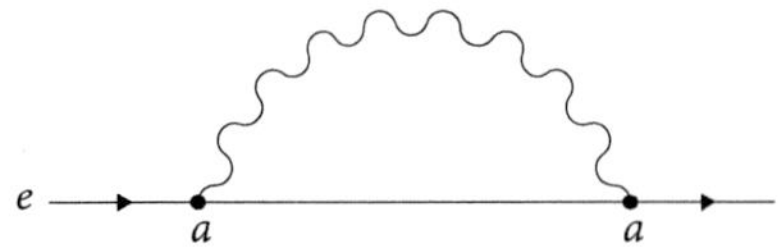	- emitting and reabsorbing a virtual photon
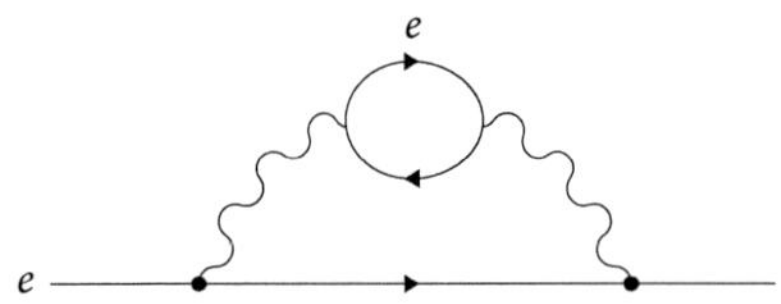	- vacuum containing the virtual e^+e^- pair, becomes polarized (**"vacuum polarization"**)

What we have discussed in the above ⇒ **Quantum electrodynamics (QED)**

- There is no restriction on momentum k of particles in such virtual state.
- spectrum of the virtual photons(ex. [IV]) of 4-momentum k in such process;

$$f(k)\,dk \Rightarrow \int \frac{dk}{k} \Rightarrow \text{logarithmically divergent}$$

to solve this,

① "bare" mass or "bare" charge × divergent integral

$\xrightarrow[\text{by}]{\text{replaced}}$ "physical" mass or "physical" charge: **"renormalization"**

※ "physical" means experimentally measurable

$\xrightarrow[\text{renormalization}]{\text{as a result of}}$

② α is running constant (i.e. depend on the energy at which measurements are being made)

c.f. $\alpha(Z_0) = \dfrac{1}{128}$

③ theory has the property of gauge invariance (invariant under arbitrary redefinitions of the global potential scale or gauge)

■ **Strong interactions** (mediating boson is a gluon, which is neutral, massless, strong: **Quantum chromodynamics (QCD)** force carrier)

↳ $J^P = 1^-$: vector particle

$\Sigma^0(1192) \rightarrow \Lambda\ \gamma$: lifetime $\tau \sim 10^{-19}$ EM interaction
$(uds)\quad (uds)$

$\Sigma^0(1385) \rightarrow \Lambda\ \pi^0$: lifetime $\tau \sim 10^{-23}$ strong interaction
$(uds)\ (uds)(u\overline{u})$

$$\left(\because\ \Gamma = 36\ \text{MeV} \Rightarrow \tau = \frac{\hbar}{\Gamma}\right)$$

In a leading-order process, the decay rate is proportional to the square of the coupling constant.

$$\Rightarrow \frac{\alpha_s}{\alpha_{EM}} \sim \sqrt{\frac{10^{-19}}{20^{-23}}} \sim 100$$

$\Rightarrow \alpha_{EM} = \frac{e^2}{4\pi} = \frac{1}{137}$ and $\alpha_s = \frac{g_s^2}{4\pi} \sim 1$, g_s is the strong charge of constituent quarks

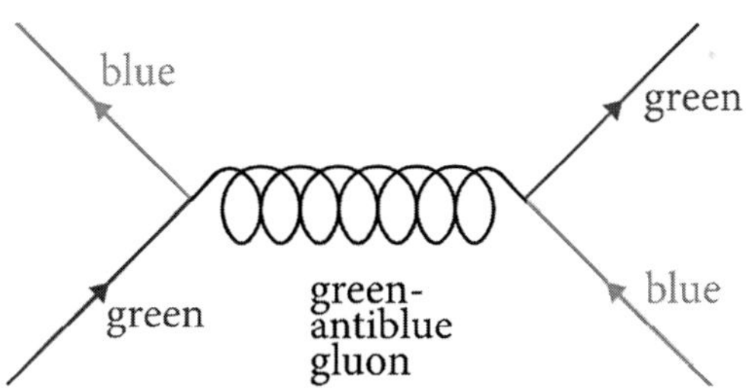

Feynman diagram for an interaction between quarks generated by a gluon.

gluon is $g\overline{b}$ $\Rightarrow$ gluon carries one color and one anti-color

$\Rightarrow$ Since there are three quark colors $(r,\ b,\ g)$, $SU(3)$; quark carries color charge

There are $3^2 = 9$ possible color combinations.

⇒ However one of these is the colorless singlet.

(※ one of these is excluded by the condition that color must be transferred: sum of three gluons would not transfer any color.)

$\frac{1}{\sqrt{3}}(r\bar{r}+b\bar{b}+g\bar{g})$: colorless singlet state

⇒ 8 active gluon states: $r\bar{b}$, $r\bar{g}$, $b\bar{g}$, $b\bar{r}$, $g\bar{r}$, $g\bar{b}$,

$$\frac{1}{\sqrt{2}}(r\bar{r}-b\bar{b}),\ \frac{1}{\sqrt{6}}(r\bar{r}+b\bar{b}-2g\bar{g})$$

- quark-antiquark potential (interquark potential)

$$U_s=-\frac{4}{3}\frac{\alpha_s}{r}+kr$$

- 1st term dominates at small r, arises from single-gluon exchange

(c.f. Coulomb potential $U_{\text{em}}=-\frac{\alpha}{r}$)

- 2nd term is that the confinement at large r
- coefficient in 1st term: $\frac{4}{3}=\frac{1}{2}\times\frac{8}{3}$, where $1/2$ is the historical reason, 8 is the color gluon states and 3 is for three quark colors

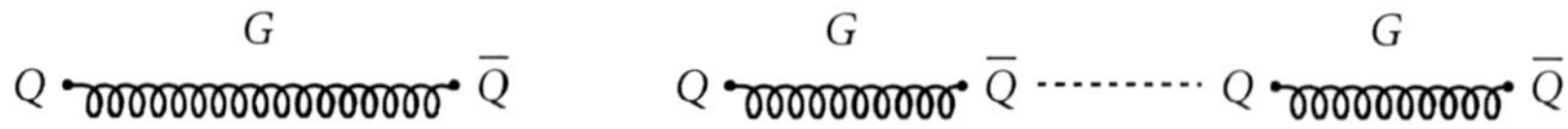

(a) (b)

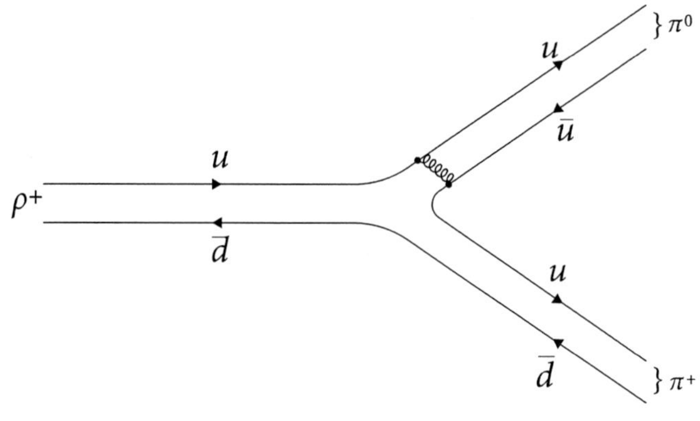

$\Rightarrow$ create $q\bar{q}$ pair involving two short string, energy proportional to string length fragmentation of the quarks to hadrons: "hadronization"

■ Weak and electroweak Interaction

$\Sigma^+(1189) \rightarrow p\pi^0$: lifetime $\sim 10^{-10}$ sec, weak interaction

$(uus)\quad (uud)(u\bar{u})$

$\Sigma^0(1192) \rightarrow \Lambda\gamma$: lifetime $\sim 10^{-19}$ sec, electromagnetic interaction

$(uds)\quad\quad (uds)$

$\Rightarrow \dfrac{\alpha_{\text{weak}}}{\alpha_{\text{EM}}} \sim \sqrt{\dfrac{10^{-19}}{10^{-10}}} \sim 10^{-5}$: weak coupling is smaller by a factor of order $\sim 10^{-5}$ than EM coupling

Observable the weak interactions involve either neutrino or quarks with a flavor change ($\Delta S = 1,\ \Delta C = 1$ *etc.*) that is forbidden for the strong interaction

ex nuclear β-decay

$n \rightarrow p + e^- + \bar{\nu}_e$

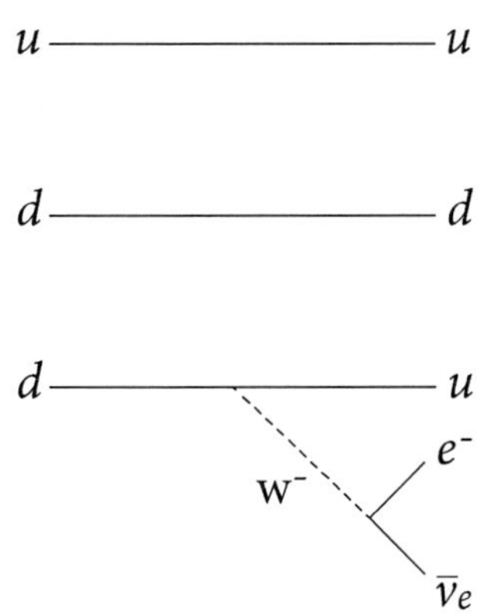

and its inverse $\bar{\nu}_e + p \rightarrow n + e^+$

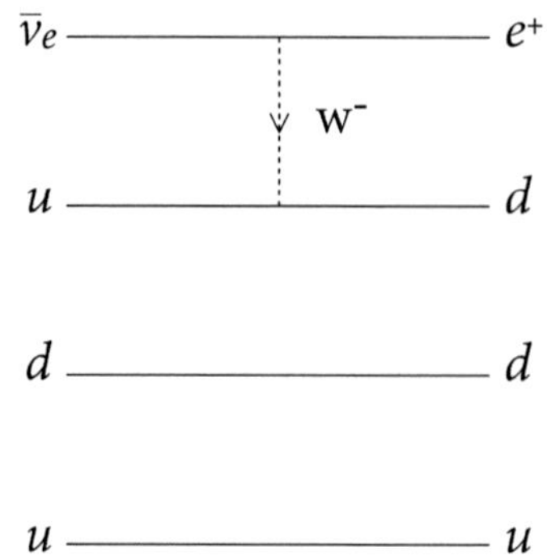

The weak interactions are weak because they are mediated by massive $W^{\pm}$ (charged-current weak interactions) and Z^0(neutral-current weak interactions) which give rise to interactions of very short range.

ex $e^- + \bar{\nu}_\mu \rightarrow e^- + \bar{\nu}_\mu$

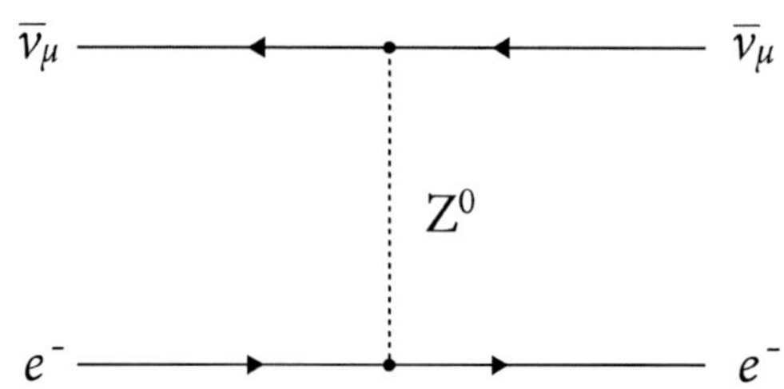

EM scattering: $f = \dfrac{e^2}{q^2}$

Weak scattering: $f(q^2) = \dfrac{g^2}{q^2 + M_{W,Z}^2}$, where g is the weak charge

for $q^2 \ll M_{W,Z}^2$, the scattering amplitude is independent on q^2, i.e. interaction is pointlike

$\Rightarrow \dfrac{g^2}{M_{W,Z}^2} \equiv G_F$ (Fermi coupling constant) $\approx 10^{-5}$ GeV^{-2} from measurement of β-decay rates

Glashow, Weinberg and Salam (GWS) proposed that

$$g = e \ \text{(in practice, } \alpha_{\mathrm{EM}} = \frac{1}{137},\ \alpha_{\mathrm{weak}} = \frac{1}{128})$$

then,

$$M_{W,Z} \sim \frac{g}{\sqrt{G_F}} = \frac{e}{\sqrt{G_F}} = \sqrt{\frac{4\pi\alpha}{G_F}} \sim 90\ \mathrm{GeV} \left(\because\ \alpha = \frac{e^2}{4\pi} \Rightarrow e = 4\pi\alpha \right)$$

⇒ agrees well with measured values

■ Gravitational Interactions

mediated by graviton $J^P = 2^+$

force between two equal point masses M: $\dfrac{G_N M^2}{r^2}$,

where $G_N = 6.672\times10^{-11}\dfrac{\mathrm{N\cdot m^2}}{\mathrm{kg^2}} = 6.7\times10^{-39}\ \hbar c(\mathrm{GeV}/c^2)^{-2}$

is the Newtonian constant (see "physical constants") and

r is the separation between two masses

by comparing with $\dfrac{e^2}{r^2}$, quantity $\dfrac{G_N M^2}{\hbar c}$ is dimensionless

※ c.f. $\alpha[\text{dimensionless}] = e^2/4\pi\hbar c$

$$\because\ [\alpha] = \left[\frac{e^2}{4\pi\hbar c}\right] \Rightarrow [e^2] = [\hbar c][\alpha] = \text{dimensionless}\cdot[\hbar c]$$

$$[e^2] = [G_N M^2] = \text{dimensionless}\cdot[\hbar c]$$

$$\Rightarrow \text{dimensionless} = \left[\frac{G_N M^2}{\hbar c}\right]$$

then

$$\frac{G_N M^2}{4\pi\hbar c} = \frac{6.7\times10^{-39}\ \hbar c\ (\mathrm{GeV})^2}{4\pi\hbar c}\frac{c^4}{c^4\quad(\mathrm{GeV})^2} \approx 5.34\times10^{-40}$$

if we take as the unit of mass,

$Mc^2 = 1\ \mathrm{GeV}$ (for the mass scales common in high energy physics)

$$\left(\leftrightarrow \frac{e^2}{4\pi\hbar c} = \frac{1}{137}\right)$$

→ It would approach unity only for hypothetical elementary particles of mass equal to the Planck mass, $\left(\frac{\hbar c}{G_N}\right)^{\frac{1}{2}} = 1.22\times 10^{19}$ GeV

(※ in the figure of Big Bang, 10^{19} GeV exists)

Planck length $\ell_p = \frac{\hbar c}{M_p c^2} \approx 1.6\times 10^{-35}$ m $\left(\because\ [\hbar c] = \text{MeV}\cdot\text{fm} \Rightarrow \text{fm} = \frac{\hbar c}{\text{MeV}}\right)$

※ It is the size of the space that we usually know no longer exists.

Two point like particles, each having the Planck mass and separated by the Planck length, gravitational potential energy is

$$\frac{G_N M_p^2}{\ell_p} = G_N \frac{\hbar c}{G_N}\left(\frac{M_P c^2}{\hbar c}\right) = M_p c^2;\ \text{rest mass}$$

⇒ quantum gravitational effects become important at the Planck scale

Summary	Relative force strength
strong	1
electromagnetic	10^{-2}
weak	10^{-7}
gravitational	10^{-39}

■ Interaction cross-section in high energy physics

$$\begin{cases} 1\ \text{barn} = 1\text{b} = 10^{-28}\ \text{m}^2 \\ 1\ \text{nb} = 10^{-9}\ \text{b} = 10^{-37}\ \text{m}^2 \\ 1\ \text{pb} = 10^{-12}\ \text{b} = 10^{-40}\ \text{m}^2 \\ 1\ \text{fb} = 10^{-15}\ \text{b} = 10^{-43}\ \text{m}^2 \end{cases}$$

$a+b \rightarrow c+d$: parallel beam of particles of type a impinges normally on a target of thickness Δx, containing n_b particles of type b per unit volume

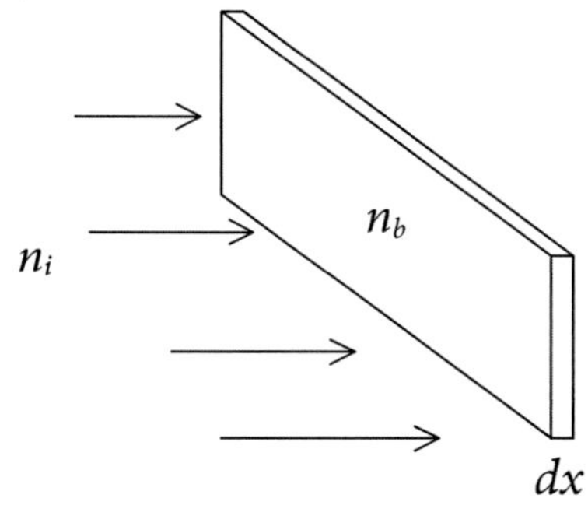

n_i: incident particle density

v_i: incident particle velocity relative to the target

(relative velocity of a and b)

$\phi = n_i v_i$ $\left([\phi] = \dfrac{\text{No. of particles}}{\text{area} \cdot \text{time}}\right)$: flux through the target

If each of the target particles has an effective cross-section σ, the probability that any particle a will hit a target particle is $\sigma n_b \Delta x$.

Then, the number of reactions per unit time per area of beam is $\phi \sigma n_b \Delta x$

($\leftarrow$ $\phi\sigma = W$: reaction rate(=[flux][cross-section]) per target particle)

The reaction rate W (known as **Fermi's Second Golden Rule**: non-rel. Q.M.)

$W = \dfrac{2\pi}{\hbar}|M_{\text{if}}|^2 \rho_f$: M_{if} is the matrix element between initial and final states

$= \int \Psi_f^* U \Psi_i dV$

ρ_f is the energy density $\dfrac{dN}{dE}$ of the final states

number of states in phase space available to a particular final-states particle, directed into solid angle $d\Omega$ and enclosed in a physical volume V is

$$dN = \frac{V}{(2\pi\hbar)^3} p^2 dp d\Omega$$

└ phase space available to particle (with neglecting its spin)

Then

$$\frac{d\sigma}{d\Omega} = \frac{W}{\phi} = \frac{W}{v_i} = \frac{2\pi}{\hbar}\frac{|M_{\rm if}|^2}{v_i}\frac{1}{(2\pi\hbar)^3}p_f^2\frac{dp_f}{dE_0} \quad \cdots\cdots (1)$$

└ by one incident particle, i.e. $n_i = 1$

($\because$ $V = 1$ (volume factor cancels with normalization of wave functions)

because arbitrary normalization volume V

cancels out in the expression for the cross-section)

here $p_f = |\overrightarrow{p_c}| = |\overrightarrow{p_d}|$ is the center-of-momentum system

$E_0 = E_c + E_d$ is the total cms energy $= \sqrt{\overrightarrow{p_f}^2 + m_c^2} + \sqrt{\overrightarrow{p_f}^2 + m_d^2}$

Let`s find $\frac{dp_f}{dE_0}$ in Eq (1);

$$dE_0 = \left(\frac{1}{2}2p_f\frac{1}{\sqrt{p_f^2+m_c^2}} + \frac{1}{2}2p_f\frac{1}{\sqrt{p_f^2+m_d^2}}\right)dp_f = p_f\left(\frac{1}{E_c}+\frac{1}{E_d}\right)dp_f$$

$$= p_f\frac{E_c+E_d}{E_cE_d}dp_f = p_f\frac{E_0}{E_cE_d}dp_f$$

$\Rightarrow \frac{dp_f}{dE_0} = \frac{E_cE_d}{E_0p_f} = \frac{1}{v_f}$; v_f is the relative velocity of c and d ($\Rightarrow v_f = v_c + v_d$)

$\left(\because \frac{p_fp_f}{Mv_f^2p_f} = \frac{Mv_f}{Mv_f^2} = \frac{1}{v_f}\right)$, most general relationship: $\vec{p} = \frac{E}{c^2}\vec{v}$

$$\therefore \frac{d\sigma}{d\Omega} = \frac{1}{4\pi^2\hbar^4}|M_{\rm if}|^2\frac{p_f^2}{v_iv_f} \quad \cdots\cdots (2)$$

Considering spin,

available spin substates for initial state is $g_i = (2S_a+1)(2S_b+1)$

for final state is $g_f = (2S_c+1)(2S_d+1)$

So, multiply $\frac{g_f}{g_i}$ factor to Eq. (2)

$$\Rightarrow \frac{d\sigma}{d\Omega}(a+b \to c+d) = \frac{1}{4\pi^2\hbar^4}|M_{\rm if}|^2\frac{p_f^2g_f}{v_iv_fg_i}$$

■ Decays and Resonances

relationship between mean lifetime and the reaction rate is

$$\begin{cases} \tau = \dfrac{1}{W} \\ \Gamma\tau = \hbar \end{cases} \Rightarrow \Gamma = \frac{\hbar}{\tau} = \hbar W = 2\pi |M|^2 \int \rho_f d\Omega$$

The decay rate for a particle A is equal to fractional number of decays of assembly of A particles per unit time and width of a decaying state,

$$\Gamma = -\hbar \frac{dN_A}{dt}\frac{1}{N_A}$$

└ The negative sign is to make Γ positive.

$$\Rightarrow \frac{dN_A}{N_A} = -\frac{\Gamma}{\hbar}dt$$

$\therefore\ N_A(t) = N_A(0)e^{-\frac{\Gamma}{\hbar}t}$: **radioactive decay law**

The wave function describing a non-stationary decaying state of central angular frequency $\omega_R = \frac{E_R}{\hbar}$ (E_R: resonance energy) is

$$\Psi(t) = \Psi(0)e^{-i\omega_R t}e^{-t/2\tau} = \Psi(0)e^{-(iE_R + \Gamma/2)t}, \text{ in } \hbar = c = 1 \text{ units}$$

So, intensity $\Psi^*(t)\Psi(t) = \Psi(0)^*\Psi(0)e^{-\Gamma t}$ satisfy radioactive decay law

Fourier transform

$$\Psi(\omega) = \int_0^\infty \Psi(t)e^{i\omega t}dt = \int_0^\infty \Psi(t)e^{iEt}dt, \text{ in } \hbar = c = 1 \text{ units}$$

The amplitude as a function of E is

$$\Psi(E) = \int \Psi(t)e^{iEt}dt = \Psi(0)\int e^{-[i(E_R - E) + \Gamma/2]t}dt = \frac{K}{(E - E_R) - i\Gamma/2},$$

K is some constant

Cross-section $\sigma(E)$ measuring the probability of two particles a and b forming a resonant state c is proportional to $\Psi^*(E)\Psi(E) = |\Psi(E)|^2$

$$\Rightarrow \sigma(E) = \sigma_{\max} \frac{\Gamma^2/4}{(E-E_R)^2 + \Gamma^2/4}$$: **Breit-Wigner(BW) resonance formula**

↳ when particles collide, they can form an unstable, broad state, known as 「resonance」. cross-section which measures probability of forming the resonant state follows the BW distribution.

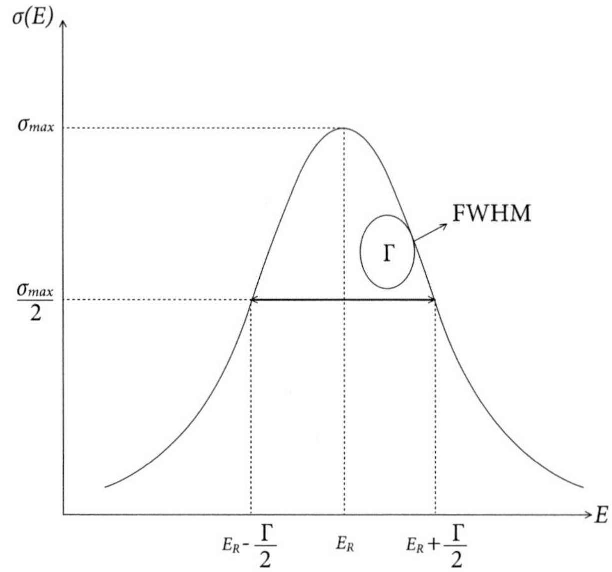

Gaussian: $f = \frac{1}{\sqrt{2\pi}\,\sigma} e^{-\frac{(x-\mu)^2}{2\sigma^2}} \xrightarrow[\text{half max}]{x\text{ giving}} x = \sqrt{2\ln 2}\,\sigma$

$\Rightarrow$ for FWHM, $\Gamma = 2\sqrt{2\ln 2}\,\sigma$

Relativistic Breit-Wigner formula

$$\sigma(S) = \sigma_{\max} \frac{M_0^2 \Gamma^2}{(S - M_0^2)^2 + M_0^2 \Gamma^2},$$

where $E = \sqrt{S}$ is the cms energy, and $M_0 (= E_R)$ is the central resonance mass.

Considering spin, a term in the right side is added $\frac{(2J+1)}{(2S_a+1)(2S_b+1)}$

When spin and angular momentum are considered

$$\sigma = \frac{4\pi\lambda^2(2J+1)}{(2S_a+1)(2S_b+1)}\frac{\Gamma^2/4}{(E-E_R)^2+\Gamma^2/4},$$

where λ is the wavelength of scattered and scattering particle in their common cms.

(※ $p = \hbar k = \frac{h}{2\pi}\frac{2\pi}{\lambda} = \frac{h}{\lambda}$ ⇒ When p is known, λ is determined.)

ex $\pi^+ p \to \Delta^{++}(1232) \to \pi^+ p$

$$J = \frac{3}{2},\ \Gamma = 120\ \text{MeV},\ S_a = \frac{1}{2},\ S_b = 0 \Rightarrow 8\pi\lambda^2 = \sigma_{max}$$

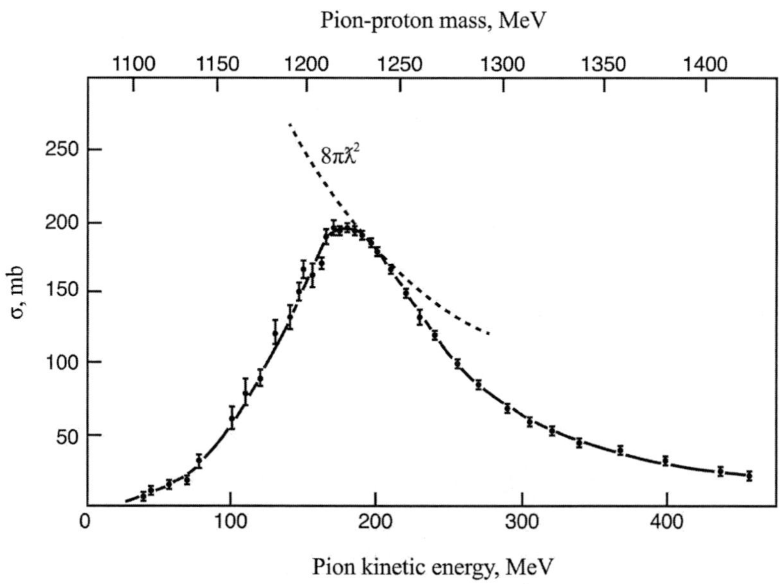

ex Z^0 resonance: more neutrino families there are, the large the cross-section and broader the resonance

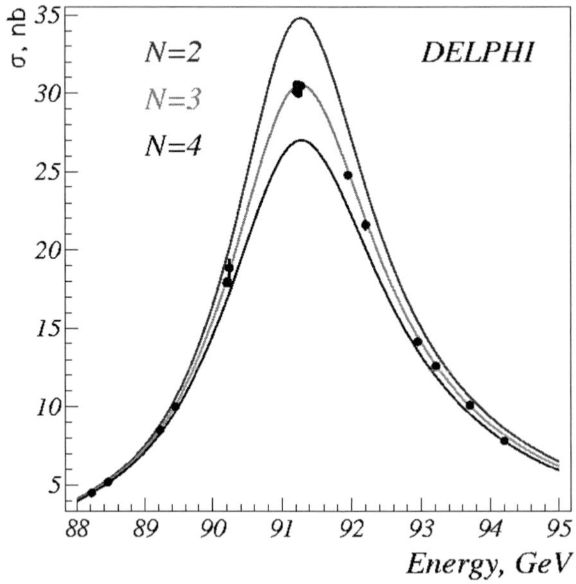

<The Z^0 peak in e^+e^- interactions as measured by DELPHI>

Large Electron-Positron Collider experiments; ALEPH, DELPHI, L3 and OPAL

$$e^+e^- \xrightarrow{Z^0} \ell^+\ell^-,\ \text{hadrons}(q\bar{q}),\ \nu\bar{\nu}$$

circumference: 27 km (total 11 years)

1989.7 ~1994

1995 upgraded for the second operation phase (~ 200 GeV)

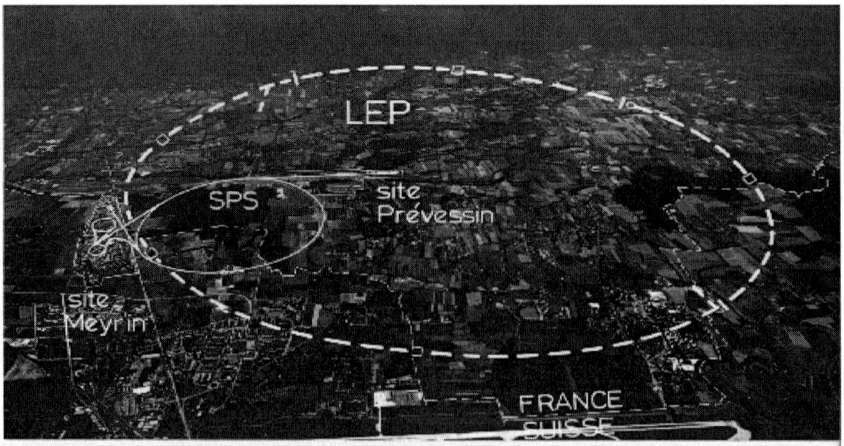

(https://www.icepp.s.u-tokyo.ac.jp/history/lep-e.html)

Chapter 3

Invariance Principles and Conservation Laws

■ Introduction

invariance $\Rightarrow$ symmetry $\begin{cases}\text{continuous} \\ \text{discrete}\end{cases}$

① Continuous symmetries lead to additive conservation laws

- energy, momentum

② Discrete symmetries lead to multiplicative conservation laws

- parity, charge conjugation

※ Noether`s theorem: The symmetry of the action(differentiable) of the physical system corresponds to one conservation law.

ex continuous transformation

energy invariance(symmetry) of equations describing a physical system under

(i) a translation in space $\xrightarrow[\text{no external force}]{\text{isolated system}}$ conservation of linear momentum

$$\llcorner \left(\frac{d\vec{p}}{dt} = \vec{F} = 0 \rightarrow \vec{p} \text{ constant} \right)$$

(ii) a rotation in space → conservation of angular momentum

(iii) translation in time → conservation of energy

(iv) gauge transformation → conservation of charge

ex discrete transformation

parity operation(spatial reflection through the origin of coordinate)

= spatial inversion of coordinates

■ Translation and Rotation operators

(i) for infinitesimal translation δr in space,

$$\Psi(r+\delta r) = \Psi' = \Psi(r) + \delta r\frac{\partial \Psi(r)}{\partial r} \equiv D\Psi(r),$$

where $D = 1 + \delta r\frac{\partial}{\partial r}$ is the infinitesimal space translation operator

$$\Rightarrow D = 1 + \delta r\frac{\partial}{\partial r} = 1 + \frac{i}{\hbar}p\delta r \quad \left(\because\ p = \frac{\hbar}{i}\frac{\partial}{\partial r}\right)$$

(ii) for a finite translation Δr, $\Delta r = n\delta r$ (where $n \to \infty$ as $\delta r \to 0$)

$$D = \lim_{n\to\infty}\left(1 + \frac{i}{\hbar}p\frac{\Delta r}{n}\right)^n \xrightarrow[\text{(binomial theorem)}]{\text{apply}} e^{\frac{i}{\hbar}p\Delta r}$$

$\Rightarrow D^* = e^{-\frac{i}{\hbar}p\Delta r} = D^{-1}$ $(\because\ DD^{-1} = 1)$: D is an unitary operator

(iii) similarly, for infinitesimal rotation about some axis, Z

$$R = 1 + \delta\phi\frac{\partial}{\partial\phi}$$

$$\left[\because\ R\Psi(x,y,z) = \Psi(x+\delta x, y+\delta y, z) = \Psi(x,y,z) + \delta x\frac{\partial\Psi}{\partial x} + \delta y\frac{\partial\Psi}{\partial y} \quad \cdots \quad (1)\right]$$

meanwhile,

$$L_z = (\vec{r}\times\vec{p})_Z = \begin{vmatrix} \hat{x} & \hat{y} & \hat{z} \\ x & y & z \\ \frac{\hbar}{i}\frac{\partial}{\partial x} & \frac{\hbar}{i}\frac{\partial}{\partial y} & \frac{\hbar}{i}\frac{\partial}{\partial z} \end{vmatrix}_z = \frac{\hbar}{i}\left(x\frac{\partial}{\partial y} - y\frac{\partial}{\partial x}\right) \quad \cdots \quad (2)$$

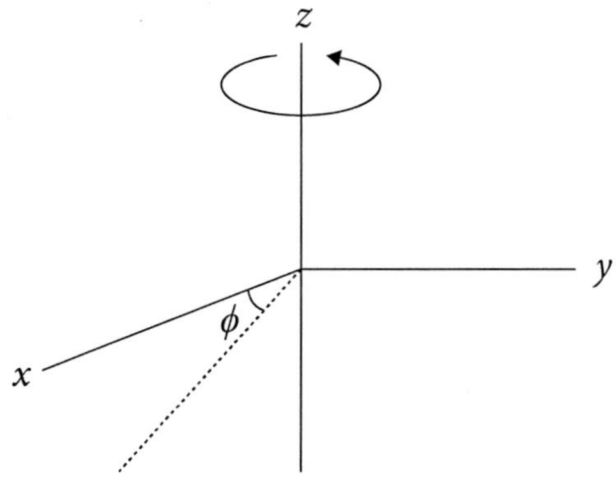

$$\begin{cases} x = r\cos\phi \Rightarrow \delta x = -r\sin\phi\,\delta\phi = -y\delta\phi \\ y = r\sin\phi \Rightarrow \delta y = r\cos\phi\,\delta\phi = x\delta\phi \end{cases}$$

Substituting these into Eq. (1),

$$R\Psi(x,y,z) = \Psi - y\delta\phi\frac{\partial\Psi}{\partial x} + x\delta\phi\frac{\partial\Psi}{\partial y}$$

$$= \left[1+\left(x\frac{\partial}{\partial y} - y\frac{\partial}{\partial x}\right)\delta\phi\right]\Psi = \left(1+\delta\phi\frac{\partial}{\partial\phi}\right)\Psi$$

$$\left[\because \frac{\partial\Psi(x,y)}{\partial\phi} = \frac{\delta\Psi}{\delta x}\frac{\delta x}{\delta\phi} + \frac{\delta\Psi}{\delta y}\frac{\delta y}{\delta\phi} = \left(-y\frac{\partial}{\partial x} + x\frac{\partial}{\partial y}\right) \Rightarrow \frac{\partial}{\partial\phi} = \left(-y\frac{\partial}{\partial x} + x\frac{\partial}{\partial y}\right)\right]$$

$$\therefore\ R = 1+\delta\phi\frac{\partial}{\partial\phi} \quad \cdots \quad (3)$$

Nother theorem:

$$\frac{\partial}{\partial t} < \Psi | A | \Psi > \ = \ < \Psi | \frac{\partial}{\partial t} A\Psi > + < \frac{\partial\Psi}{\partial t} | A\Psi >$$

$$= -i\hbar < \Psi | HA | \Psi > + i\hbar < \Psi | AH | \Psi >$$

$$= -i\hbar < \Psi | [A, H] | \Psi >$$

i.e. if a Hamiltonian is symmetrical with respect to an operator, the measure associated to such operator is conserved, vice versa.

Group name	Matrices in Group
$U(n)$	$n\times n$ unitary $U^{+}=U^{-1}$
$SU(n)$	$n\times n$ unitary with determinant 1
$O(n)$	$n\times n$ orthogonal $\tilde{O}O=1$
$SO(n)$	$n\times n$ orthogonal with $\tilde{O}O=1$

From Eq. (2),

$$\frac{\partial}{\partial\phi}=\frac{i}{\hbar}L_Z \cdots \quad (4)$$

Substituting Eq. (4) into Eq. (3),

$$R=1+\delta\phi\frac{i}{\hbar}L_Z=1+\frac{i}{\hbar}L_Z\delta\phi$$

Since a finite rotation $\Delta\phi$ is obtained by repeating the infinitesimal rotation n times, where $n\rightarrow\infty$ as $\delta\phi\rightarrow 0$, $(\Delta\phi=n\delta\phi)$

$$R=\lim_{n\rightarrow\infty}\left(1+\frac{i}{\hbar}L_z\frac{\Delta\phi}{n}\right)^n=e^{\frac{i}{\hbar}L_Z\Delta\phi}$$

$\Rightarrow R^{*}=R^{-1}=e^{-\frac{i}{\hbar}L_Z\Delta\phi}$; R is the unitary operator

The parity operation(P): discrete symmetry $\Rightarrow$ multiplicative quantum number

= spatial inversion of coordinates: $(x,y,z)\rightarrow(-x,-y,-z)$

$$P\Psi(\vec{r})=\Psi(-\vec{r})$$

$$\Rightarrow PP\Psi(\vec{r})=P\Psi(-\vec{r})=\Psi(\vec{r})$$

$\Rightarrow P^2=1$: P is an unitary operator and the eigenvalue of the operator is ± 1

$P=+1$: even wave function under parity operation

$P=-1$: odd wave function under parity operation

ex for $\Psi=\cos x$

$P\Psi=\cos(-x)=\cos x=+\Psi\Rightarrow\Psi$ is even$(P=+1)$

for $\Psi = \sin x$

$P\Psi = \sin(-x) = -\sin x = -\Psi \Rightarrow \Psi$ is odd$(P = -1)$

for $\Psi = \cos x + \sin x$

$P\Psi = \cos(-x) + \sin(-x) = \cos x - \sin x \neq \pm\Psi$

$\rightarrow$ therefore, Ψ has no definite parity eigenvalue

For hydrogen atom wave function,

wave function=radial part $\times$ angular part

$\Rightarrow \Psi(r, \theta, \phi) = u(r) Y_\ell^m(\theta,\phi)$, where $Y_\ell^m(\theta,\phi)$ is the spherical harmonics

※ The spherical harmonics describe a state in a spherically symmetric potential with definite angular momentum

$$= u(r)\sqrt{\frac{(2\ell+1)(\ell-m)!}{4\pi(\ell+m)!}}\, P_\ell^m(\cos\theta)e^{im\phi},$$

where P_ℓ^m is the associated Legendre function

spatial inversion $\vec{r} \rightarrow -\vec{r}$ is equivalent to

$$\theta \rightarrow \pi - \theta \text{ and } \phi \rightarrow \pi + \phi$$

so, under spatial inversion

$e^{im\phi} \rightarrow e^{im\pi}e^{im\phi} = (-1)^m e^{im\phi}$ and

$P_\ell^m(\cos\theta) \rightarrow P_\ell^m[\cos(\pi-\theta)] = (-1)^{\ell+m} P_\ell^m(\cos\theta)$

$\Rightarrow Y_\ell^m(\theta,\phi) \rightarrow Y_\ell^m(\pi-\theta, \pi+\theta) = (-1)^\ell Y_\ell^m(\theta,\phi)$:

spherical harmonic functions have parity $(-1)^\ell$, i.e. $P = (-1)^\ell$

$$\Rightarrow \begin{cases} S,\ D,\ G,\ \cdots,\cdots: \text{atomic states have even parity} \\ P,\ F,\ N,\ \cdots,\cdots: \text{atomic states have odd parity} \end{cases}$$

ex parity is equal to product of parts

└ two pions in angular momentum ℓ state $P = P(\pi_1) \cdot P(\pi_2) \cdot (-1)^\ell$

■ Photon parity

Electric dipole(dominant effect of interaction of electron in atom with electromagnetic field) transitions between states have angular momentum selection rule $\Delta\ell = \pm 1$

⇒ So, as a result of transition, parity atomic state must change

⇒ parity of the electromagnetic($E1$) radiation(photons) emitted in this case must be -1, so the parity of the whole system(atom+photon) is conserved.

└ photon: $J^P = 1^-$

(※ The transitions between atomic levels with a single photon emission are of the electric dipole type. For them, the rule $\Delta\ell = \pm 1$ applies.

From a property of spherical harmonics, two levels have opposite parity.

⇒ For parity conservations, parity of photon is negative)

For strong interaction,

(i) $p + p \rightarrow \pi^+ + p + n$

by convention(definition), parties of neutron and proton are $+1$

└ Since baryon is preserved, the parity of protons and neutrons can be the same value on the left and right sides, and conventionally, it is chosen as $+$.

⇒ $P_\pi = -1$

(ii) $p + p \rightarrow K^+ + \Lambda + p$

$(\bar{s}u)$ (sud)

$S = +1$ $S = -1$

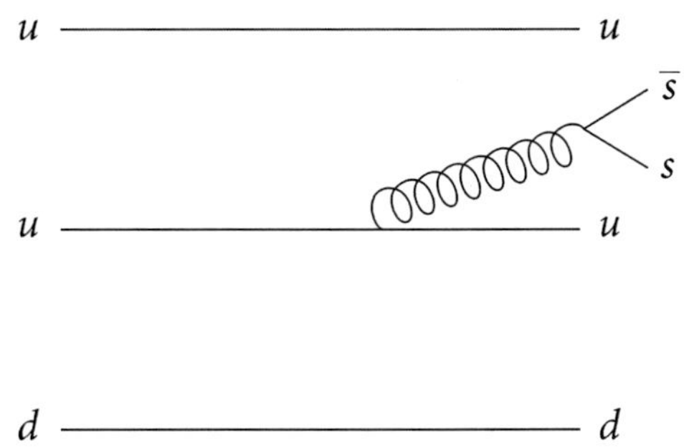

strange particles must be created in association

└ parity of ΛK^+ pair is measured

⇒ odd parity

⇒ hyperon Λ is assigned even parity as the nucleon

⇒ parity of K^+ is odd

■ Pion spin and parity($J^P = 0^-$)

Let\`s determine spin and parity of the charged and neutral pions.

(i) Spin

reversible reaction: invariant under time reversal and space inversion

$$p + p \rightleftarrows \pi^+ + d$$

$\Rightarrow |M_{\mathrm{if}}|^2 = |M_{\mathrm{fi}}|^2$

From $\dfrac{d\sigma}{d\Omega}(a + b \rightarrow c + d) = \dfrac{1}{4\pi^2\hbar^4}|M_{\mathrm{if}}|^2 \dfrac{p_f^2}{v_i v_f}$,

total number of spin substates available is $g_f = (2S_c + 1)(2S_d + 1)$ for unpolarized a and b

$$\sigma_{pp \rightarrow \pi^+ d} \propto (2S_\pi + 1)(2S_d + 1)p_\pi^2,$$

where $p_\pi = p_d$ in cms momentum and $S_d = 1$(deutron spin is 1)

$$\sigma_{\pi^+ d \rightarrow pp} \propto \frac{1}{2}(2S_p + 1)^2 p_p^2,$$

└ two protons in the final state are identical

※ for $K_\pi \approx 20$ MeV, predicted $\sigma \approx 3$ mb for $S_\pi = 0$

$\sigma \approx 1$ mb for $S_\pi = 1$

experimentally measured value is $\sigma \approx 3-4$ mb

⇒ From experiment of $\dfrac{\sigma_{pp \to \pi^+ d}}{\sigma_{\pi^+ d \to pp}}$, $S_{\pi^+} = 0$ was obtained.

(ii) for neutral pion,

$\pi^0 \to 2\gamma$, where $S_\gamma = 1$ ⇒ pion spin should be integral

total photon`s z-component: $S_z = 0$ or 2

(※ along the flight path of γs in the π^0 rest frame)

ⓐ If $S_{\pi^0} = 1$, $S_z = 0$ is only possible

└ (※ two-photon amplitude must behave under spatial rotations like $P_1^m(\cos\theta)$ with $m = 0$)

⇒ situation corresponds to two right(left)-circularly polarized photons travelling in opposite directions

⇒ under a 180° rotation about an axis normal to z, $\theta \to \pi - \theta$, $P_1^0 \propto \cos\theta$ changes sign

⇒ contradict with two identical bosons must be symmetric by Bose statistics

ⓑ therefore, $S_{\pi^0} = 0$ or $S_{\pi^0} \geq 2$

└ ruled out by pion production statistics

In high energy interactions, positive, negative and neutral pions are produced in equal numbers.

⇒ $S_{\pi^0} = 0$; with same spin multiplicity as its charged counterparts

(iii) Parity

ⓐ slow negative pions captured in deuterium, capture takes place from an atomic S-state of the pion

(※ deuterium(D or ^2H): one proton and one neutron = heavy hydrogen)

$\pi^- + d \rightarrow n + n,$

where $S_d = 1$, because of inside deutron, $P_d = (-1)^{\ell_d} P_n P_p = +1$

initial state has angular momentum $J = 1$

$$[\because J = L + S = 0 + (0+1) = 1] \quad S_\pi \; S_d$$

wave function of two neutrons,

(※ $L+S$ must be even by antisymmetric requirement on identical fermions

$\Rightarrow {}^{2S+1}J_L = {}^3P_1$

initial state is $P_d P_\pi (-1)^\ell = (+1) P_d P_\pi = -1$)

$\Psi = \phi_{(space)} \alpha_{(spin)}$ and total spin $S = 0$ or 1

then, there are four possible combinations of two neutrons:

spin triplet	$\alpha(1,1) = \uparrow\uparrow$	symmetric under spin exchange
	$\alpha(1,0) = \frac{1}{\sqrt{2}}(\uparrow\downarrow + \downarrow\uparrow)$	
	$\alpha(1,-1) = \downarrow\downarrow$	
spin singlet	$\alpha(0,0) = \frac{1}{\sqrt{2}}(\uparrow\downarrow - \downarrow\uparrow)$	antisymmetric under spin exchange

$\Rightarrow$ symmetry of spin function $(-1)^{S+1}$

Meanwhile, for spatial function ϕ,

symmetry under interchange of space coordinates is $(-1)^L$

$\Rightarrow$ overall symmetry under total interchange (space and spin) $= (-1)^{L+S+1}$

$\Rightarrow$ Since we are dealing with identical fermions, Ψ must be antisymmetric under interchange.

$\Rightarrow$ $L+S$ must be even and J takes on values $L+S$, $L+S-1$, ……, $|L-S|$

$\Rightarrow$ $J=1$ can be form by combining state with

	L	S
$J=1$	0	1
	1	0
	1	1
	2	1

$\Rightarrow$ state can be $\left[{}^1S_0,\ {}^3P_{0,1,2},\ {}^1D_2,\ {}^3F_{2,3,4},\ \cdots\right]$

	ℓ
S: Sharp	0
P: Principal	1
D: Diffuse	2
F: Fundamental	3

$\Rightarrow$ only possible state $L=1$ and $S=1$

$\Rightarrow$ two neutrons are ${}^3P_1$ state with parity $(-1)^L=-1$

($\therefore$ Since both neutrons and deuteron have intrinsic parity $+1$, only way to obtain negative parity in the initial state is $P_\pi=-1$)

$$J^P=\begin{cases}0^+ & : \text{scalar particle}\\ 0^- & : \text{pseodoscalar}\\ 1^- & : \text{vector}\\ 1^+ & : \text{axial vector}\end{cases}$$

ⓑ $\pi^0 \rightarrow 2\gamma$ for the parity of the neutral pion

initial state π^0 has $J=0$, and the final state consists of identical bosons:

(※ simplest wave functions describing two-photon system are written in terms of $\vec{k},\ \vec{\epsilon}_1,\ \vec{\epsilon}_2$)

$\Psi_1(2\gamma) = A(\vec{\epsilon_1} \cdot \vec{\epsilon_2}) \propto \cos\phi$,

where A is the constant, $\vec{\epsilon_1}$ and $\vec{\epsilon_2}$ are the polarization vectors, and ϕ is the angle between polarization planes.

Ψ_1 is the scalar $\Rightarrow$ even under the space inversion

$\Rightarrow$ positive parity for the pion ($P = +1$)

$\Psi_2(2\gamma) = B(\vec{\epsilon_1} \times \vec{\epsilon_2}) \cdot \vec{k} \propto \sin\phi$,

where B is the constant, $\vec{k}$ is the momentum vector of the photon in the pion rest frame.

Ψ_2 is scalar $\Rightarrow$ odd under the space inversion

$\Rightarrow$ "pseudoscalar" quantity (O^-)

$\Rightarrow$ negative parity for the pion ($P = -1$)

Intensities or rates $\sim |\Psi|^2$

$$|\Psi_1|^2 \sim \cos^2\phi = \frac{1}{2}(1 + \cos 2\phi) \text{ or}$$

$$|\Psi_2|^2 \sim \sin^2\phi = \frac{1}{2}(1 - \cos 2\phi)$$

Experiment; rare decay of pion $[\pi^0 \rightarrow (e^+ + e^-) + (e^+ + e^-)]$

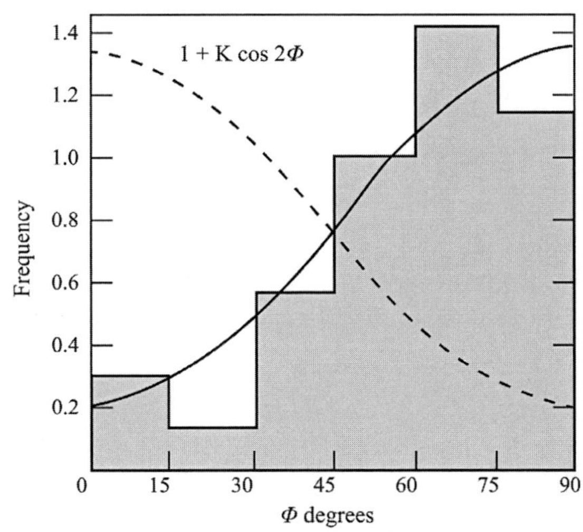

$\pi^0 \to (e^+ + e^-) + (e^+ + e^-)$: $BR = \alpha^2 \simeq 10^{-4}$: pion is pseudoscalar

└ planes of pairs follow E vectors of internally converted photons($\pi^0 \to \gamma\gamma$)

$\Rightarrow$ pion is pseudoscalar

■ Parity of particles and antiparticles

Dirac theory of fermions requires that particle(fermions) and antiparticle (antifermions) to have opposite intrinsic parity.

This is checked in the decay $e^+e^-({}^1S_0) \to 2\gamma$:

spin-singlet ground state of positronium decays to two photons

by Dirac theory, e^+e^- has opposite parity

$\Rightarrow$ two photon state has negative parity

$\Rightarrow$ preferentially orthogonal polarization of photons (such as $\pi^0 \to 2\gamma$)

$\Rightarrow$ $\sin^2\phi$ dependence (ϕ is the azimuthal angle between polarization vectors)

If orbital angular momentum L is involved in the two-particle system

① parity of the fermion-antifermion pair is $(-1)^{L+1}$

$\Leftarrow (-1)^L(-1)$

└ due to space inversion

due to opposite intrinsic parity

ex for $\pi^+(u\bar{d})$

for S-state combination, intrinsic parity $= -1$

Meanwhile,

② boson and antiboson have the same parity

$\Rightarrow$ $\pi^+\pi^-$ system in a state of angular momentum L has $(-1)^L$ parity

so, $J = L + S$

$\Rightarrow 1 = L + 0 \Rightarrow L = 1$

└ spin of two pion system

$\Rightarrow \rho^0 \rightarrow \pi^+\pi^-$, total angular momentum of $\rho = 1$ ($J = 1$)

ρ meson, of mass 770 MeV and $J = 1$ (ρ is meson, so spin is integer)

$\Rightarrow J^p = 1^-$

This is exactly the case of π^0 decay

- the photon polarizations must have $\sin^2\phi$ form

$\frac{\text{Rate}(90^\circ)}{\text{Rate}(0^\circ)} = 2.04 \pm 0.08$, expect 2.00 $\Rightarrow$ conclude $P_{(e^+)} = -P_{(e^-)}$

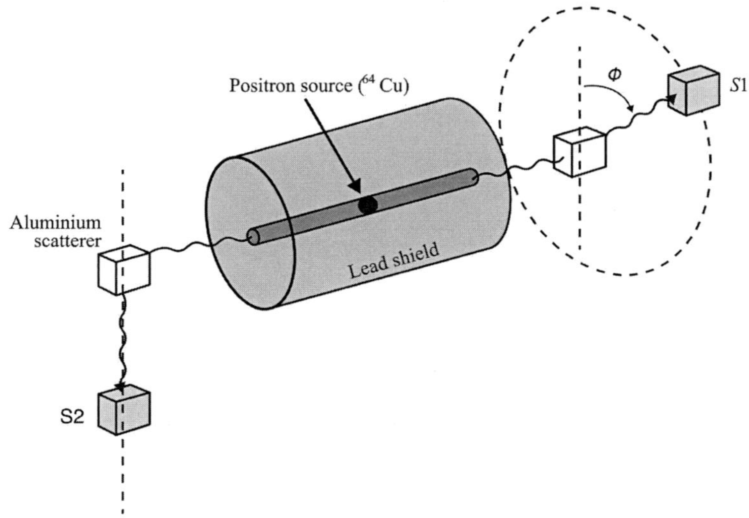

■ Tests of parity conservation

- parity is conserved in the strong and electromagnetic interactions but is not conserved in the weak interaction.

※ parity conservation is not a universal law of physics

※ in experiment, tiny degrees of parity violation are observed, because Hamiltonian also contains contributions from weak interactions between particles involved.

- for neutrino, only LH neutrino is observed in nature

※ this is a maximally violated symmetry

upon the spatial inversion, $\vec{p}$ is reversed but $\vec{\sigma}$ is unchanged

⇒ RH neutrino state under parity operation

⇒ weak interaction is not invariant under the spatial inversion and do not conserve parity

■ Summary of Parity: talk of parity of state only if it is eigenstate of Parity.

Vacuum is such a state and its parity is set positive by definition.

- under parity operation

$\vec{r} \to -\vec{r}$: coordinates

$t \to t$: time is not changed

$\vec{p} \to -\vec{p}$: $\because \vec{p} = m\frac{d\vec{r}}{dt}$

$\vec{L} \to \vec{L}$: $\because \vec{L} = \vec{r} \times \vec{p}$, (similarly $\vec{\sigma} \to \vec{\sigma}$)

$\vec{S} \to \vec{S}$: spin is not changed

More generally,

- scalar quantities remain unchanged (0^+)
- pseudoscalars change their sign (0^-)
- vectors change sign (1^-)
- axial vectors do not (1^+)

Quantum Field theory requires fermions and their antifermions to have opposite parities, and requires bosons and their antibosons to have the same parity.

$\Rightarrow$ general rule at the quark level: all quarks have positive parity, antiquarks have negative parity

Parity conservation is not an universal law of physics.

$\Rightarrow$ The weak interaction does not conserve parity.

$A \rightarrow B + C$

J assumes spin of decaying particle

$\Rightarrow$ conservation of the angular momentum requires that total angular momentum of final state must be J

$\Rightarrow$ if two decay products are the spinless, their relative angular momentum $(\ell) = J$ (i.e. equal spin of A) ($\because$ $J = L + S \Rightarrow J = L + 0$)

$\Rightarrow \ell = J$

$\Rightarrow$ conservation of parity in the decay implies

$$P_A = P_B P_C (-1)^{\ell} = P_B P_C (-1)^J = P_B P_C$$

(if decaying particle has spin zero, $J = 0$)

$\Rightarrow P_A = P_B P_C$

$$\Rightarrow \text{allowed decays: } \begin{cases} 0^+ \rightarrow 0^+ + 0^+ \\ 0^+ \rightarrow 0^- + 0^- \\ 0^- \rightarrow 0^+ + 0^- \end{cases}$$

■ Charge conjugation invariance

charge conjugation(reverses the signs of charge and magnetic moment of a particle, leaving other coord. unchanged) is invariant in the strong and electromagnetic interactions but is not invariant in the weak interaction.

$$C|\pi^0> \ = \eta|\pi^0>,$$

where C is the charge conjugation operator and η is the constant.

$(\Rightarrow C^2|\pi^0> \ = \eta C|\pi^0> \ = \eta^2|\pi^0> \ \Rightarrow \eta^2 = 1)$

so,

$$C|\pi^0> = \pm|\pi^0>$$

※ only neutral bosons are the eigenstates of C

$\because$ $C|\pi^+> = |\pi^->$, $C|\pi^-> = |\pi^+>$

To find the sign,

Note that electromagnetic fields are produced by moving charges, that change sign under C.

⇒ photons has $C=-1$: charge conjugation eigenvalue of the γ is -1

and

$$\pi^0 \rightarrow 2\gamma$$

⇒ $C|\pi^0> = +|\pi^0>$: remember that this topic is related with G-parity

(※ $C|\eta^0> = +|\eta^0>$: η also decays into two photons)

In experiment, $\frac{\pi^0 \rightarrow 3\gamma}{\pi^0 \rightarrow 2\gamma} < 3\times10^{-8}$ (see PDG book)

and $\frac{\eta \rightarrow 3\gamma}{\eta \rightarrow 2\gamma} < 4\times10^{-5}$

Under C operation, LH neutrino → LH antineutrino, which is not found in nature.

⇒ Charge conjugation is not invariant in the weak interaction.

However, under the combined operation CP

LH neutrino → RH antineutrino

⇒ CP is (approximately) conserved in the weak interaction

⇒ CP violation in the weak interaction occurs at 10^{-4} level

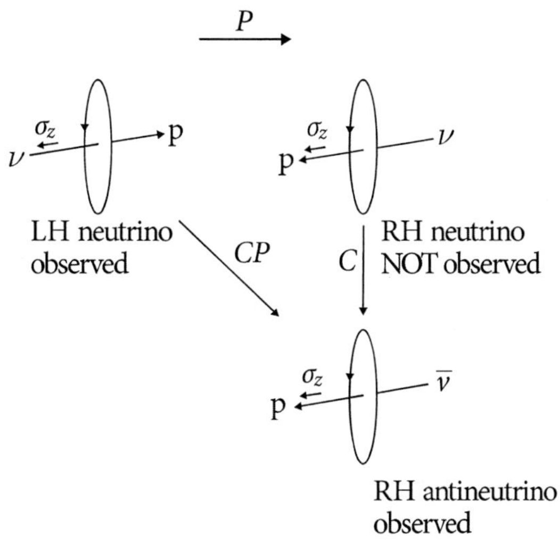

Charge conservation and gauge invariance

experimental limit for change non-conservation,

$$\frac{n \rightarrow p\nu_e\bar{\nu}_e}{n \rightarrow pe^-\bar{\nu}_e} < 9\times10^{-24}$$

conservation quantity	↔	invariance principle (gauge transformation)
linear momentum		translation in space
angular momentum		rotation
charge		electromagnetic field (long-range field)
energy		time

In a field theory with gauge in variance, there is a high degree of symmetry, and this results in the systematic cancellation of divergence between different diagram, order by order

⇒ such a renormalizable theory predicts cross-sections and decay rates that are finite and calculable to all orders in the coupling constant.

■ Baryon and lepton conservation

※ Baryon conservation is established to a very low level of violation experimental limits for baryon number B and lepton number L non-conservation,

$\tau(p \rightarrow e^{+}\pi^{0}) > 8\times10^{33}$ yr: limits on the mean lifetime

$\tau({}^{76}Ge \rightarrow {}^{76}Se + 0\nu + e^{-} + e^{-}) > 10^{26}$ yr: neutrinoless double beta decay

└ Selenium

In our present cold universe, B and L numbers are conserved but there are no deep theoretical reasons to suppose that this conservation rule is absolute.

(ref. PDG "Baryon Conservation")

(ref. PDG "Lepton Conservation")

■ CPT invariance (ref. PDG "CPT Invariance":)

(※ consequency is that masses, lifetime, moments, etc, of particles and antiparticles must be identical)

⇒ search possible mass difference between proton and antiproton

$$\frac{|m_p - m_{\bar{p}}|}{m_p} \leq 10^{-8}$$

This theorem states that all interactions are invariant under the successive operation of C(charge conjugation), P(parity operation), and T(time reversal) taken in any order.

▶ CP violation and T violation

└ long-lived neutral kaon violates CP with a rate of 2×10^{-3}

$$K_L^o = \frac{1}{\sqrt{2}}(|\,K^o> - |\,\overline{K^o}>)$$

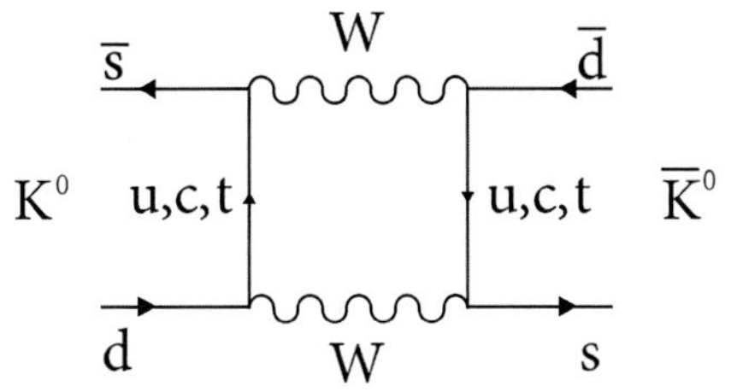

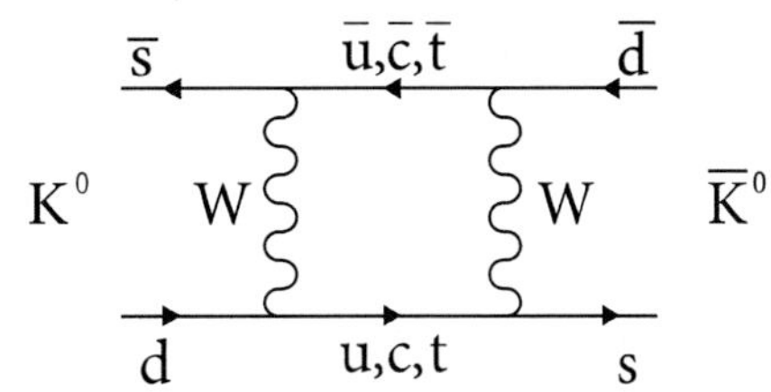

$K_L^o \rightarrow \pi\pi\pi$ $(> 99\%)$

CP -1 -1

$K_L^o \rightarrow \pi\pi$ (0.2%)

CP -1 +1

Until 1964, it was believed that, although the weak interactions were known to violate C and P invariance separately, CP symmetry was respected.

$K^o \rightarrow 3\pi$ with CP eigenvalue –1

CP -1 -1

$\rightarrow 2\pi$ with CP eigenvalue +1, with probability 2×10^{-3} CP violation

+1

- SM of CP violation explains its origin from CP-violating phase in mixing matrix of six quark flavors
- CP violation implies T violation (PDG book) through the CPT theorem (CPT can be invariant)

Experimental limits on T violation $< 10^{-3}$

ex $\mu^+ \rightarrow e^+ + \nu_e + \bar{\nu}_\mu$

$n \rightarrow p + e^- + \bar{\nu}_e$

Observation of the polarization transverse to a plane defined by the momentum vectors of the secondary particles is a signal for T violation.

■ **Neutron electric dipole moment (EDM):** the most sensitive tests of T violation come from searches for the neutron and electron dipole moments

① electron $(0.07 \pm 0.07) \times 10^{-26}$ e · cm

② neutron $< 0.29 \times 10^{-25}$ e · cm

electric dipole moment $\vec{\sigma} \cdot \vec{E}$

operator	$\vec{\sigma}$	$\vec{E}(= -\vec{\nabla} U)$	$\vec{\sigma} \cdot \vec{E}$
P	$\vec{\sigma}$	$-\vec{E}$	$-\vec{\sigma} \cdot \vec{E}$
T	$-\vec{\sigma}$	$\vec{E}$	$\vec{\sigma} \cdot \vec{E}$

⇒ EDM violates both P and T symmetries

Existence of these electric dipole moments would also violate parity.

On dimensional grounds, (How big might we expect EDM to be?)

EDM=charge(e) × length(ℓ) × T violating parameter(f)

$$\ell \sim \frac{e^2 M}{M_W^2} = \frac{4\pi\alpha M}{M_W^2} = 1.4 \times 10^{-5}\ \mathrm{GeV}^{-1} \simeq 10^{-19}\ \mathrm{cm},$$

where $\alpha = 1/137$, $M = 0.94$ GeV, $M_W = 80$ GeV

$f = 10^{-3}$, level of CP violation in K^0 decay

Then,

EDM $\sim e(10^{-19}\mathrm{cm})(10^{-3}) = 10^{-22}$ e · cm, the length is very short compared with the radius(10^{-15} m) of the neutron

		Effect of T	Effect of P
position	r	r	-r
momentun	p	-p	-p
spin	$\sigma = (r \times p)$	$-\sigma$	σ
electric field	E $(= -\nabla V)$	E	-E
magnetic field	B	-B	B
mag. dip. mom.	$\sigma \cdot B$	$\sigma \cdot B$	$\sigma \cdot B$
el. dipl. mom.	$\sigma \cdot E$	$-\sigma \cdot E$	$-\sigma \cdot E$
long. pol.	$\sigma \cdot p$	$\sigma \cdot p$	$-\sigma \cdot p$
trans. pol.	$\sigma \cdot (p_1 \times p_2)$	$-\sigma \cdot (p_1 \times p_2)$	$\sigma \cdot (p_1 \times p_2)$

$\vec{\nabla} \cdot \vec{E} = 4\pi\rho_e$ (in here, ρ_e is the even under time reversal)

$\vec{\nabla} \cdot \vec{B} = \dfrac{\partial \vec{E}}{\partial t} + \vec{J}$ (in here, $\vec{J}$ is the odd under time reversal)

■ Young's tableau(타블로)

$\boxed{N}$: fundamental representation of SU(N)

$\overline{N}$ =

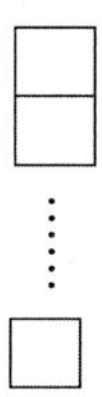

$\Rightarrow$ column of length $N-1$ represent $\overline{N}$

$\Rightarrow$ $SU(3)$ $\square = 3$

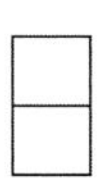
$= \overline{3}$

$\Rightarrow$ $SU(2)$ $\square = 2$

※ A way to number a box

N	N+1	N+2
N-1	N	N+1
N-2	N-1	N

- a value in the denominator: a number of boxes passing per hook
- a value of a numerator: a number of in boxes
- $\text{dim} = \dfrac{\text{numerator}}{\text{denominator}}$

ex-1

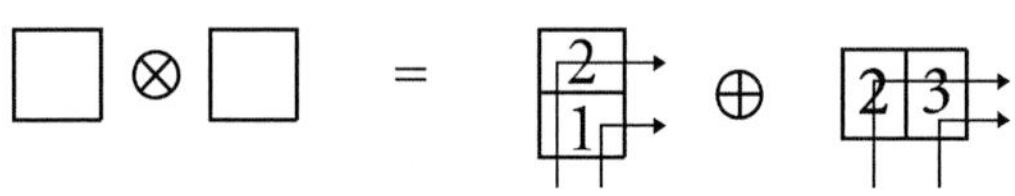

$$2 \otimes 2 \qquad \frac{2}{1 \cdot 2} \qquad \frac{2 \cdot 3}{1 \cdot 2}$$

$$1 \oplus 3$$

singlet triplet

ex-2

$3 \otimes \bar{3} \Rightarrow$

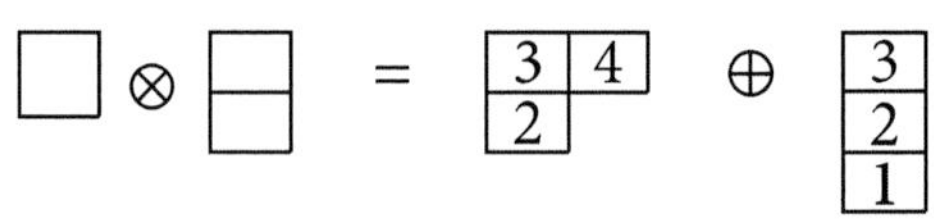

$$\frac{3 \cdot 4 \cdot 2}{1 \cdot 1 \cdot 3} \qquad \frac{1 \cdot 2 \cdot 3}{1 \cdot 2 \cdot 3}$$

$$8 \oplus 1$$

octet singlet

■ **Isospin symmetry** (1932, Heisenberg, think of proton and neutron as two charge states of nucleon)

- Isospin is conserved quantum number in the strong interactions but is not conserved in the electromagnetic interactions

A nucleon(proton, neutron) is ascribed a quantum number, isospin(I), with $I=\frac{1}{2}$

Then,

$$\text{charge}=\frac{Q}{e}=\frac{1}{2}+I_3 \quad \text{if we assign } I_3=+\frac{1}{2} \text{ for proton}$$

$$I_3=-\frac{1}{2} \text{ for neutron}$$

Isospin conservation corresponds to invariance of the length of this vector under rotation of the coordinate axes in isospin space.

for the pion (an isospin triplet $I=1$)

$$\pi^+ = u\bar{d} \quad (I_3=+1)$$

$$\pi^- = d\bar{u} \quad (I_3=-1)$$

$$\pi^0 = \frac{1}{\sqrt{2}}(d\bar{d}-u\bar{u}) \quad (I_3=0)$$

- isospin symmetry applies between all baryons and mesons which transform one to another by interchange of u and d quarks.
- within quark model, we can think of this symmetry as being a symmetry between u and d quarks, $I_3(u)=\frac{1}{2}$, $I_3(d)=-\frac{1}{2}$ because $\begin{cases} p=duu \\ n=ddu \end{cases}$
- masses of pions are similar, ($m_{\pi^\pm}=140$ MeV, $m_{\pi^0}=135$ MeV) reflecting the similar masses of u and d quarks

Isospin in two-nucleon and pion-nucleon systems

(I) for two nucleons, (this is completely analogous to combination of two spin $\frac{1}{2}$ states)

$$\begin{cases} \chi(1,1)=p(1)p(2) \\ \chi(1,0)=\frac{1}{\sqrt{2}}[p(1)n(2)+n(1)p(2)] \\ \chi(1,-1)=n(1)n(2) \end{cases}$$

$\Rightarrow I=1$ triplet, symmetric under interchange $1\leftrightarrow 2$

$$\chi(0,0) = \frac{1}{\sqrt{2}}[p(1)n(2) - n(1)p(2)]$$

$\Rightarrow I = 0$ singlet, antisymmetric under interchange

$\xrightarrow[\text{theory}]{\text{group}}$ $2 \otimes 2 = 1 \otimes 3$; Combining these doublets yields a triplet plus a singlet

└ two component doublet

RHS: 1(← singlet), 3(← triplet)

Total wave function for two-nucleon state:

$$\Psi_{(\text{total})} = \phi_{(\text{space})}\alpha_{(\text{spin})}\chi_{(\text{isospin})}$$

for deuteron(pn), which has spin 1

α is symmetric under interchange of two nucleons

ϕ has $(-1)^{\ell}$ symmetry $\Rightarrow$ for deuteron, $\ell = 0$

$\Rightarrow \phi$ is symmetric

$\Psi_{(\text{total})}$ should be antisymmetric $\Rightarrow \chi$ must be antisymmetric

$\Rightarrow$ deuteron is a isosinglet (i.e. $I = 0$)

① $p + p \rightarrow d + \pi^{+}$

$I =$ 1 0 1

$p(1)p(2) = \chi(1, 1)$

② $p + n \rightarrow d + \pi^{0}$

$I =$ 1(50%) 0 1

or 0(50%)

$\Leftarrow$ 1(50%) from isospin conservation

reaction proceeds only through the $I = 1$ channel because the isospin conserved in the strong interaction.

$$\Rightarrow \therefore \frac{\sigma_{②}}{\sigma_{①}} = \frac{1}{2}$$

※ cross-section is proportional to the square of the matrix element connecting initial and final states

$$\sigma \propto < \Psi_f | H\Psi_i > ^2 = M_{if}^2 \Rightarrow M_1 = < \Psi_f(1) | H_1 | \Psi_i(1) >,$$

where 1 is the isospin value.

(II) pion-nucleon scattering: $I_{total} = I_\pi + I_N = 1 + \frac{1}{2} = \frac{3}{2}$ or $\frac{1}{2}$

Clebsch-Gordon coefficients in pion-nucleon scattering

	$I=\frac{3}{2}$				$I=\frac{1}{2}$	
	$I_3=\frac{3}{2}$	$I_3=\frac{1}{2}$	$I_3=-\frac{1}{2}$	$I_3=-\frac{3}{2}$	$I_3=\frac{1}{2}$	$I_3=-\frac{1}{2}$
$\pi^+ p$	1					
$\pi^+ n$		$\sqrt{\frac{1}{3}}$			$\sqrt{\frac{2}{3}}$	
$\pi^0 p$		$\sqrt{\frac{2}{3}}$			$-\sqrt{\frac{1}{3}}$	
$\pi^0 n$			$\sqrt{\frac{2}{3}}$			$\sqrt{\frac{1}{3}}$
$\pi^- p$			$\sqrt{\frac{1}{3}}$			$-\sqrt{\frac{2}{3}}$
$\pi^- n$				1		

※ Apply the lowering operator to $\left|1 \ \frac{1}{2}\right\rangle \Rightarrow \left|1 \ -\frac{1}{2}\right\rangle$ and $\left|0 \ \frac{1}{2}\right\rangle$

Consider an application of isospin conservation:

(a) $p+p \rightarrow d+\pi^0$ (isospin of final state is 1) $\Rightarrow$ initial state $I=1$

(b) $p+n \rightarrow d+\pi^0$ (isospin of final state is 1) $\Rightarrow$ initial state $I=0$ or 1

($\because$ CG coeff. say 50%, 50%)

$$\Rightarrow \therefore \ \frac{\sigma_b}{\sigma_a} = \frac{1}{2}$$

Cross-section is proportional to the square of the matrix element connecting initial and final states

$\Rightarrow \sigma \propto \ <\Psi_f | H | \Psi_i>^2 = M_{\text{if}}^2$, where H is the isospin operator

$$M_1 = <\Psi_f(1/2) | H_1 | \Psi_i(1/2)> \quad (\because \text{ Possible } I \text{ value is } 3/2 \text{ or } 1/2)$$

$$M_3 = <\Psi_f(3/2) | H_3 | \Psi_i(3/2)>$$

Consider pion-nucleon elastic scattering:

for ⓐ $\pi^+ p \rightarrow \pi^+ p$, $\sigma_a = K| M_3 |^2$ is purely $I = \frac{3}{2}$, K is some constant

$$|\Psi_i> = |\Psi_f> = |\chi\left(\frac{3}{2}, \frac{3}{2}\right)>$$

for ⓑ $\pi^- p \rightarrow \pi^- p$, mixture of $I = \frac{1}{2}$ and $I = \frac{3}{2}$

$$|\Psi_i> = |\Psi_f> = \sqrt{\frac{1}{3}}\, |\chi\left(\frac{3}{2}, -\frac{1}{2}\right)> - \sqrt{\frac{2}{3}}\, |\chi\left(\frac{1}{2}, -\frac{1}{2}\right)>$$

$$\text{so, } \sigma_b = K\,|<\Psi_f|\,(H_1 + H_3)\,|\Psi_i>|^2 = K\left|\frac{1}{3}M_3 + \frac{2}{3}M_1\right|^2$$

($\because$ A channel with equal I value is only allowed.)

for ⓒ $\pi^- p \rightarrow \pi^- n$ (charge exchange), mixture of $I = \frac{1}{2}$ and $I = \frac{3}{2}$

$$|\Psi_i> = \sqrt{\frac{1}{3}}\, |\chi\left(\frac{3}{2}, -\frac{1}{2}\right)> - \sqrt{\frac{2}{3}}\, |\chi\left(\frac{1}{2}, -\frac{1}{2}\right)>$$

$$|\Psi_f> = \sqrt{\frac{2}{3}}\, |\chi\left(\frac{3}{2}, -\frac{1}{2}\right)> + \sqrt{\frac{1}{3}}\, |\chi\left(\frac{1}{2}, -\frac{1}{2}\right)>$$

$$\text{so, } \sigma_c = K\left|\sqrt{\frac{2}{9}}\,M_3 - \sqrt{\frac{2}{9}}\,M_1\right|^2$$

$$\Rightarrow \therefore\ \sigma_a : \sigma_b : \sigma_c = |M_3|^3 : \frac{1}{9}|M_3 + 2M_1|^2 : \frac{2}{9}|M_3 - M_1|^2$$

If one or other isospin amplitude dominates under experimental conditions,

ex if $M_3 \gg M_1$, $\sigma_a : \sigma_b : \sigma_c = 9 : 1 : 2$

if $M_1 \gg M_3$, $\sigma_a : \sigma_b : \sigma_c = 0 : 2 : 1$

Δ^+ ($M_{\Delta^+} \approx 1232$ MeV, $I(J^P) = \frac{3}{2}\left(\frac{3}{2}^+\right)$) decays to p or n: strong interaction dominates

$$|\Delta^+> = |\frac{3}{2}, \frac{1}{2}> \qquad |\Delta^{++}> = |\frac{3}{2}, \frac{3}{2}>$$

$$|p> = |\frac{1}{2}, \frac{1}{2}> \qquad |\Delta^+> = |\frac{3}{2}, \frac{1}{2}>$$

$$|n> = |\frac{1}{2}, -\frac{1}{2}> \qquad |\Delta^0> = |\frac{3}{2}, -\frac{1}{2}>$$

$$|\pi^+> = |1, 1> \qquad |\Delta^-> = |\frac{3}{2}, -\frac{3}{2}>$$

$$|\pi^0> = |1, 0>$$

$$|\pi^-> = |1, -1>$$

(reference ※ PDG C-G coeff.)

$$|\frac{3}{2}, \frac{1}{2}> = \underbrace{\frac{1}{\sqrt{3}} |\frac{1}{2}, -\frac{1}{2}> |1,1>}_{\pi^+ n} + \underbrace{\sqrt{\frac{2}{3}} |\frac{1}{2}, \frac{1}{2}> |1,0>}_{\pi^0 p}$$

$$\Rightarrow \therefore \ \frac{\Delta^+ \to \pi^0 p}{\Delta^+ \to \pi^+ n} = \frac{|<\pi^0 p | \Delta^+>|^2}{|<\pi^+ n | \Delta^+>|^2} = \frac{(\sqrt{2/3})^2}{(\sqrt{1/3})^2} = 2$$

43. CLEBSCH-GORDAN COEFFICIENTS, SPHERICAL HARMONICS, AND d FUNCTIONS

Note: A square-root sign is to be understood over *every* coefficient, *e.g.*, for $-8/15$ read $-\sqrt{8/15}$.

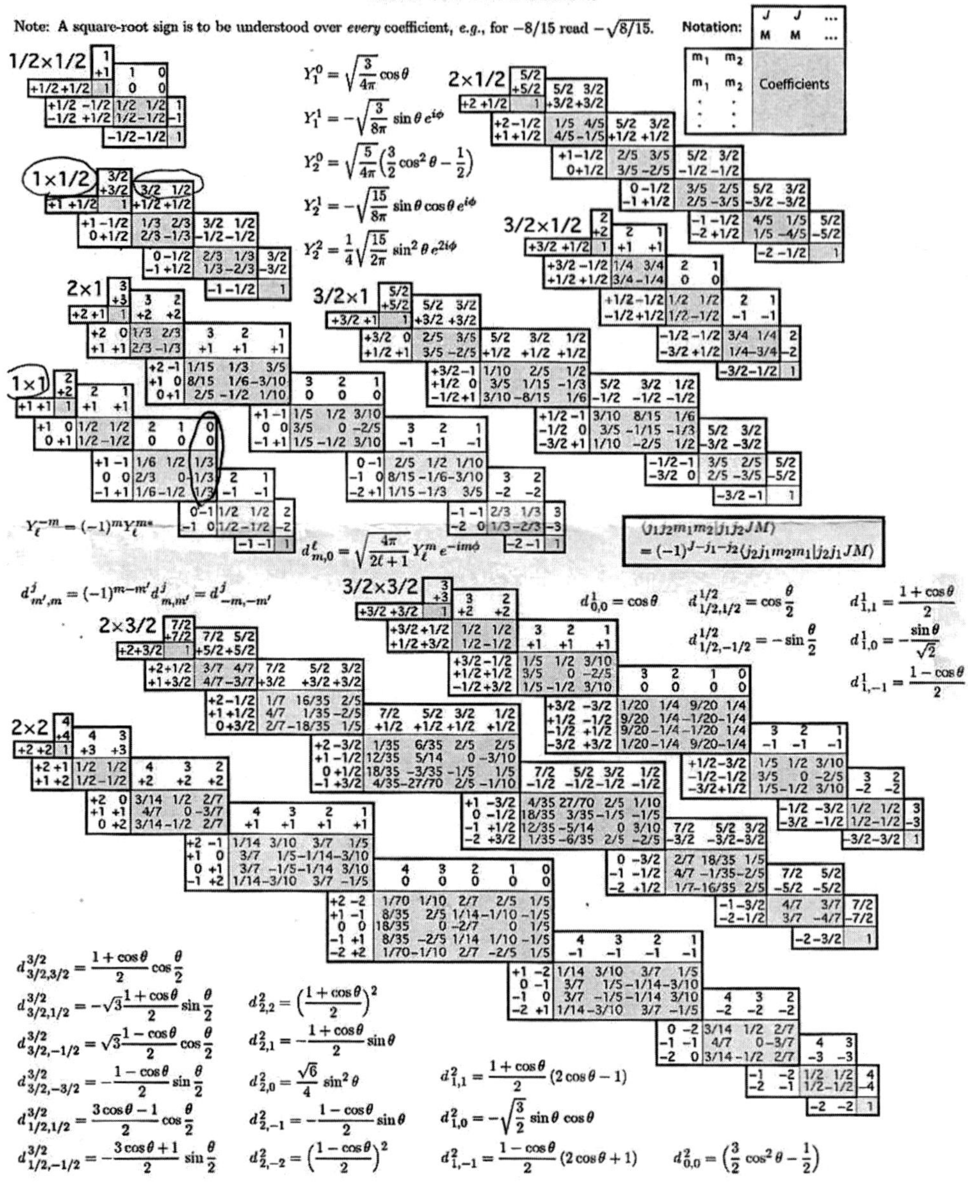

Figure 43.1: The sign convention is that of Wigner (*Group Theory*, Academic Press, New York, 1959), also used by Condon and Shortley (*The Theory of Atomic Spectra*, Cambridge Univ. Press, New York, 1953), Rose (*Elementary Theory of Angular Momentum*, Wiley, New York, 1957), and Cohen (*Tables of the Clebsch-Gordan Coefficients*, North American Rockwell Science Center, Thousand Oaks, Calif., 1974).

(source: particle data group)

■ Isospin(I), strangeness(S) and hypercharge(Y)

$$\frac{Q}{|e|} = I_3 + \frac{B}{2} \Rightarrow \text{for } I_3 = \pm\frac{1}{2} \text{ and } B = 1,\ \frac{Q}{|e|} = 1 \text{ or } 0$$

(1: proton, 0: neutron)

let`s extend to include strangeness quantum number S,

$$\frac{Q}{|e|} = I_3 + \frac{B+S}{2} = I_3 + \frac{Y}{2}, \text{ where } Y \text{ is the hypercharge}$$

above equation is the original form of Gell-Mann-Nishijima formula in 1955 based on empirical experiments

$$Y = B + S - \frac{C - B^* + T}{3}$$

$\Rightarrow \frac{1}{3}$ for u and d quark

$-\frac{2}{3}$ for s quark

0 for all other quarks

ex u quark and d quark

$$u : I_3 = \frac{1}{2},\ B = \frac{1}{3},\ S = 0 \Rightarrow Q = \frac{2}{3}e$$

$$d : I_3 = -\frac{1}{2},\ B = \frac{1}{3},\ S = 0 \Rightarrow Q = -\frac{1}{3}e$$

ex for kaon of $S = +1,\ B = 0$ and $I_3 = \pm\frac{1}{2}\left(K^{\pm} = I = \frac{1}{2}\right)$,

$$\frac{Q}{|e|} = 1 \text{ and } 0 \text{ for } K^{+}(\bar{s}u) \text{ and } K^{0}(d\bar{s})$$

for $\Lambda(uds)$ hyperon, $B = +1,\ S = -1$ and $I_3 = 0$

$$\frac{Q}{|e|} = 0$$

- quark quantum numbers

$$\frac{Q}{|e|} = I_3 + \frac{1}{2}(B + S + C + B^* + T),$$

where B is $\frac{1}{3}$ for all quarks and B^* is the beauty quantum number

flavor	I	I_3	S	C	B^*	T	$Q/\lvert e\rvert$
u	$\frac{1}{2}$	$\frac{1}{2}$					$\frac{2}{3}$
d	$\frac{1}{2}$	$-\frac{1}{2}$					$-\frac{1}{3}$
s			-1				$-\frac{1}{3}$
c				$+1$			$\frac{2}{3}$
b					-1		$-\frac{1}{3}$
t						$+1$	$\frac{2}{3}$

※ s, c, b, t : flavor of quark has same sign as its charge:

ex strangeness of $K^+ = +1$,

bottomness of $B^+(u\bar{b}) = +1$, charm of $D_S^- = -1$

Conservation Rules

Conserved quantity	Strong	electromagnetic	weak
energy-momentum	O	O	O
charge			
baryon number			
lepton number			
CPT	O	O	O
Parity	O	O	X
C (charge conjugation)	O	O	X
CP (or T)	O	O	$\sim 10^{-3}$ violation
I (isospin)	O	X	X
G-parity	O	X	X

※ G-parity is not a fundamental quantum number, however, it is convenient when dealing with non-strange states with zero baryonic number.

$\Rightarrow$ typically, these states, pion system, are eigenstate of G

(G is defined as C followed by $180°$ rotation around y-axis in isotopic space(iso-vector) $\Rightarrow G = e^{+i\pi I_y}C$

$$I_3 \xrightarrow{e^{i\pi I_2}} -I_3 \xrightarrow{C} I_3$$

rotation around y-axis implies $\theta \to \pi - \theta,\ \phi = \pi - \phi$

for $I_3 = 0,\ Y_\ell^0 \to (-1)^\ell Y_\ell^0$ i.e. isospin state $\chi(I, 0) \to (-1)^I \chi(I, 0)$

π^0 is the eigenstate of C and

$$C\,|\,\pi^+> \;=\; +\,|\,\pi^->,\quad C\,|\,\pi^-> \;=\; +\,|\,\pi^+>$$

Relationship between Cartesian components and charged states:

$$|\,\pi^+> \;=\; \frac{1}{\sqrt{2}}(\,|\,\pi_x> + i\,|\,\pi_y>)$$

$$|\,\pi^0> \;=\; |\,\pi_z>$$

$$|\,\pi^-> \;=\; \frac{1}{\sqrt{2}}(\,|\,\pi_x> - i\,|\,\pi_y>)$$

apply C and then rotate

$$|\,\pi^+> \;=\; \frac{1}{\sqrt{2}}(\,|\,\pi_x> + i\,|\,\pi_y>) \xrightarrow{C} |\,\pi^-> \;=\; \frac{1}{\sqrt{2}}(\,|\,\pi_x> - i\,|\,\pi_y>)$$

$$\xrightarrow{e^{i\pi I_y}} = \frac{1}{\sqrt{2}}(-\,|\,\pi_x> - i\,|\,\pi_y>) = -\,|\,\pi^+>$$

$$|\,\pi^0> \;=\; |\,\pi_z> \xrightarrow{C} |\,\pi^0> \;=\; |\,\pi_z> \xrightarrow{e^{i\pi I_y}} -\,|\,\pi_z> \;=\; -\,|\,\pi^0>$$

$$|\,\pi^-> \;=\; \frac{1}{\sqrt{2}}(\,|\,\pi_x> - i\,|\,\pi_y>) \xrightarrow{C} |\,\pi^+> \;=\; \frac{1}{\sqrt{2}}(\,|\,\pi_x> + i\,|\,\pi_y>)$$

$$\xrightarrow{e^{i\pi I_y}} = \frac{1}{\sqrt{2}}(-\,|\,\pi_x> + i\,|\,\pi_y>) = -\,|\,\pi^->$$

$$\Rightarrow G\,|\,\pi> \;=\; -\,|\,\pi>$$

For a system of n_π pions, $G = (-1)^{n}\pi$: G symmetry is useful for multiple pion states

Only strong interaction conserves G-parity because the EM and weak interactions violate the isospin.

$G = e^{+i\pi I_y}C$ is the composition of the isospin and charge conjugation

$$\begin{cases} G \mid \pi^0 > \ = e^{+i\pi I_y} \mid \pi^0 > \ = - \mid \pi^0 > \\ G \mid \pi^\pm > \ =- e^{+i\pi I_y} \mid \pi^\mp > \ = - \mid \pi^\pm > \end{cases} \Rightarrow G \mid \pi > \ = (-1) \mid \pi >$$

$$\therefore \text{ in Q.M.,} \quad e^{iJ_y\pi} \mid J, J_z > \ = (-1)^{J-J_z} \mid J, -J_z >$$

$$\text{similarly, } e^{iI_2\pi} \mid I, I_3 > \ = (-1)^{I-I_3} \mid I, -I_3 >$$

$$\Rightarrow e^{iI_2\pi} \mid \pi^\pm > \ = + \mid \pi^\mp >, \quad e^{iI_2\pi} \mid \pi^0 > \ = - \mid \pi^0 >$$

$$G \mid \pi^0 > \ = - \mid \pi^0 >$$

since π is isotriplet

$$C \mid \pi^+ > \ = - \mid \pi^- >, \ C \mid \pi^- > \ = - \mid \pi^+ >$$

EM field is produced by charge

$\Rightarrow$ photon has $=-1$

$\Rightarrow \pi^0 \rightarrow 2\gamma \Rightarrow \pi^0 = +1$ charge conjugation

[I] Relativistic mechanics

(i) $E^2 = p^2c^2 + m^2c^4$ (Einstein equation)

(ii) Total Energy: $E = \gamma mc^2 = \dfrac{mc^2}{\sqrt{1-\beta^2}}$ [GeV]

$$\left(\because \gamma^2 = \frac{1}{1-\beta^2} \Rightarrow \beta = \sqrt{1-\frac{1}{\gamma^2}} \right)$$

(iii) Linear momentum: $\vec{p} = \gamma m \vec{v}$ [GeV/c]

(iv) $\frac{\vec{p}}{E} = \frac{\gamma m\vec{v}}{\gamma mc^2} = \frac{\vec{v}}{c} = \vec{\beta}$ (unit of $\vec{p}$ is $[\mathrm{GeV}/c]$, unit of E is GeV)

$$\left(\because \ \gamma = \frac{E}{mc^2}\right)$$

(v) Kinetic energy: $K = E - mc^2 = (\gamma - 1)mc^2$

(vi) Lorentz factor: $\gamma = \frac{E}{mc^2} = \frac{K + mc^2}{mc^2} = 1 + \frac{K}{mc^2}$

$$\gamma\beta = \left(\frac{E}{mc^2}\right)\frac{p}{E} \xrightarrow{\hbar = c = 1} \frac{1}{m}\sqrt{E^2 - m^2} = \sqrt{\left(\frac{E}{m}\right)^2 - 1}$$

[II] **Average:** weighted average of selected data attempt to use uncorrected variables as much as possible, but correlations are taken into account

$$\bar{x} \pm \sigma_{\bar{x}} = \frac{\sum w_i x_i}{\sum w_i} \pm \frac{1}{\sqrt{\sum_i w_i}}, \quad \text{here } w_i = \frac{1}{(\sigma_i)^2}$$

$$= \frac{1}{w}\sum_{i=1}^{n} w_i x_i \pm \frac{1}{\sqrt{w}}$$

each normally distributed with mean μ_i and σ_i^2

If $\frac{\chi^2}{N-1}$ $\left(\chi^2 = \sum_i \frac{(x_i - \mu_i)^2}{\sigma_i^2}\right)$ is greater than 1, but not greatly so, we average data but then also do the following:

① We increase our quoted error, $\delta\bar{x}$, by a scale factor S defined as

$$S = \sqrt{\frac{\chi^2}{N-1}} \quad \text{(scaling factor does not effect central values)}$$

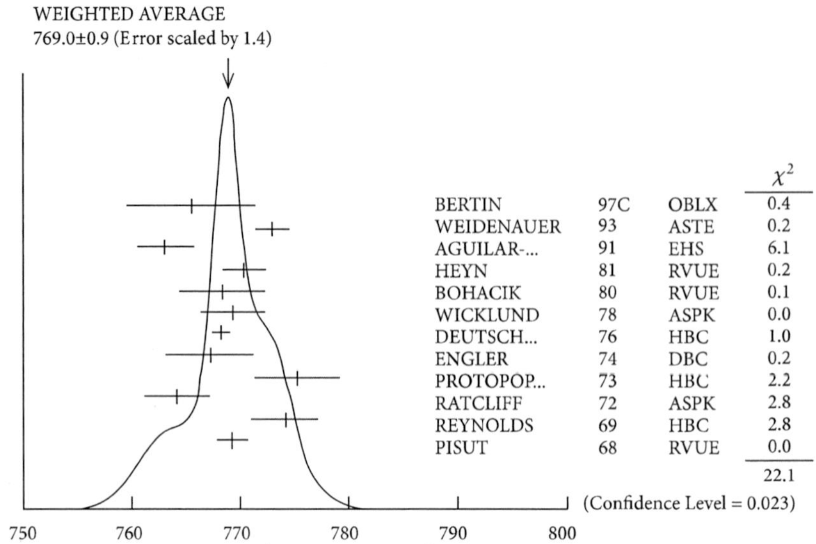

(SOURCE: particle data group)

NEUTRAL ONLY, OTHER REACTION

VALUE(MeV)	EVTS	DOCUMENT	TECN	COMMENT
769.0±0.9	OUR AVERAGE	Error includes scale factor of 1.4		See the ideogram above.
765±6		BERTIN 97C	OBLX	$0.0\bar{p}p \rightarrow \pi^+\pi^-\pi^0$
773±1.6		WEIDENAUER 93	ASTE	$\bar{p}p \rightarrow \pi^+\pi^-\omega$
762.6±2.6		AGUILAR··· 91	EHS	$400pp$
770±2		HEYN 81	RVUE	Pion form factor
768±4		BOHACIK 80	RVUE	
769±3		WICKLUND 78	ASPK	$3,4,6\pi^{\pm}N$
768±1	76000	DEUTSCH··· 76	HBC	$16\pi^+p$
767±4	4100	ENGLER 74	DBC	$6\pi^+n \rightarrow \pi^+\pi^-p$
775±4	32000	PROTOPOP··· 73	HBC	$7.1\pi^+p,\ t < 0.4$
764±3	6800	RATCLIFF 72	ASPK	$15\pi^-p,\ t < 0.3$
774±3	1700	REYNOLDS 69	HBC	$2.26\pi^-p$
769.2±1.5	13300	PISUT 68	RVUE	$1.7\text{–}3.2\pi^-p,\ t < 10$

(source: particle data group)

$$\frac{\chi^2}{N-1} \approx 2.0$$

$$\text{scale factor } S = \sqrt{\frac{\chi^2}{N-1}} \approx \sqrt{2.0} \approx 1.4$$

Chapter 4

Quarks in Hadrons

Patterns of observed particle led to proposed in early 1960`s that hadrons were composed of quarks ($u$, $d$, $s$ at that time). During 1970`s two more quarks were discovered: charm(1974), beauty(1977) and top quark in 1995.

■ Charm and Beauty: heavy quarkonium states

charmonium states, $\Psi(c\bar{c}$ bound states), were observed in the Stanford Positron Electron Accelerating Ring)

• two complementary experiments

① 1974 e^+e^- collides SPEAR at SLAC

$e^+e^- \rightarrow \Psi(3100) \rightarrow$ hadrons

└ e^+e^-, $\mu^+\mu^-$ [$\Psi'(3.7\ \mathrm{GeV})$ was also found]

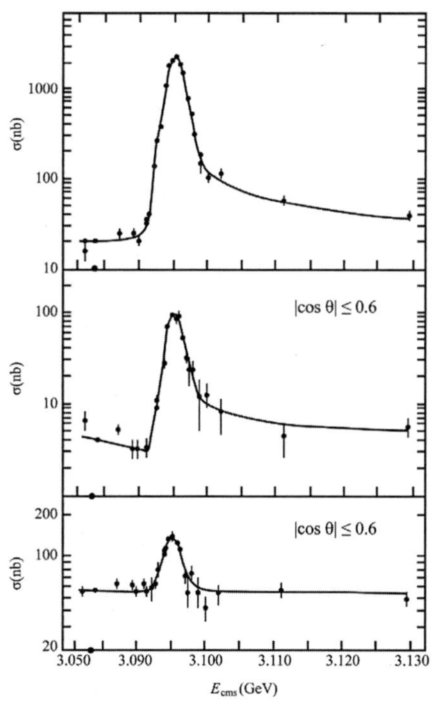

($\Psi \rightarrow$ (top: hadrons, middle: $\pi^+\pi^-$, $\mu^+\mu^-$ and K^+K^-, bottom: e^+e^-)

② 1974 AGS(Alternating Gradient Synchrotron) working at 30 GeV in BNL

$P + B_e \rightarrow J/\Psi$ + anything

$\quad\quad\quad\quad\quad\quad \llcorner\ e^+e^-$

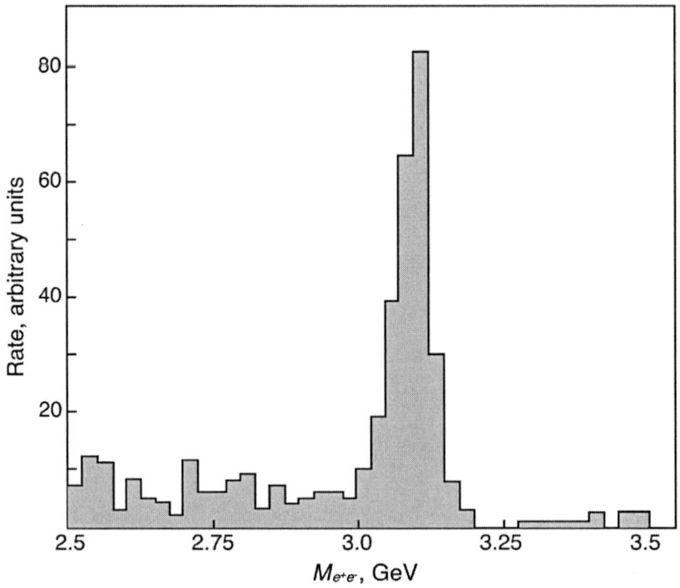

- observed widths (of order of several MeV) of the peaks are dominated by the experimental resolution and true width of the J/Ψ is much smaller, which is determined from the total reaction rate and the leptonic branching ratio from Breit-Wigner formula:

  total width from integral of the cross-section

$$\sigma = \frac{4\pi\lambda\!\!\!^{-2}(2J+1)}{(2S_a+1)(2S_b+1)} \frac{\Gamma^4/4}{\left[(E-E_R)^2+\Gamma^4/4\right]}, \text{ where } \Gamma = \Gamma_{\text{initial}}\Gamma_{\text{final}} = \Gamma_{ee}^2$$

and $\lambda\!\!\!^{-}$ is the de Broglie wavelength of e^+e^- in the cms.

$$\Rightarrow \sigma(E)_{e^+e^- \rightarrow J/\Psi \rightarrow e^+e^-} = \frac{4\pi\lambda\!\!\!^{-2}(2J+1)}{(2S_1+1)(2S_2+1)} \frac{\Gamma_{e^+e^-}^4/4}{\left[(E-E_R)^2+\Gamma^4/4\right]},$$

where E is the cms energy, E_R is the energy at the resonance peak, Γ is the total width of resonance and $\Gamma_{e^+e^-}$ is the partial width for $\Psi \to e^+e^-$ with $S_1 = S_2 = \frac{1}{2}$ and assuming that spin of resonance state is $J = 1(J^{pc} = 1^{--})$

let, $\tan\theta = \dfrac{2(E - E_R)}{\Gamma} \Rightarrow dE = \dfrac{\Gamma}{2}\dfrac{d\theta}{\cos^2\theta}$

then, $\displaystyle\int_0^{\infty} \sigma(E)dE = \frac{3\pi^2}{2}\lambda\!\!\!^-{}^2\left(\frac{\Gamma_{e^+e^-}}{\Gamma}\right)^2\Gamma \quad \left(\because \int \frac{dx}{1+x^2} = \pi,\ x = \tan\theta\right)$

here, $\dfrac{\Gamma_{e^+e^-}}{\Gamma} = 0.06,\quad \lambda\!\!\!^- = \dfrac{\lambda}{2\pi} = \dfrac{1}{2\pi}\dfrac{h}{p} = \dfrac{\hbar c}{pc} = \dfrac{197\ \text{Mev}\cdot\text{fm}}{1550\ \text{MeV}}$

($\because$ 1550 MeV is equal to half of J/Ψ invariant mass)

State	Mass (MeV)	J^P, I	Γ, (MeV)	Branching ratio	
$J/\Psi(3100)$ $=\Psi(1S)$	3097.88 ± 0.04	$1^-,0$	0.087	hadrons	88%
				e^+e^-	6%
				e^+e^-	6%
$J/\Psi(3700)$ $=\Psi(2S)$	3686 ± 0.1	$1^-,0$	0.28	$\Psi + 2\pi$	50%
				$X + \gamma$	24%
				e^+e^-	0.9%
				$\mu^+\mu^-$	0.8%

and $\displaystyle\int \sigma(E)dE = 800\ \text{nb MeV}$

$\Rightarrow \Gamma = 0.087$ MeV: true width of $J/\Psi \Rightarrow$ very narrow width

$\Rightarrow \Gamma_{e^+e^-} = 5$ KeV

(※ c.f. $\Gamma_{\rho(770\text{MeV})} = 6.8$ KeV, $\Gamma_{\omega(784\text{MeV})} = 0.6$ KeV, $\Gamma_{\phi(1020\text{MeV})} = 1.4$ KeV)

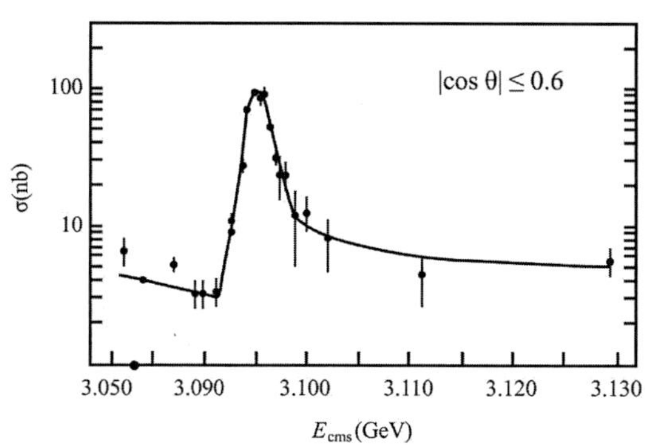

$\Leftarrow$

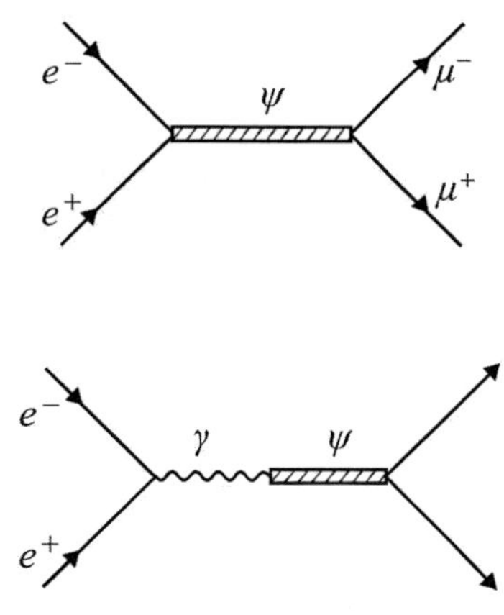

(interference of direct and virtual photon channels: (top) amplitude for production of Ψ via direct channel (bottom) amplitude for production of Ψ via intermediate virtual photon.)

So, Ψ must have the same quantum numbers as photon

$\Rightarrow J^{PC} = 1^{--}$ for Ψ ($\because$ charge conjugation of photon is -1)

and

$I = 0$ for Ψ ($\because$ $\Psi \to \rho\pi$, $I(e) = 1$, $I(\pi) = 1 \Rightarrow \rho^+\pi^-$, $\rho^0\pi^0$, $\rho^-\pi^+$ are found to be equally populated $\Rightarrow$ from Clebsch-Gordan coefficient, $I = 0$)

Narrowness of J/Ψ and Ψ' is explained by QCD, 3G exchange required.

extreme narrowness of J/Ψ states

$\Rightarrow$ introduce new quantum number C (by GIM [Glashow, Iliopoulous, Maiani 1970]) and large mass of J/Ψ

$\Rightarrow$ charm quark must be massive

$\Rightarrow$ postulated that J/Ψ(vector particle) is $c\bar{c}$ bound state, "charmonium". Other combinations with a net charm number, $c\bar{d}$, called as charmed meson.

[$c\bar{d} = D^+(1870\ \text{MeV})$]

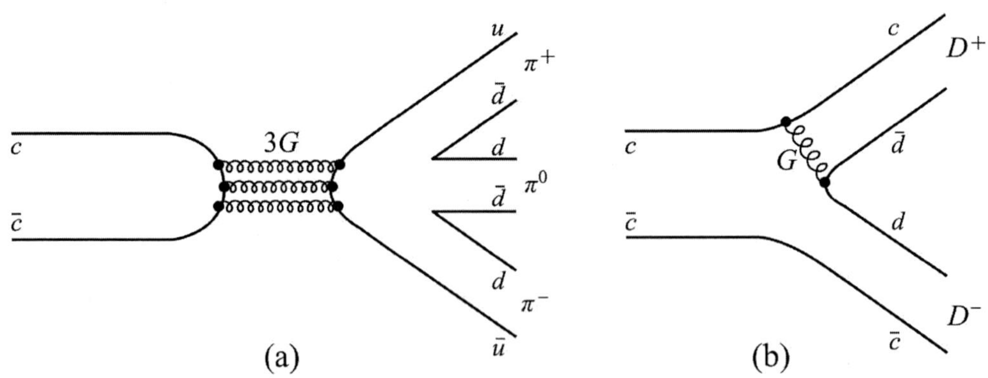

(a) "OZI(Okubo, Zweig, Iizuka)" forbidden diagram

$\Psi(3100) \rightarrow 3\pi$ with $\Gamma = 0.076$ MeV

(b) favored but forbidden by energy conservation ($D^{\pm}$ mass $= 1.869$ GeV) for Ψ and Ψ'

$\Psi(3770) \rightarrow D\overline{D}$ with $\Gamma = 24$ MeV

(Since $2M_D = 3.75$ GeV, $\Psi(3.1)$ and $\Psi(3.7)$ are forbidden by energy conservation)

$\Rightarrow$ i.e. "OZI" forbidden diagram is the only one allowed for hadronic decay of low-mass state

Bottomonium states

Upsilon states, $Y(b\bar{b}$ bound state) were observed by 1977 at FNAL

$P + Be,\ Cu,\ Pt \rightarrow \mu^{+} + \mu^{-} +$ anything: 400GeV proton-nucleus collisions

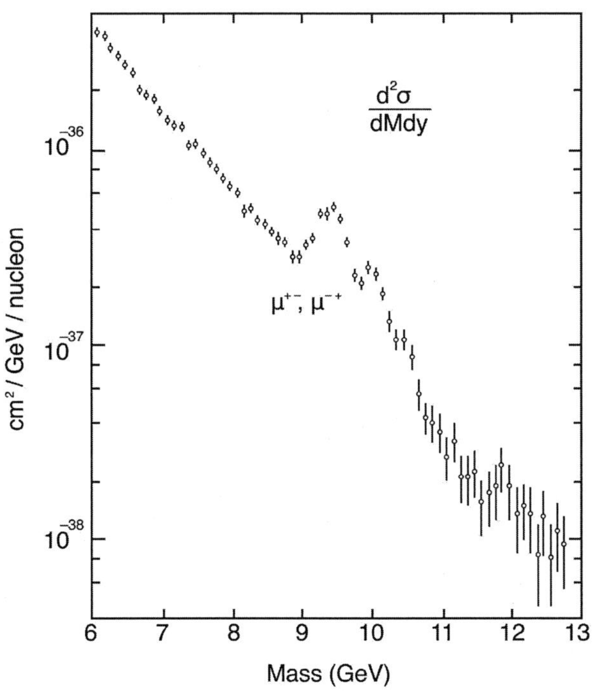

In this experiment, a broad peak with resonance combined is shown.

⇒ electron-positron collides refine measurements: Later, through e^+e^- experiments at DORIS in Hamburg, Germany and CESR in Cornell, $\Upsilon(1S)$, $\Upsilon(2S)$, $\Upsilon(3S)$, …… were discovered.

	$\Upsilon(1S)$	$\Upsilon(2S)$	$\Upsilon(3S)$	$\Upsilon(4S)$
Mass(MeV)	9460.4±0.2	10023.3±0.3	10355.3±0.5	10580.0±3.5
$\Gamma_{e^+e^-}$(KeV)	1.32	0.52		0.25
Γ_{tot}(MeV)	0.053	0.044	0.026	10

(※ in the table, 0.2, 0.3, 0.5: narrow because of the OZI rule,
3.5: above threshold (10558 MeV) for decay into $B\overline{B}$ pairs)

Summary	I	I_3	S	C	B^*	T	Q/e
u	1/2	1/2	0	0	0	0	2/3
d	1/2	-1/2	0	0	0	0	-1/3
s	0	0	-1	0	0	0	-1/3
c	0	0	0	1	0	0	2/3
b	0	0	0	0	-1	0	-1/3
t	0	0	0	0	0	1	2/3

[I] Comparison of quarkonium and positronium levels

an electron and position orbiting around their common center of mass

⇒ bound quantum state as positronium

Positronium states: $e^+e^- \rightarrow \gamma\gamma$

- $\tau = 1.25 \times 10^{-10}$ sec
- singlet state
- even angular momentum $J=0 \Rightarrow C=(-1)^{L+S}=(-1)^J=+1$
- $C=(-1)^{n\gamma} \xrightarrow{n=2} C=+1$

C and P of e^+e^- system under particle exchange:

$(-1)^{S+1}$ for spin function

$(-1)^L$ for space function

$\Rightarrow$ $(-1)^{L+1}$ after taking account of opposite intrinsic parities of particle and antiparticle

$\Rightarrow$ $(-1)^{L+S}$ is total symmetry

$\Rightarrow$ equivalent to the C operation $(e^+ \leftrightarrow e^-)$

Bose symmetry of the two-photon system: two gamma rays with 511 KeV (Positron-Emission Tomography; PET)

$\Rightarrow$ $(-1)^{0+1}(-1)^{S+1} = +1 \Rightarrow S=0$ (spin singlet states)

for a system of n photons, $C=(-1)^n$: probability quickly decreases with number

3γ decays $\Rightarrow$ $C=-1 \Rightarrow S=1$ (spin triplet state)

$[e^+e^- \to \gamma\gamma\gamma]$

- $\tau = 1.4\times 10^{-7}$ sec
- triplet state $\Rightarrow$ $J=1 \to C=(-1)^{L+S}=(-1)^{0+1}=-1$
- $C=(-1)^{n\gamma} \to C=-1$

Principal energy levels from non-relativistic Schrödinger equation in a Coulomb potential:

energy level for hydrogen atom,

$$E_n = -\frac{\alpha^2 \mu c^2}{2n^2}\left(=\frac{13.6}{n^2}\ \mathrm{eV}\right),\ \alpha = \frac{e^2}{4\pi\hbar c}$$

where n is the principal quantum number,

μ is the reduced mass($=\frac{m}{2}$ for positronium)

$$\left(\because \frac{1}{\mu} = \frac{1}{m_{e^+}} + \frac{1}{m_{e^-}} = \frac{2}{m} \Rightarrow \mu = \frac{m}{2} \right)$$

$$\xrightarrow[\text{positronium}]{\text{for}} E_n = -\frac{\alpha^2 mc^2}{4n^2}$$

• relativistic Q.M.

$$\text{relativistic correction}\begin{cases} \text{spin}-\text{orbit coupling} \rightarrow \text{fine structure} \\ \text{spin}-\text{spin coupling} \rightarrow \text{hyperfine structure} \end{cases}$$

$$\xrightarrow[\text{a given n}]{\text{for}} \begin{cases} \text{into S, P, D, } \cdots\cdots \text{ states} \\ \text{into triplet}(^3S_1) \text{ and singlet}(^1S_0) \end{cases}$$

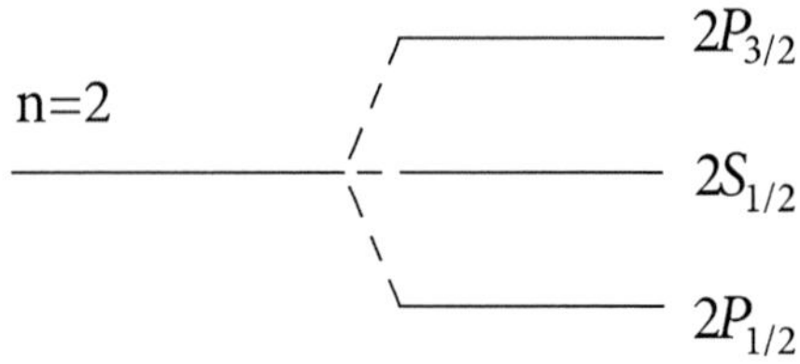

in positronium, these are about the same size in positronium

$$\Delta E(\text{fine structure}) \sim \frac{\alpha^4 mc^2}{n^3}$$

and triplet and singlet states are called ortho-positronium(o-Ps) and para-positronium (p-Ps), respectively.

(※ "ortho"="straight", "para"="against")

Excellent agreement on theory and experimental results for lifetimes and energy levels

① Lifetime of singlet state

ⓐ decay rate $\sim$ α^2 since two photons are coupled.

ⓑ $\sim |\Psi_{(0)}|^2$: square of e^+e^- wave function at the origin.

$$\left(\Psi_{100} = \left(\frac{1}{\pi a^3} \right)^{1/2} e^{-r/a} \Rightarrow |\Psi(0)|^2 = \frac{1}{\pi a^3} \right),$$

$$\Rightarrow |\Psi(0)|^2 = \frac{1}{\pi a^3},$$

where Bohr radius in positronium is double that in hydrogen

$$a = 2\times\frac{\hbar^2}{me^2} = \frac{2\hbar^2}{m}\frac{1}{\alpha\hbar c} = \frac{2\hbar}{mc\alpha} \qquad \left(\therefore\ \alpha = \frac{e^2}{\hbar c}\right)$$

<dimensional analysis>

decay width Γ in energy units is obtained by

$$\frac{\alpha^2|\Psi_{(0)}|^2}{m^2} \text{ (in natural units)} \Rightarrow \frac{[\text{energy}]^3}{[\text{energy}]^2} = [\text{energy}]$$

($\because$ α: dimensionless, m^2:[energy]^2,

$$|\Psi_{(0)}|^2 \propto \frac{1}{a^3} = \left(\frac{1}{1/\text{energy}}\right)^3 = [\text{energy}]^3)$$

$\Rightarrow \Gamma \sim \alpha^2(m^3\alpha^3)m^{-2} = \alpha^5 m$ (actual calculation $= \frac{\alpha^5 m}{2}$)

② lifetimes of triplet state is slower by a factor of order α

actual calculation $= \frac{2(\pi^2-9)}{9\pi}\alpha^6 m$

$\Rightarrow$ comparison between calculation and experimental result are in good agreement

[II] Quarkonium levels

Striking similarities with positronium $\Rightarrow$ quantum states of a bound fermion-antifermion ($q\bar{q}$) pair

※ different potential:

- positronium $\rightarrow$ Coulomb

  $-\frac{e^2}{r} \rightarrow -\frac{\alpha}{r}$

- quarkonium $\rightarrow$ potential from QCD

QCD potential in the heavy $q\bar{q}$ system is,

$$U_{\text{QCD}} = -\frac{4}{3}\frac{\alpha_s}{r} + kr,$$

where $-4/3$ is the color factor, α_s is the quark-gluon coupling($\simeq 0.2$ from studies of deep inelastic lepton scattering) , $-\frac{4}{3}\frac{\alpha_s}{r}$ is the Coulomb-like term, kr($1\ \text{GeV}\ \text{fm}^{-1}$ from studies of deep inelastic lepton scattering) is the linear term.

For heavy quarks, non-relativistic approximation is valid:

$$\begin{cases} E = \dfrac{p^2}{2\mu} = \dfrac{p^2}{2 \cdot \dfrac{1}{2}m} = \dfrac{p^2}{m} \\ E = \dfrac{\alpha^2 mc^2}{4n^2} \end{cases} \Rightarrow p^2 = \left(\frac{\alpha}{2n}\right)^2 m^2 c^2$$

$$\Rightarrow \frac{p}{mc} = \left(\frac{\alpha}{2n}\right)\frac{mc}{mc} = \frac{\alpha}{2n}$$

$$\Rightarrow \frac{p}{mc} = \frac{\alpha_s}{2n} = \frac{\alpha_s}{2} \quad (\text{for } n = 1)$$

$$\approx 0.1$$

<running of the strong coupling constant>

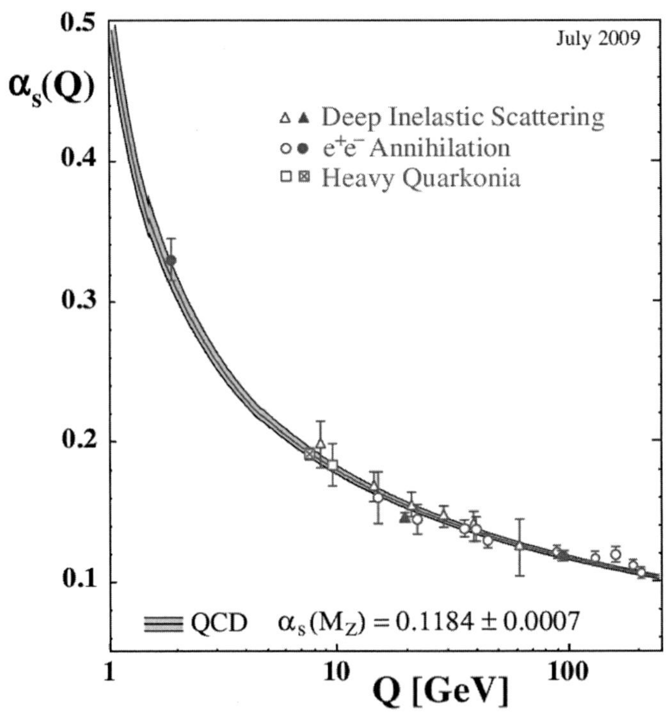

$\Rightarrow$ non-relativistic approximation for heavy quark systems can be justified.

but, light $q\bar{q}$ system, where α_s could even be of order unity

$\Rightarrow$ non-relativistic approximation is not so obvious.

<for Quarkonium, Similar energy levels to positronium>

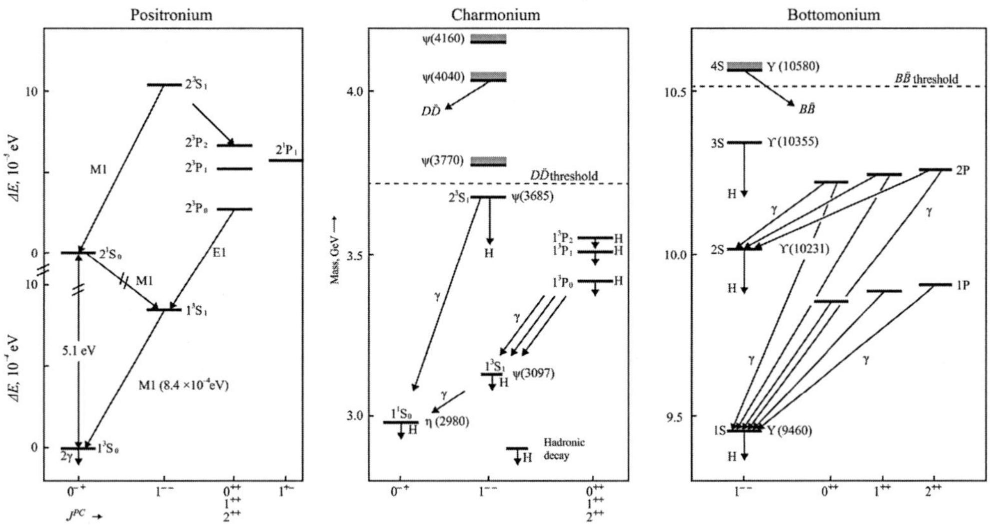

fine structure is of order α_s, and therefore coarser than for positronium, as observed

■ **The baryon decuplet** (lightest spin $\frac{3}{2}$ baryons)

quark model as a model of hadron structure in deep inelastic lepton scattering by nucleons, the nucleon appeared to be built from "partons"(light, quasi-free point like constituents).

$\Rightarrow$ partons turned out to be quarks, they were confined inside nucleon by special confining potential

<Quark quantum numbers>

flavor	B	J	I	I_3	S	Q/e
u	1/3	1/2	1/2	+1/2	0	+2/3
d	1/3	1/2	1/2	-1/2	0	-1/3
s	1/3	1/2	0	0	-1	-1/3

$3\otimes3\otimes3=10_S\oplus8_M\oplus8_M\oplus1_A$: $3\times3\times3=27$ combinations are possible

1 is anti-symmetric under interchange of two quarks
10 is symmetric under interchange of two quarks
8's are mixed under interchange of two quarks

10 baryon states in strangeness(S) and third component(I_3) of isospin

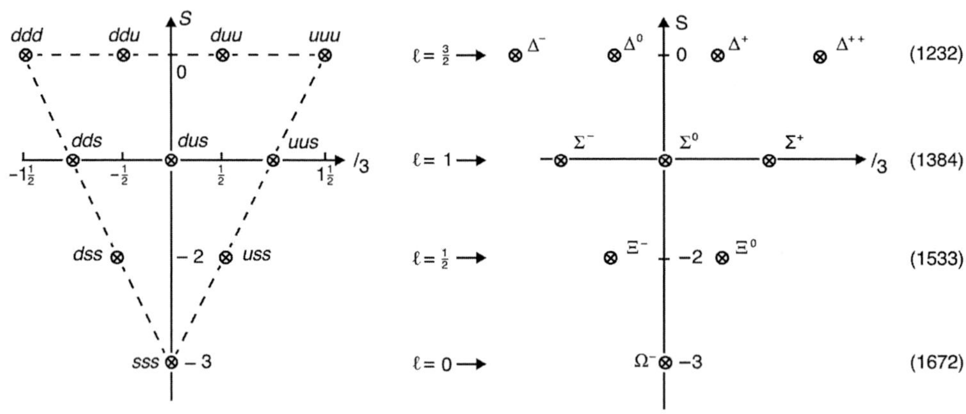

(a) (b)

(b) Meson mass of each isospin multiplet; differing only by a few MeV between members of each isospin.

$\Delta(1232)$: existing in charge substates Δ^{++}, Δ^+, Δ^0, Δ^-
with $S=0$, $I=\frac{3}{2}$ isospin quadruplet

$\Sigma(1384)$: Σ^+, Σ^0, Σ^- with $S=-1$, $I=1$ isospin triplet

$\Xi(1533)$: Ξ^0, Ξ^- with $S=-2$, $I=\frac{1}{2}$ isospin doublet

$\Omega^-(1672)$: with $S=-3$, $I=0$ isospin singlet

⇒ mass difference for each increment of strangeness is roughly the same.

(※ We see an orderly increase of mass with number of strange quarks)

⇒ Ω^- baryon was predicted on this basis three years before it was observed at BNL in 1964 (see the next page):

$K^- + p \to \Omega^- + K^+ + K^0$

↳ $\Xi^0 + \pi^-$ ($\Delta S = 1$, weak decay)

↳ $\pi^0 + \Lambda$ ($\Delta S = 1$, weak decay, $\Lambda = (uds)$)

↳ $p + \pi^-$ ($\Delta S = 1$, weak decay)

→ $\gamma + \gamma$ (electromagnetic decay)

↳ e^+e^-

→ e^+e^-

Discovery of Ω^-

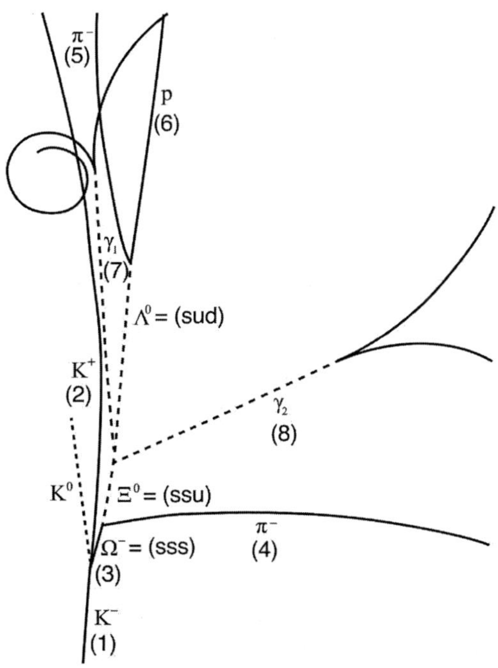

(V. E. Barnes et al., Observation of a Hyperon with Strangeness Minus Three, Phys. Rev. Lett. 12, 204)

baryon consists of three quarks

⇒ 27 combinations are possible (3×3×3)

⇒ what so special about 10 of them?

⇒ introduce some symmetry principle, concerning quark spin as well as quark flavor

corner states, uuu, ddd and sss are symmetric under interchange.

⇒ require same symmetry for the other states

⇒ **ex** full form of the ddu state is

$\frac{1}{\sqrt{3}}(ddu+udd+dud)$, which is symmetric under interchange of any two quarks

for dus is

$$\frac{1}{\sqrt{6}}(dsu+uds+sud+sdu+dus+usd)$$

※ The different states in the multiplet are obtained by applying the isospin shift operators $I_{\pm}$ successively.

i.e.

$$I_{-}(uuu) = [I_{-}(u)](uu)+(u)[I_{-}(u)](u)+(uu)[I_{-}(u)] = duu+udu+uud$$

⇒ These 10 states are the only completely symmetric combinations we can make.

⇒ of the remaining 17, one(singlet) is completely antisymmetric(changing sign under interchange of any two quarks),

$$dsu+uds+sud-usd-sdu-dus$$

⇒ Other 16 states consist of two octets of mixed symmetry

⇒ $3\otimes3\otimes3 = 27 = 1\oplus8\oplus8\oplus10$

(1 and 10 is for antisym. and sym. under interchange of two quarks, respectively, and $8\oplus8$: mixed under interchange of two quarks)

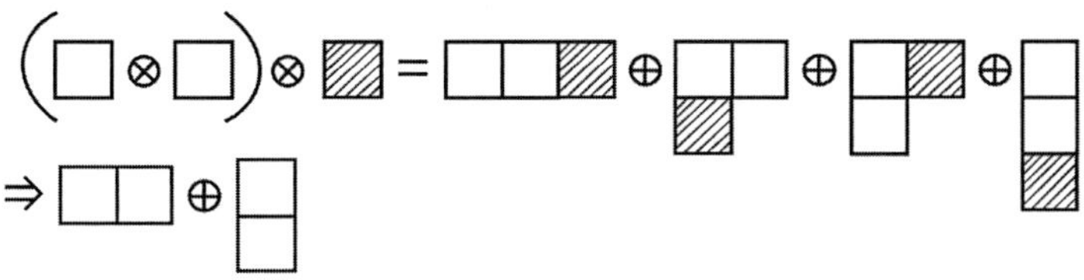

For combining any two of the basic quark states,

$3 \otimes 3 = 6 \oplus 3$ (6 and 3 is for sym. and antisym.)

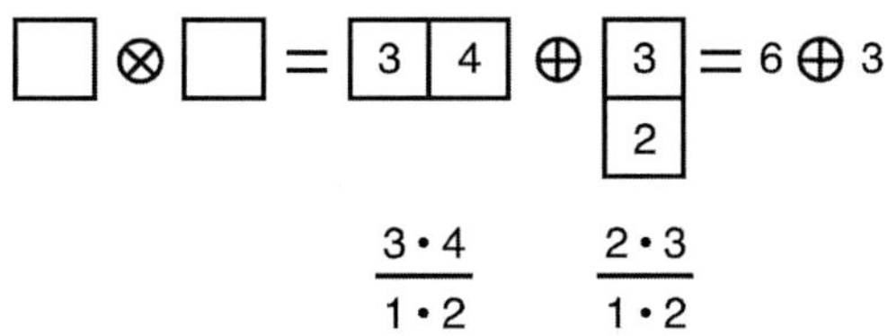

※ quarks are fundamental representations of group $SU(3)$: $3 \otimes \overline{3} = 8 \oplus 1$

■ Quark spin and color

in decuplet, quarks sit in the spatially symmetric ground state $\Rightarrow \ell = 0$

$\Rightarrow J = \frac{3}{2}$(spin $\frac{3}{2}$ baryon, therefore $u\uparrow u\uparrow u\uparrow$) is obtained by quarks in symmetric spin state

$\Rightarrow \Delta^{++} = \uparrow\uparrow\uparrow$, violates the Pauli principle, which is two or more identical fermions can't exist in the same quantum state

$\Rightarrow$ another degree of freedom is needed

$\Rightarrow$ "color"(quarks exists in three colors: red, green, blue)

※ $u\uparrow$ (red)$u\uparrow$ (green)$u\uparrow$ (blue)

baryons and mesons are zero net color (colorless) $\Rightarrow$ they are color singlets.

The interquark force is independent of the colors involved.

$\Rightarrow \Delta^{++}$ consists of one red, one green, and one blue u quark

$\Rightarrow$ and we have to anti-symmetrize color:

$$u\uparrow u\uparrow u\uparrow (rgb - rbg + brg - bgr + gbr - grb)$$

[experiment]: Validity of color quantum number has been demonstrated.

① $R = \dfrac{\sigma_{(e^+e^- \to hadrons)}}{\sigma_{(e^+e^- \to \mu^+\mu^-)}}$

$$= N_c\left[\left(\frac{1}{3}\right)^2 + \left(\frac{1}{3}\right)^2 + \left(\frac{2}{3}\right)^2\right] = 3 \cdot \frac{2}{3} = 2 \text{ for } \sqrt{S} < 3 \text{ GeV}$$

$$= N_c\left[\left(\frac{1}{3}\right)^2 + \left(\frac{1}{3}\right)^2 + \left(\frac{1}{3}\right)^2 + \left(\frac{2}{3}\right)^2 + \left(\frac{2}{3}\right)^2\right] = 3 \cdot \frac{11}{9} = \frac{11}{3} \text{ for } \sqrt{S} > 10 \text{ GeV}$$

($\because$ only u, d, s quarks are involved for $\sqrt{S} < 3$ GeV

u, d, s, c, b quarks are involved for $\sqrt{S} > 10$ GeV)

<R as a function of the total e^+e^- center-of-mass energy>

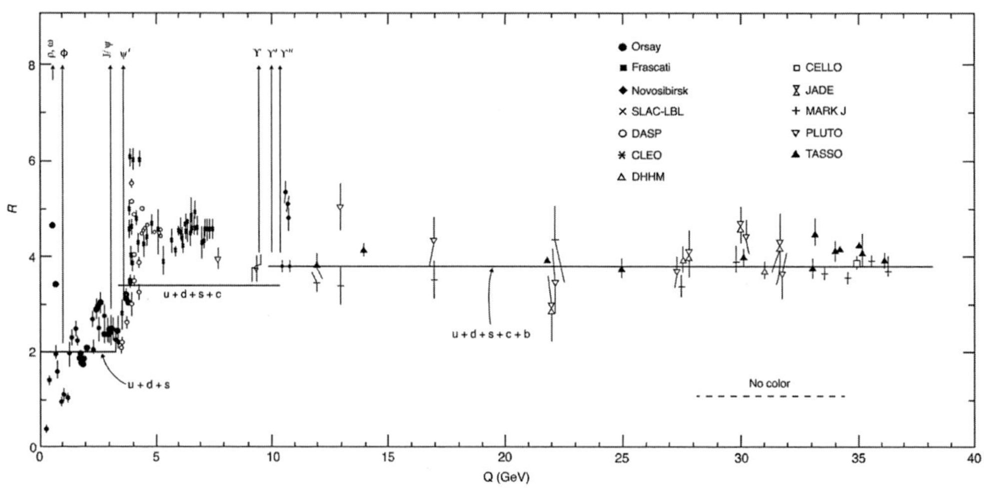

② $\pi^0 \to 2\gamma$

We also know from the rate of decays of π^0 that there are three colors.

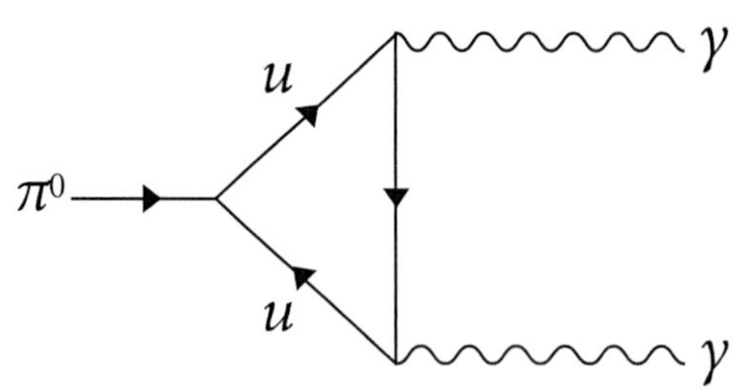

triangle diagram: colors involved are not distinguishable

$\Rightarrow$ three amplitudes must be added coherently

$\Rightarrow$ $N_c^2 = 9$

(※ triangle diagram is the source of "anomalies", which spoil renormalizability of the SM.

$\Rightarrow$ sum of quarks and leptons charge equal to 0, anomalies disappear

(When $N_c = 3$, the sum of all charges becomes 0.)

($\because$ The sum of the charges of (e, μ, τ) is -3, that of (u, d, s, c, b, t) is $+1$, that times 3 is $+3$. Then, the sum of all charges is 0)

$$\begin{cases} \Gamma_{(\pi^0 \to 2\gamma)} = \left(\dfrac{N_c}{3}\right)^2 \dfrac{\alpha^2 m_\pi^3}{64\pi^3 f_\pi^2} = 7.73\left(\dfrac{N_c}{3}\right)^2 \text{ eV} \\ \Gamma_{(\text{obs})} = 7.7 \pm 0.6 \text{ eV} \end{cases}$$

where f_π is pion decay constant, 92.4 MeV)

$\Rightarrow$ $N_c = 2.99 \pm 0.12$

$$※\ \tau = \frac{\hbar}{\Gamma} = \frac{197 \text{ Mev} \cdot \text{fm}}{3 \times 10^{23} \text{ fm/sec} \cdot \Gamma} = \frac{6.6 \times 10^{-16} \text{ sec}}{\Gamma(\text{eV})} = 8.5 \times 10^{-17} \text{sec} \left(\frac{3}{N_c}\right)^2$$

$$d = \gamma\beta c\tau = \left(\frac{3}{N_c}\right)^2 \times 2.6 \times 10^{-2} \,\mu\text{m} \left(\frac{p}{m}\right)$$

Suppose $p = 56$ eV, then $d \approx 1\,\mu\text{m} \left(\dfrac{3}{N_c}\right)^2$

■ π^0 lifetime experiment

$p + A \rightarrow \pi^0 + X$

$\quad\quad\quad\quad\ \ \llcorner \gamma\gamma$

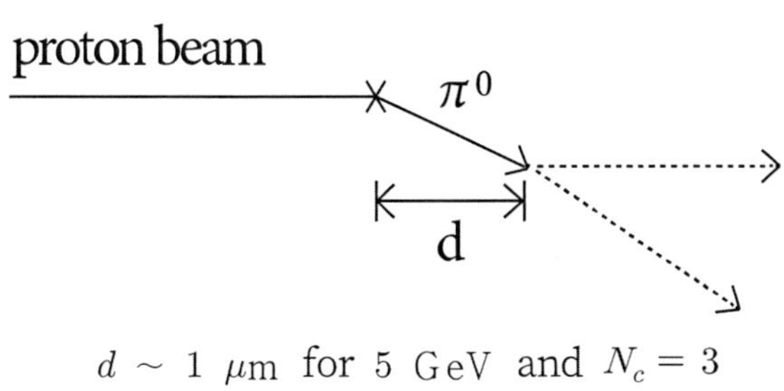

$d \sim 1\ \mu\text{m}$ for 5 GeV and $N_c = 3$

Technique: use thin foils for target, convert photons into e^+e^- pairs, and count as function of target thickness(t)

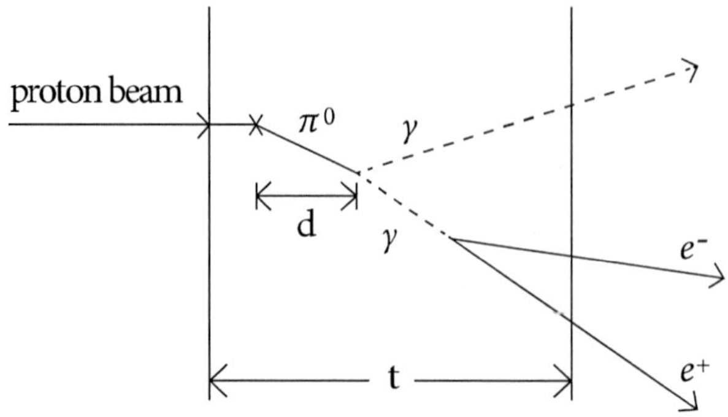

(i) production rate of π^0 is Kdx

(ii) then, π^0s decay with decay length λ

(iii) Pairs are produced immediately as Dalitz pairs in fraction B of decays, or appear in conversion with probability $\dfrac{dy}{X_0}$

Pair Production Rate

$$R(t) = kt\left\{B + \frac{1}{X_0}\left[\frac{t}{2} - \lambda + \frac{\lambda^2}{t(1-e^{t/\lambda})}\right]\right\}$$

thin foil $\Rightarrow t \ll \lambda$

$$\Rightarrow R(t) \approx kt\left(B + \frac{t^2}{6\lambda X_0}\right)$$

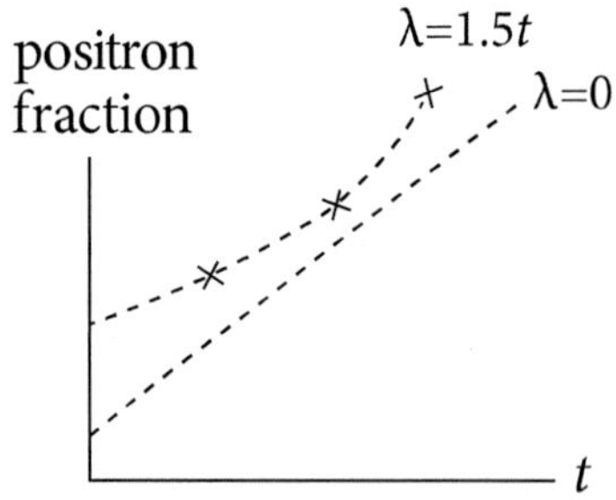

($t = 3,\ 4,\ 18,\ 58\ \mu$m, $p = 5$ GeV/c)

PDG $\Rightarrow \tau = (8.4 \pm 0.4) \times 10^{-17}$ sec $\Rightarrow N_c = 3$

■ The baryon octet

decuplet: Multiplet including the neutron and proton, lightest spin 1/2 baryons

wave functions are symmetric in spin, flavor, in space

(i.e. under each separately)

We can also form three-quark states that are symmetric under simultaneous interchange of flavor and spin of any quark pair

⇒ These states are identified with members $J^P = \frac{1}{2}^+$,

baryon octet that includes the proton and neutron as members

※ total function must be anti-symmetric

· (flavor)(spin)(color)(space)

· color is antisymmetric for all hadrons because they are color-neutral, singlet states

· space is symmetric because $L = 0 \Rightarrow \therefore$ (flavor)(spin) is symmetric

Let`s construct baryon octet wave functions

put two quarks in the spin singlet state: $\frac{1}{\sqrt{2}}(\uparrow\downarrow - \downarrow\uparrow)$ antisymmetric

⇒ We need a flavor antisymmetric combination of u and d to make the overall state symmetric:

$$\frac{1}{\sqrt{2}}(ud - du)$$

⇒ then, add the third quark u, with spin up,

$$(u_\uparrow d_\downarrow - u_\downarrow d_\uparrow - d_\uparrow u_\downarrow + d_\downarrow u_\uparrow)u_\uparrow$$

⇒ sym. under interchange of the flavor and spin, for whole expression be symmetric, (Now symmetrize by making cyclic permutation)

$$\begin{aligned}\phi_{(P, J_z = +\frac{1}{2})} = \frac{1}{\sqrt{18}}(&2u_\uparrow u_\uparrow d_\downarrow + 2d_\downarrow u_\uparrow u_\uparrow + 2u_\uparrow d_\downarrow u_\uparrow \\ &- u_\downarrow d_\uparrow u_\uparrow - u_\uparrow u_\downarrow d_\uparrow - u_\downarrow u_\uparrow d_\uparrow \\ &- d_\uparrow u_\downarrow u_\uparrow - u_\uparrow d_\uparrow u_\downarrow - d_\uparrow u_\uparrow u_\downarrow)\end{aligned}$$

The other members of the $J^P = \frac{1}{2}^+$ baryon octet can be worked out in similar fashion.

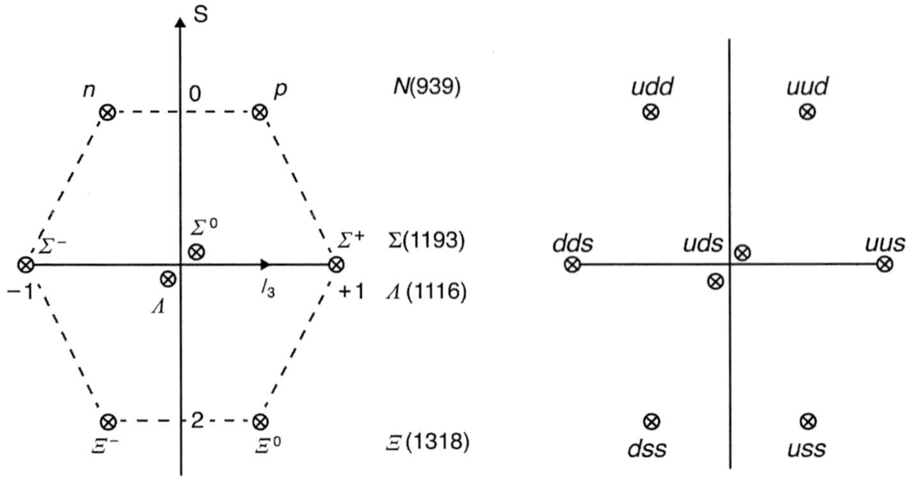

Notice the masses:

$$m_N = 939,\ m_\Sigma = 1193 = 939 + 254,\ m_\Lambda = 1116 = 939 + 177,$$

$$m_\Xi = 1318 = m_\Sigma + 125 = m_\Lambda + 202$$

Pattern is more complicated than for decuplet.

eight members are:
- n and $p(939)$ nucleon isospin doublet $\left(I = \frac{1}{2},\ S = 0\right)$
- $\Sigma(1193)$ isotriplet $(I = 1,\ S = -1)$
- $\Xi(1318)$ isodoublet $\left(I = \frac{1}{2},\ S = -2\right)$
- $\Lambda(1116)$ isosinglet $(I = 0,\ S = -1)$

in baryon decuplet, baryon masses differ because of differences in the strange quark content

⇒ But, in baryon octet

$$\underset{1193\,\text{MeV}}{m_\Sigma} = \underset{1116\,\text{MeV}}{m_\Lambda} \qquad \underset{177\,\text{MeV}}{m_\Lambda - m_N} = \underset{203\,\text{MeV}}{m_\Xi - m_\Lambda}$$

⇒ rather poor agreement: can be described by "hyperfine splitting" effects of the interquark interactions.

■ Quark-antiquark combinations ($3 \otimes \bar{3} = 8 \oplus 1$): the light pseudoscalar mesons

① for three flavors

number of meson families: $3 \times 3 = 9$ states (nonets)

Since spins of quarks and antiquarks are $\frac{1}{2}$,

$J = 1$ (spin-triplet state) and $J = 0$ (spin-singlet state) exists:

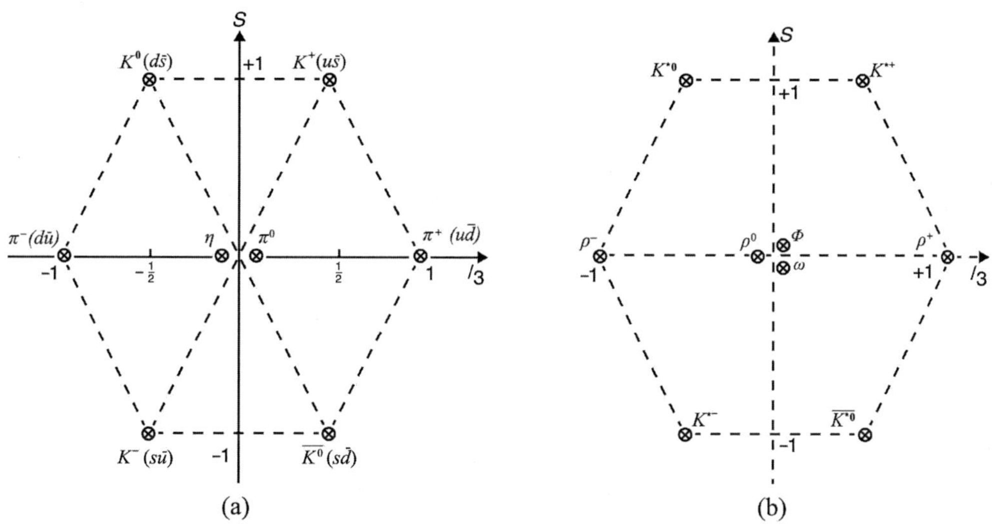

(a) lowest-lying pseudoscalar meson state ($J^P = 0^-$)

(b) vector meson nonet ($J^P = 1^-$)

Under charge conjugation operation (Condon-Shortly convention)

	nucleons		quarks	
I_3	particle	antiparticle	particle	antiparticle
+1/2	p	$+\bar{n}$	u	$+\bar{d}$
-1/2	n	$-\bar{p}$	d	$-\bar{u}$

for $\ell = 0$ and spin-singlet,

Since fermion and antifermion have opposite intrinsic parity,

$J^P = 0^- \Rightarrow$ pseudoscalar mesons

for u and d quarks and antiquarks (this table ignores the strange quark)

	I	I_3	wave function	Q/e
$I=1$ triplet	1	1	$u\bar{d} = \pi^+$	+1
	1	-1	$-\bar{u}d = \pi^-$	-1
	1	0	$\frac{1}{\sqrt{2}}(d\bar{d} - u\bar{u}) = \pi^0$	0
$I=0$ singlet	0	0	$\frac{1}{\sqrt{2}}(d\bar{d} + u\bar{u}) = \eta(550)$	0

($\because$ by applying isospin shift operator $I_\pm$, $I_+ d = u$, $I_+ \bar{u} = -\bar{d}$, $I_+ u = I_+ \bar{d} = 0$)

② introduction of s quark, possible state are $3^2 = 9$

	I	I	S	Meson	Quark combination	Decay	Mass (MeV)
octet	1	1	0	π^+	$u\bar{d}$	$\pi^\pm \to \mu\nu$	140
	1	−1	0	π^-	$d\bar{u}$	$\pi^0 \to 2\gamma$	135
	1	0	0	π^0	$(d\bar{d} - u\bar{u})/\sqrt{2}$	$K^+ \to \mu\nu$	494
	1/2	1/2	+1	K^+	$u\bar{s}$	$K^0 \to \pi^+\pi^-$	498
	1/2	−1/2	+1	K^0	$d\bar{s}$	$K^- \to \mu\nu$	494
	1/2	−1/2	−1	K^-	$\bar{u}s$	$\bar{K}^0 \to \pi^+\pi^-$	498
	1/2	1/2	−1	$\bar{K}^0$	$\bar{d}s$		
	0	0	0	η_8	$(d\bar{d} + u\bar{u} - 2s\bar{s})/\sqrt{6}$	$\eta \to 2\gamma$	549
singlet	0	0	0	η_0	$(d\bar{d} + u\bar{u} + s\bar{s})/\sqrt{3}$	$\eta \to \eta\pi\pi$ $\to 2\gamma$	958

actual state η and η' in nature appear to be linear combinations of the wave functions η_8 and η_0

Gell-Mann Okubo Mass Formula:

$$4(M_{K^0})^2 = M_{\pi^0}^2 + 3M_\eta^2$$
$$0.988\ \mathrm{GeV}^2 \qquad 0.924\ \mathrm{GeV}^2$$

Mass(MeV)	Dominant decay mode
770	$\rho \to 2\pi$
892	$K^* \to K\pi$
782	$\omega \to 3\pi$
1020	$\phi \to K\overline{K}$

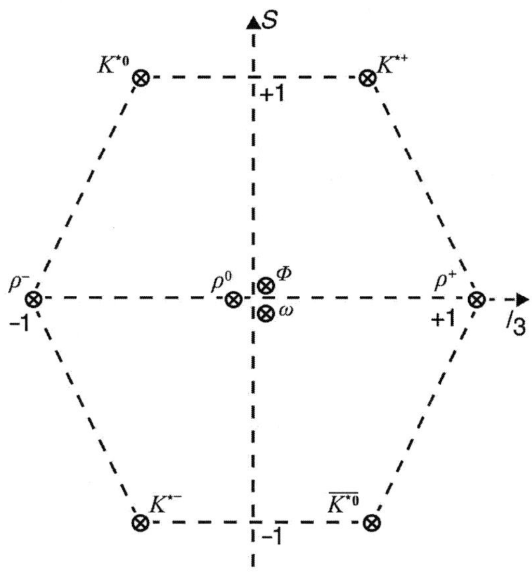

■ The light vector mesons

for $\ell = 0$ and spin-triplet $\Rightarrow$ $J^P = 1^-$ $\Rightarrow$ vector mesons

assume the octet and singlet states are mixed and θ be mixing angle,

$$\phi = \phi_0 \sin\theta - \phi_8 \cos\theta$$

$\updownarrow$ orthogonal

$$\omega = \phi_8 \sin\theta + \phi_0 \cos\theta,$$

where ϕ and ω are the physical vector mesons and

ϕ_0 and ϕ_8 are the singlet and octet states, respectively,

with $I = S = 0$

assuming the matrix element of energy operator between states yields the squared mass

$$\Rightarrow \begin{cases} M_\phi^2 = M_0^2 \sin^2\theta + M_8^2 \cos^2\theta - 2M_{08}^2 \sin\theta\cos\theta \cdots\cdots (1) \\ M_\omega^2 = M_8^2 \sin^2\theta + M_0^2 \cos^2\theta + 2M_{08}^2 \sin\theta\cos\theta \cdots\cdots (2) \end{cases}$$

since ϕ and ω are orthogonal,

$$(\text{※ } M_{\phi\omega}^2 = M_{08}^2 \sin^2\theta + M_0^2 \sin\theta\cos\theta - M_8^2 \sin\theta\cos\theta - M_{08}^2 \cos^2\theta)$$

$$M_{\phi\omega}^2 = 0 = (M_0^2 - M_8^2)\sin\theta\cos\theta + M_{08}^2(\sin^2\theta - \cos^2\theta) \cdots\cdots (3)$$

Eliminating M_{08} and M_0 between these three equations gives

$$\tan^2\theta = \frac{M_\phi^2 - M_8^2}{M_8^2 - M_\omega^2} \cdots\cdots (4)$$

from empirical mass formula: Gell-Mann and OKubo mass formula

$$M_\rho^2 + 3M_8^2 = 4M_{K^*}^2 \Rightarrow M_8^2 = \frac{1}{3}\left(4M_{K^*}^2 - M_\rho^2\right) \cdots\cdots (5)$$

$\xrightarrow{\text{Eq. (4),(5)}}$ observed mass gives $\theta \simeq 40^\circ$

for particular case, $\sin\theta = \frac{1}{\sqrt{3}}$ $(\theta \simeq 35^\circ)$ "ideal mixing"

→ almost true in practice

$$\begin{cases} \phi = \frac{1}{\sqrt{3}}(\phi_0 - \sqrt{2}\,\phi_8) \\ \omega = \frac{1}{\sqrt{3}}(\phi_8 + \sqrt{2}\,\phi_0) \end{cases} \cdots\cdots (6)$$

	I	I	S	Meson	Quark combination	Decay	Mass (MeV)
octet	1	1	0	π^+	$u\bar{d}$	$\pi^{\pm}\to\mu\nu$	140
	1	−1	0	π^-	$d\bar{u}$		
	1	0	0	π^0	$(d\bar{d}-u\bar{u})/\sqrt{2}$	$\pi^0\to 2\gamma$	135
	1/2	1/2	+1	K^+	$u\bar{s}$	$K^+\to\mu\nu$	494
	1/2	−1/2	+1	K^0	$d\bar{s}$	$K^0\to\pi^+\pi^-$	498
	1/2	−1/2	−1	K^-	$\bar{u}s$	$K^-\to\mu\nu$	494
	1/2	1/2	−1	$\bar{K}^0$	$\bar{d}s$	$\bar{K}^0\to\pi^+\pi^-$	498
	0	0	0	η_8	$(d\bar{d}+u\bar{u}-2s\bar{s})/\sqrt{6}$	$\eta\to 2\gamma$	549
singlet	0	0	0	η_0	$(d\bar{d}+u\bar{u}+s\bar{s})/\sqrt{3}$	$\eta\to\eta\pi\pi$ $\to 2\gamma$	958

$$\text{“ideal mixing”}\Rightarrow \begin{cases} \phi_0 = \dfrac{1}{\sqrt{3}}(d\bar{d}+u\bar{u}+s\bar{s}) \\ \phi_8 = \dfrac{1}{\sqrt{6}}(d\bar{d}+u\bar{u}-2s\bar{s}) \end{cases} \quad \cdots\cdots (7)$$

Substituting Eq. (7) into (6),

$$\Rightarrow \begin{cases} \phi = s\bar{s} \\ \omega = \dfrac{1}{\sqrt{2}}(u\bar{u}+d\bar{d}) \end{cases}$$

$\therefore$ ϕ is composed of s quarks only, and ω is composed of u and d quarks.

$\phi(1020)\to$ K^+K^-, $K^0\bar{K}^0$ 84%; $\pi^+\pi^-\pi^0$ 15%

$\omega(783)\to$ $\pi^+\pi^-\pi^0$ 90%; $\pi^+\pi^-$, $\pi^0\gamma$ 10%

phase-space factors favor 3π decay

since Q-value is 600 MeV compared with Q= 24 MeV

for $K\bar{K}$ decay, but $K\bar{K}$ is dominant.

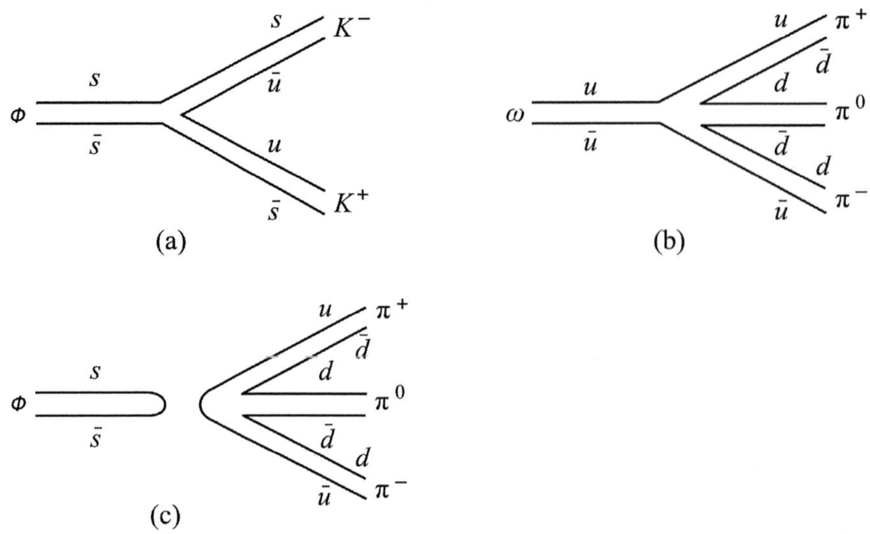

■ Other tests of quark model: quark model can be verified experimentally

- pion-nucleon cross-sections

high energy hadron-hadron cross-sections can be interpreted as due to additive effects of scattering amplitudes in collisions between individual constituent quarks.

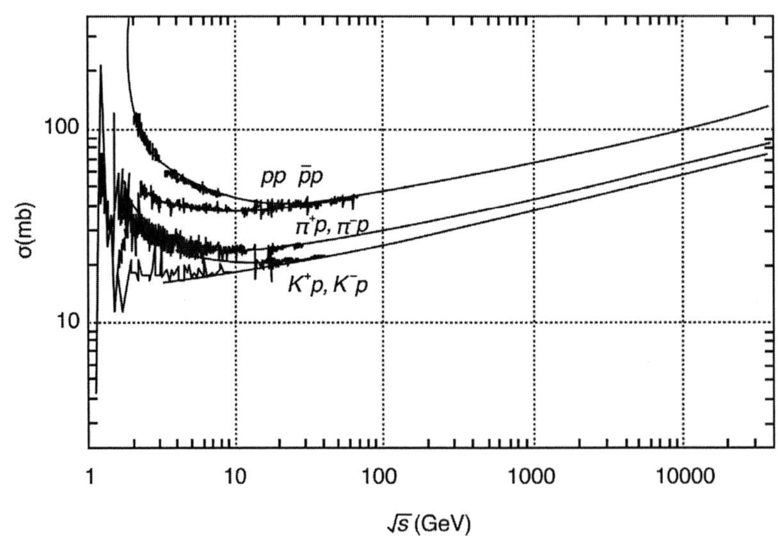

from simple quark counting

$$\frac{\sigma(\pi N)}{\sigma(NN)} = \frac{2}{3}$$

(※ at $\sqrt{S} = 60$ GeV, $\sigma(\pi N) \sim 25$ mb, $\sigma(NN) \sim 38$ mb $\Rightarrow$ ratio $\sim \frac{2}{3}$)

lepton pair production on isoscalar targets(^{12}C)

This provides a test of quark charge assignments.

6 proton(uud), 6 neutron (ddu) $\Rightarrow$ number of $u = 6\times2+6=18$,

number of $d = 6+6\times2=18$

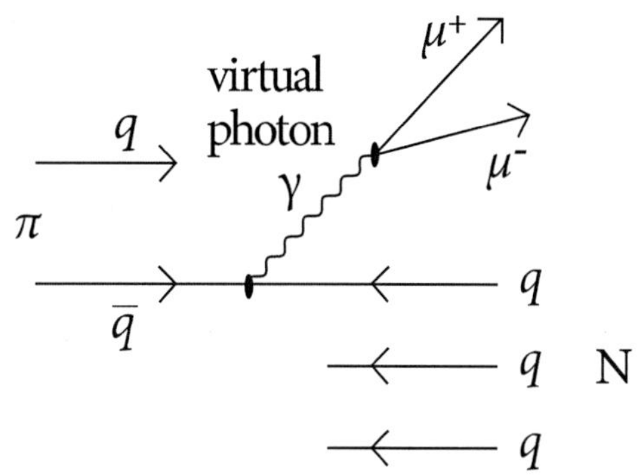

<Drell-Yan process>

cross-section for a virtual photon(EM process) is proportional to square of quark charge:

$$\sigma(\pi^- C \rightarrow \mu^+\mu^- + \cdots) \sim 18Q_u^2 = 18\left(\frac{4}{9}\right)$$

$$\sigma(\pi^+ C \rightarrow \mu^+\mu^- + \cdots) \sim 18Q_d^2 = 18\left(\frac{1}{9}\right)$$

$$(\leftarrow \pi^- = \bar{u}d,\ \pi^+ = \bar{d}u,\ ^{12}C = 18u + 18d)$$

$\Rightarrow \dfrac{\sigma(\pi^- C)}{\sigma(\pi^+ C)} = 4:1$, verified by experiment

• vector meson decays to leptons

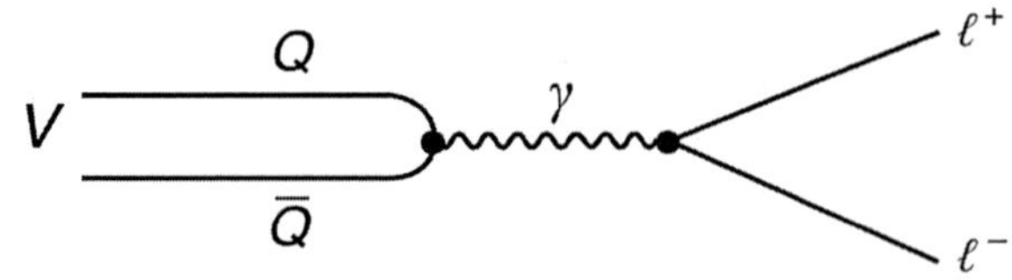

Meson	Quark wave function	$\lvert\Sigma\alpha_i Q_i\rvert^2$	$\Gamma_{e^+e^-}$ (KeV)	$\Gamma_{e^+e^-}/\lvert\Sigma\alpha_i Q_i\rvert^2$
ρ	$(u\bar{u}-d\bar{d})/\sqrt{2}$	1/2	6.8 ± 0.3	13.6 ± 0.6
ω	$(u\bar{u}-d\bar{d})/\sqrt{2}$	1/18	0.60 ± 0.02	10.8 ± 0.4
Φ	$s\bar{s}$	1/9	1.37 ± 0.05	12.3 ± 0.5
ψ	$c\bar{c}$	4/9	5.3 ± 0.4	11.9 ± 0.9
Y	$b\bar{b}$	1/9	1.32 ± 0.05	11.9 ± 0.5

$$\left[\frac{1}{\sqrt{2}}(Q_u - Q_d)\right]^2 = \frac{1}{2}\left(Q_u^2 + Q_d^2 - 2Q_u Q_d\right) = \frac{1}{2}\left[\frac{4}{9} + \frac{1}{9} - 2 \cdot \frac{2}{3}\left(-\frac{1}{3}\right)\right] = \frac{1}{2}$$

similarly,

$$\frac{1}{2}\left[\frac{4}{9} + \frac{1}{9} + 2 \cdot \frac{2}{3}\left(-\frac{1}{3}\right)\right] = \frac{1}{18}$$

- leptonic decay width $\Gamma_{e^+e^-}$ is proportional to square of quark charges.
- ratios $\dfrac{\Gamma_{e^+e^-}}{\lvert\Sigma a_i Q_i\rvert^2}$ are closely similar for ρ, ω, ϕ, Ψ and Y

  $\Rightarrow$ $\lvert\Sigma a_i Q_i\rvert^2$ is squared sum of charges of quarks in the meson.
- absolute values of leptonic widths by Van Royen-Weisskopf formula,

  $\Gamma_{(V \to \ell^+\ell^-)} = \dfrac{16\pi\alpha^2 Q^2}{M_V^2}\lvert\Psi_{(0)}\rvert^2$, $\lvert\Psi_{(0)}\rvert^2$ is $q\bar{q}$ wave function at origin

  M_V: vector meson mass

  $\lvert\Psi_{(0)}\rvert^2 = V^{-1}$, when V is volume occupied by meson

  propagator's square $\sim \dfrac{1}{q^4}$, where $\lvert q \rvert^2 = M_V^2$

  phase-space factor for two bodies introduces q^2 $\Rightarrow$ gives $\dfrac{1}{M_V^2}$

  coupling to the photon $\sqrt{\alpha}\, Q_i \Rightarrow \left(\sqrt{\alpha}\, Q_i\right)^2 = \alpha Q_i$

coupling of photon to lepton $\sqrt{\alpha}\sqrt{\alpha} \leftarrow (\sqrt{\alpha})^2 = \alpha \Rightarrow$ gives $\alpha^2 Q_i^2$

■ Mass relations and hyperfine splitting

※ spin-orbit coupling → fine structure

spin-spin coupling → hyperfine structure

The observed mass differences between baryon octet and decuplet members having the same quark constitution, due to effects of quark-quark interactions (quantum chromodynamics, QCD)

hyperfine splitting of hadron energy levels by color force between quarks (due to color magnetic moment interaction)

Consider two charged point fermions with magnetic dipole moment $\vec{\mu}_i$ and $\vec{\mu}_j$.

The field at the centre of a uniformly magnetized sphere is

$B = \frac{2}{3}M$, where magnetization $M = \frac{\mu}{V}$ is the magnetic moment per unit volume($|\Psi_{(0)}|^2 = V^{-1}$)

⇒ spin-spin magnetic interaction energy is

$$\Delta E = \vec{\mu} \cdot \vec{B} = \frac{2}{3}\vec{\mu}_i \cdot \vec{\mu}_j |\Psi_{(0)}|^2$$

$$= \frac{2\pi}{3}\frac{\alpha}{m_i m_j}|\Psi_{(0)}|^2 \vec{\sigma}_i \cdot \vec{\sigma}_j \text{ (in } \hbar = c = 1) \cdots\cdots (1)$$

$$(\because \vec{\mu}_i = \frac{e_i}{2m_i}\vec{\sigma}_i, \quad e_i e_j = 4\pi\alpha \text{ with } \hbar = c = 1 \text{ units}$$

$$\mu = 2\mu_B S = 2\frac{e\hbar}{2mc}\frac{\hbar}{2}\sigma, \quad \alpha = \frac{e^2}{4\pi\hbar c})$$

[1] Now turning to quarks,

quarks carry a strong color charge

⇒ associated with color charges of spining quarks is color magnetic interaction

⇒ same form as equation (1) but electric charges is replaced by color charge

$$\Rightarrow \Delta E(q\bar{q}) = \frac{4}{3}\frac{2\pi}{3}\frac{\alpha_s}{m_i m_j}|\Psi_{(0)}|^2 \vec{\sigma}_i \cdot \vec{\sigma}_j$$

$$(\because \ U_{QCD} = -\frac{4}{3}\frac{\alpha_s}{r} + kr \ , \ U_{em} = -\frac{\alpha}{r})$$

$$\Delta E(qq) = \frac{1}{2}\frac{4}{3}\frac{2\pi}{3}\frac{\alpha_s}{m_i m_j}|\Psi_{(0)}|^2 \vec{\sigma}_i \cdot \vec{\sigma}_j$$

in here, $\vec{\sigma}_i \cdot \vec{\sigma}_j$ depends in magnitude and sign on the relative quark spin orientations

$$\vec{\alpha}_i \cdot \vec{\alpha}_j = 4\vec{S}_i \cdot \vec{S}_j \quad \left(\because \ \vec{S} = \frac{\hbar}{2}\vec{\sigma} \xrightarrow{\hbar = c = 1} \vec{S} = \frac{1}{2}\vec{\sigma}\right)$$

$$= 2[S(S+1) - S_i(S_i+1) - S_j(S_j+1)]$$

$$= \begin{cases} +1 \ \text{for} \ S=1 & \because \ 2\left[1(1+1) - \frac{1}{2}\left(\frac{1}{2}+1\right) - \frac{1}{2}\left(\frac{1}{2}+1\right)\right] \\ -3 \ \text{for} \ S=0 & \because \ 2\left[0(0+1) - \frac{1}{2}\left(\frac{1}{2}+1\right) - \frac{1}{2}\left(\frac{1}{2}+1\right)\right] \end{cases}$$

$$※ \ \vec{S} = \vec{S}_i + \vec{S}_j \Rightarrow \vec{S}_i \cdot \vec{S}_j = \frac{1}{2}\left(S^2 - S_i^2 - S_j^2\right)$$

[2] Turning to baryons

$$\Sigma \vec{\alpha}_i \cdot \vec{\alpha}_j = 4\Sigma \vec{S}_i \cdot \vec{S}_j = 4\frac{1}{2}[S(S+1) - s_i(s_i+1) - s_j(s_j+1) - s_k(s_k+1)]$$

$$= 2[S(S+1) - 3S(S+1)] \quad (\because \ S = s_i + s_j + s_k)$$

$$= \begin{cases} +3 \ \text{for} \ S = \frac{3}{2} \\ -3 \ \text{for} \ S = \frac{1}{2} \end{cases}$$

Then,

$$(\Delta E)_\Delta = \frac{4\pi\alpha_s}{9m_u^2}|\Psi_{(0)}|^2 \vec{\sigma}_i \cdot \vec{\sigma}_j = \frac{4\pi\alpha_s}{9m_u^2}|\Psi_{(0)}|^2 \times 3$$

$$(\because \ m_u \approx m_d, \ \vec{\sigma}_i \cdot \vec{\sigma}_j = +3)$$

$$= \frac{4\pi\alpha_s}{3m_u^2}|\Psi_{(0)}|^2 = \frac{3}{m_u^2}K, \text{ where } K = \frac{4\pi\alpha_s}{9}|\Psi_{(0)}|^2$$

$$(\Delta E)_N = \frac{4\pi\alpha_s}{9m_u^2}|\Psi_{(0)}|^2 \vec{\sigma}_i \cdot \vec{\sigma}_j = \frac{4\pi\alpha_s}{9m_u^2}|\Psi_{(0)}|^2 \times (-3),$$

$$= -\frac{3}{m_u^2}K, \text{ where } K = \frac{4\pi\alpha_s}{9}|\Psi_{(0)}|^2$$

$$(\Delta E)_\Sigma = \frac{K}{m_u^2} - \frac{4}{m_u m_s}K = \left(\frac{1}{m_u^2} - \frac{4}{m_u m_s}\right)K, \text{ where } K = \frac{4\pi\alpha_s}{9}|\Psi_{(0)}|^2$$

in here, $\Sigma = \Sigma^+(1193) = uus$, $\Sigma^+ = I(J^P) = 1\left(\frac{1}{2}^+\right)$

($\because$ like pair is $u_\uparrow u_\uparrow \Rightarrow \vec{\sigma}_u \cdot \vec{\sigma}_u = 1$

$2\vec{\sigma}_u \cdot \vec{\sigma}_s = \sum \vec{\sigma}_i \cdot \vec{\sigma}_j - \vec{\sigma}_u \cdot \vec{\sigma}_u = -4$ ($\because$ $-3-1=-4$))

($\because$ $\sigma_u \cdot \sigma_u + \sigma_u \cdot \sigma_s + \sigma_u \cdot \sigma_s = \sigma_u \cdot \sigma_u + 2\sigma_u \cdot \sigma_s$)

from the eight isomultiplets of baryons in the decuplet and octet, we are able to fit the unknown parameters K, m_u and m_s

$$\Rightarrow \begin{cases} m_n\ (= m_u = m_d) = 363 \text{ MeV} \\ \text{m}_s = 538 \text{ MeV} \\ \text{K}/\text{m}_n^2 = 50 \text{ MeV} \end{cases}$$

$\Rightarrow$ observed and expected masses are well agreed as shown in the below table.

Baryon and Mass (MeV)	Quark combination ($n = u$ or d)	$\Delta E/K$	Predicted Mass (MeV)
N(939)	$3n$	$-3/m_n^2$	939
Λ(1116)	$2n$, $1s$	$-3/m_n^2$	1114
Σ(1193)	$2n$, $1s$	$1/m_n^2 - 4/(m_n m_s)$	1179
Ξ(1318)	$1n$, $2s$	$1/m_n^2 - 4/(m_n m_s)$	1327
Δ(1232)	$3n$	$3/m_n^2$	1239
Σ(1384)	$2n$, $1s$	$1/m_n^2 - 4/(m_n m_s)$	1381
Ξ(1533)	$1n$, $2s$	$1/m_s^2 + 4/(m_n m_s)$	1529
Ω(1672)	$3s$	$3/m_s^2$	1682

$\Rightarrow$ hyperfine splitting due to color magnetic moment interaction accounts for $\sim$ 300 MeV mass difference between Δ(1232) and N(939)

$$\because \ (\Delta E)_\Delta = \frac{3K}{m_u^2} = 3\times 50 = 150 \text{ MeV}$$

$$(\Delta E)_N = -\frac{3K}{m_u^2} = -3\times 50 = -150 \text{ MeV}$$

$\Rightarrow$ mass difference in 300 MeV

■ Electromagnetic mass differences and isospin symmetry

actual mass of a charged hadron

$$m_{hadron} = m_{bare} + \Delta m_{em} \quad \cdots\cdots (1)$$

← ① quark constituents and their strong mutual interactions(≡"bare" mass m)

※ described in the previous section

② contribution due to electric charge of hadron(≡Δm)

= equal to work required to put charge on the previously uncharged particle

In Eq. (1),

m_{bare}: constituents plus strong interaction

Δm_{em}: expect the Δm_{em} for multiplet to be same for like charge particle

If all baryons in $\frac{1}{2}^+$ octet have similar charge distribution,

$$\underset{}{\Delta m_p} = \underset{(uus)}{\Delta m_{\Sigma^+}}, \quad \underset{(dds)}{\Delta m_{\Sigma^-}} = \underset{(dss)}{\Delta m_{\Xi^-}}, \quad \underset{(uss)}{\Delta m_{\Xi^0}} = \Delta m_n$$

Adding "bare" masses to each side and summing these equations, ($m_d \approx m_u$)

$$m_p + m_{\Sigma^-} + m_{\Xi^0} = m_{\Sigma^+} + m_{\Xi^-} + m_n$$

$$\Rightarrow (m_p - m_n) = (m_{\Sigma^+} - m_{\Sigma^-}) + (m_{\Xi^-} - m_{\Xi^0}) \cdots\cdots (2)$$

-1.3MeV -8.0MeV +6.4MeV

└ -1.6MeV

individual mass differences are associated with isospin symmetry breaking

In the context of the quark model, there are several distinct effects to consider in accounting for mass differences.

$\Rightarrow$ ① $m_d > m_u$ (difference in masses of u and d quarks), sign of each-term in Eq. (2) indicates $m_d > m_u$)

② Coulomb energy difference associated with electrical energy between pairs of quarks.

$\Rightarrow$ this will be of order $\frac{e^2}{R_0} = \frac{e^2}{\hbar c}\frac{\hbar c}{R_0} \approx 2\ \text{MeV}$,

where R_0 is the baryon size$\simeq 0.8\ \text{fm}$, $\frac{e^2}{\hbar c} = 4\pi\alpha \Rightarrow$ order $\frac{1}{137}$,

$\hbar c = 197\ \text{MeV fm}$

③ Magnetic energy difference associated with magnetic-moment (hyperfine) interaction between quark pairs.

$\Rightarrow$ this will be of order $\left(\frac{e\hbar}{mc}\right)^2\left(\frac{1}{R_0}\right)^3$,

where m is quark mass, $|\Psi_{(0)}|^2 \simeq R_0^{-3}$ and $\frac{\hbar}{mc} \sim R_0$

$\Rightarrow$ magnetic energy is in the region of $\frac{e^2}{R_0}$

$\Rightarrow 1 \sim 2\ \text{MeV}$

$\Rightarrow$ Fitting the exact forms of these three terms to the numbers in Eq. (2), it is found that $m_d - m_u = 2\ \text{MeV}$

Mass differences in isospin multiplets

	$\Delta m(\text{MeV}/c^2)$	$m_{av}(\text{MeV}/c^2)$	$10^3\ \Delta m/m$
$n-p$	1.3	939	1.4
$\Sigma^0 - \Sigma^+$	3.1	1190	2.6
$\Sigma^- - \Sigma^0$	4.9	1195	4.1
$\Xi^- - \Xi^0$	6.4	1318	4.9
$K^0 - K^\pm$	4.0	495	8.1
$\pi^\pm - \pi^0$	4.6	140	33

■ Magnetic moments of baryons

• magnetic moments of the nucleons

- proton: 2.793 nuclear magnetons
- neutron: –1.913 nuclear magnetons

[nuclear magneton= $\frac{e\hbar}{2m_p c}$]= μ_N , Bohr magneton= $\mu_B = \frac{e\hbar}{2m_e}$

for proton(uud),

two u quarks is in spin-triplet state χ: $J_\chi = 1$, $m = 0, \pm 1$

d quark state ϕ can be $J_\phi = \frac{1}{2}$, $m = \pm\frac{1}{2}$

① total angular momentum function for a spin-up proton is $\Psi(J=\frac{1}{2}, m=\frac{1}{2})$

from Clebsch-Gordan coefficients for combinning: $J_\chi = 1$, $J_\phi = \frac{1}{2}$,

$$\Psi(\frac{1}{2},\frac{1}{2}) = \sqrt{\frac{2}{3}}\chi(1,1)\phi(\frac{1}{2},-\frac{1}{2}) - \sqrt{\frac{1}{3}}\chi(1,0)\phi(\frac{1}{2},\frac{1}{2})$$

then

proton moment $\mu_p = \frac{2}{3}(2\mu_u - \mu_d) + \frac{1}{3}\mu_d = \frac{4}{3}\mu_u - \frac{1}{3}\mu_d$

② for neutron, $u \to d$ in the above wave function

so, neutron moment $\mu_N = \frac{4}{3}\mu_d - \frac{1}{3}\mu_u$ (labels u and d interchanged)

③ for $\Sigma^+ = uus$

$\Rightarrow$ replaces μ_d by μ_s in μ_p

$\Rightarrow \mu_{\Sigma^+} = \frac{4}{3}\mu_u - \frac{1}{3}\mu_s$

④ for $\Sigma^- = dds$

$\Rightarrow$ replaces μ_u by μ_d in the above μ_{Σ^+} (or replaces μ_u by μ_s in μ_n)

$\Rightarrow \mu_{\Sigma^-} = \frac{4}{3}\mu_d - \frac{1}{3}\mu_s$

⑤ for $\Lambda = uds\left(\left[I(J^P) = 0\left(\frac{1}{2}^+\right)\right]\right)$ with $I = 0$

$\Rightarrow$ pair u and d must be in an $I = 0$ isospin state

$\Rightarrow$ they must be in an antisymmetric spin state $J = 0$

$\Rightarrow$ u and d make no contribution to the moment

so, $\mu_\Lambda = \mu_s$

⑥ for $\Xi^0 = uss$

$\Rightarrow$ replaces μ_d by μ_s in μ_n

$\Rightarrow \mu_{\Xi^0} = \frac{4}{3}\mu_s - \frac{1}{3}\mu_n$

⑦ for $\Xi^- = dss$

$\Rightarrow$ replaces μ_u by μ_s in μ_p (or replaces μ_u by μ_d in the above μ_{Ξ^0})

$\Rightarrow \mu_{\Xi^-} = \frac{4}{3}\mu_s - \frac{1}{3}\mu_d$

⑧ for $\Omega^- = sss$

$\Rightarrow \mu_{\Omega^-} = 3\mu_s$

$m_u \approx m_d$ and quark charge assignments in $\vec{\mu}_i = \left(\frac{e_i}{2m_i}\right)\vec{\sigma}_i$

and

$$\begin{cases} \mu_p = \frac{4}{3}\mu_u - \frac{1}{3}\mu_d \\ \mu_n = \frac{4}{3}\mu_d - \frac{1}{3}\mu_u \end{cases} \Rightarrow \mu_n = -\frac{2}{3}\mu_p \text{ (experiment result, } \frac{\mu_n}{\mu_p} = -0.685 \text{ well agree)}$$

$$\Rightarrow \frac{\mu_n}{\mu_p} = -\frac{2}{3}$$

$$※\ \mu_p \Rightarrow \frac{4}{3}\left(\frac{2}{3}\right) - \frac{1}{3}\left(-\frac{1}{3}\right) = \frac{8}{9} + \frac{1}{9} = 1$$

$$\mu_n \Rightarrow \frac{4}{3}\left(-\frac{1}{3}\right) - \frac{1}{3}\left(\frac{2}{3}\right) = -\frac{4}{9} - \frac{2}{9} = -\frac{2}{3}$$

measured moments of proton and Λ hyperon $\Rightarrow$ gives m_n and m_s

$$\Rightarrow \begin{cases} m_n (= m_u = m_d) = 336 \text{ MeV} \\ \mathrm{m}_s = 509 \text{ MeV} \end{cases}$$

support basic correctness of quark model of hadrons

Baryon	Magnetic moment in quark model	Predicted magnetic moment	Observed magnetic moment
p	$4\mu_u/3-\mu_d/3$	+2.79	+2.793
n	$4\mu_d/3-\mu_u/3$	−1.86	−1.913
Λ	μ_s	−0.61	-0.614 ± 0.005
Σ^+	$4\mu_u/3-\mu_s/3$	+2.68	$+2.46\pm0.01$
Σ^-	$4\mu_d/3-\mu_s/3$	−1.04	-1.16 ± 0.03
Ξ^0	$4\mu_s/3-\mu_u/3$	−1.44	-1.25 ± 0.014
Ξ^-	$4\mu_s/3-\mu_d/3$	−0.51	-0.65 ± 0.01
Ω^-	$3\mu_s$	−1.84	-2.02 ± 0.05

■ Mesons built of light and heavy quarks

charmed pseudoscalar ($J^P=0^-$) D meson

$$\begin{cases} D^+ = c\bar{d} \\ D^0 = c\bar{u} \xrightarrow[3.88\%]{} K^-\pi^+ \\ D_s^+ = c\bar{s} \\ \quad\vdots \end{cases}$$

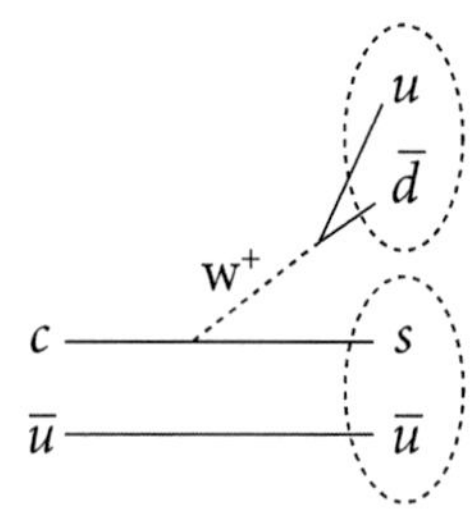

decay by weak interaction with $\tau \simeq 10^{-12}$ sec

$J^P=2^+$: radial excitations

charmed vector ($J^P=1^-$) meson

$$\begin{cases} D^{*0} = c\bar{u} \\ D_s^{*+} = c\bar{s} \\ \quad\vdots \end{cases}$$

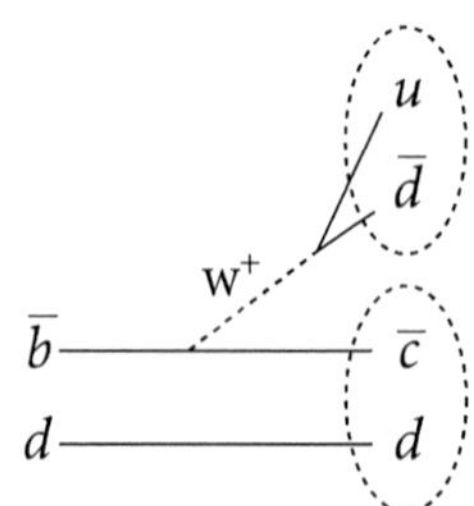

pseudoscalar B meson and vector B^* mesons

$$B^+ = u\bar{b}$$

$$B^0 = d\bar{b} \quad \rightarrow D^-\pi^+ \ (D^-\text{: charmed meson})$$

$$B_s^0 = s\bar{b}$$

$$\vdots$$

$$M_{D_s(c\bar{s})} - M_{D^0(c\bar{u})} = 99 \pm 1 \ \text{MeV}$$

$$M_{B_s(s\bar{b})} - M_{B(u\bar{b})} = 90 \pm 3 \ \text{MeV}$$

$\Rightarrow$ mass difference depends on light quark flavors and hardly at all the heavy quark mass.

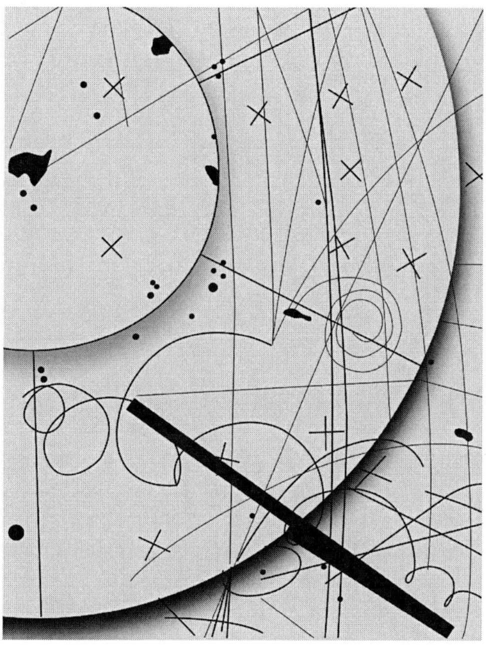

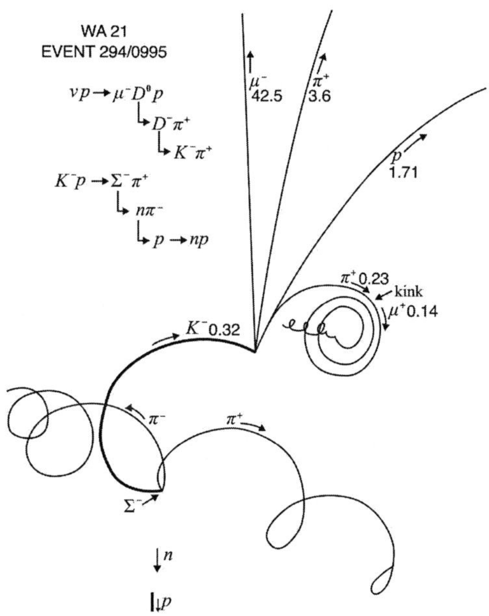

Heavy quark mass	Triplet - singlet difference	Product
$M_c \simeq 1.86$ GeV	$M_{D^*} - M_D = 0.14$ GeV	$M_c(M_{D^*} - M_D) = 0.26\ \text{GeV}^2$
$M_b \simeq 5.28$ GeV	$M_{B^*} - M_B = 0.046$ GeV	$M_b(M_{B^*} - M_B) = 0.24\ \text{GeV}^2$

■ The top quark (discovered in 1995 at FNAL, $m_t = 175\ \mathrm{GeV}/c^2$)

└ $t\bar{t}$ system is so short-lived that no discrete bound state

$t\bar{t}$ production

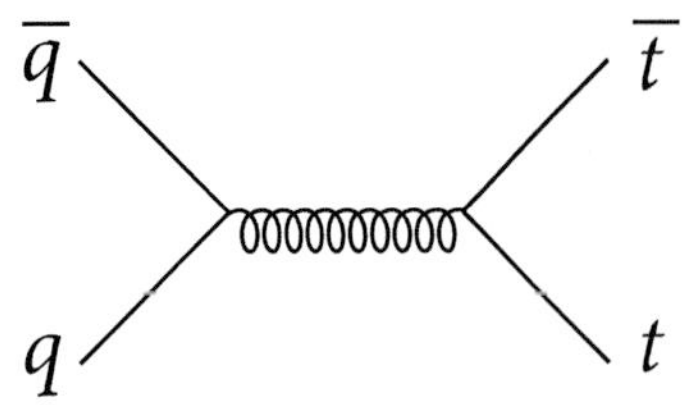

(a) via singlet-gluon exchange

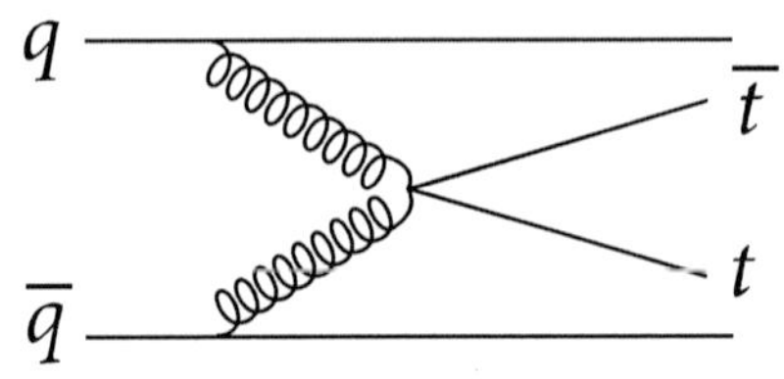

(b) via gluon fusion (less important)

then,

$t\bar{t} \to W^+ b + W^- \bar{b}$ weak decay (※ $p\bar{p} \to t\bar{t} + X$: $1/10^{10}$ collisions)

$\hookrightarrow \begin{cases} e\nu,\ \mu\nu,\ \tau\nu \\ q\bar{q} \end{cases}$

※ lifetime of B mesons : 1.6×10^{-12} sec $\Rightarrow c\tau_B = 0.48$ mm

$\Rightarrow$ The distance between B vertex and primary collision in a direction perpendicular to the beam axis is

$$\Delta z = \left(\frac{p_T}{m_B c}\right) c\tau_B\ ,\quad p_T \text{ is the transverse momentum of } B \text{ meson}$$

$\hookrightarrow \beta\gamma$

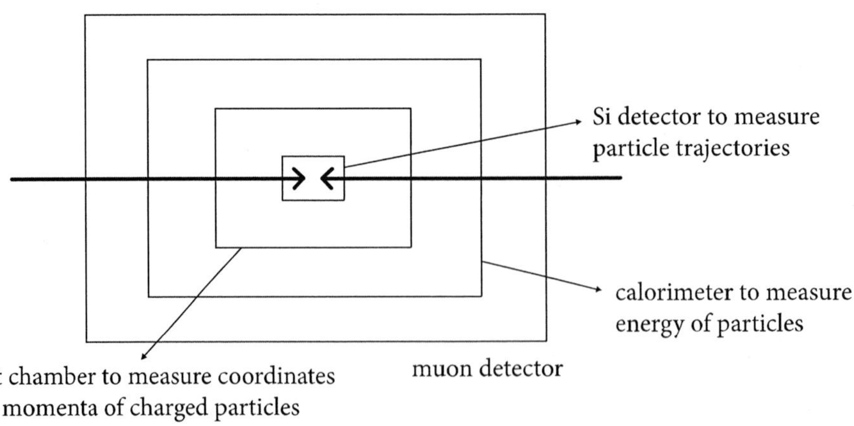

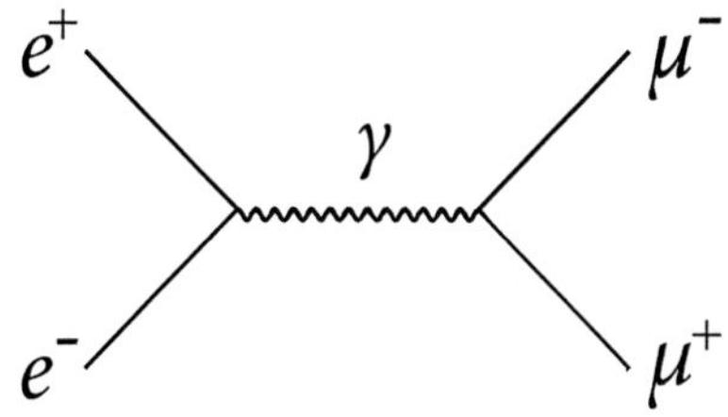

$$\sigma(e^+e^- \rightarrow \mu^+\mu^-) = \frac{4\pi\alpha^2}{3s}, \text{ where } \sqrt{s} = E_{cms}$$

$\Rightarrow$ We expect

$$\sigma(q\bar{q} \rightarrow t\bar{t}) = \frac{4\pi\alpha_s^2}{3\hat{s}}\frac{2}{9},$$

where $\hat{s}$ is squared quark-antiquark cms energy

α_s is the strong quark-gluon coupling

$\frac{2}{9}$ is the factor due to color

Chapter 5

Lepton and Quark Scattering

Introduction

Deep inelastic lepton-nucleon scattering revealed dynamical understanding of quark substructure.

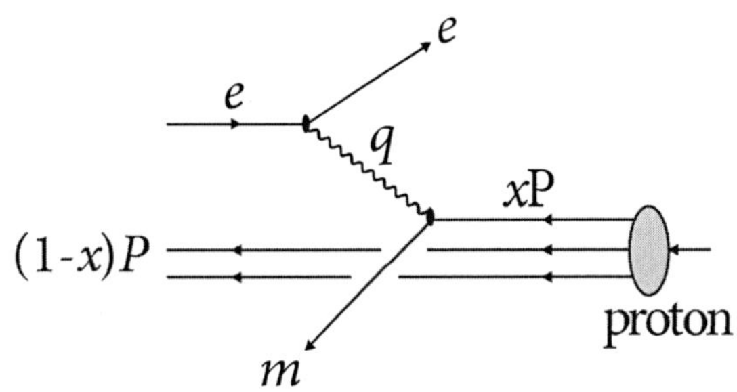

- deep inelastic lepton-nucleon scattering experiments

  ⇒ ┌① evidence for quarks

  → identified as a quark

  └② information on the nature of quark-quark interactions

- interpreted as electric scattering of lepton by a pointlike constituent of nucleon-theory of scattering of two spin $\frac{1}{2}$, point like particles required

① and ② in the above are reinforced and greatly strengthened and enlarged by e^+e^- annihilation to hadron studies at high energies.

⇒ start with simpler process $e^+e^- \to \mu^+\mu^-$ (annihilation process)

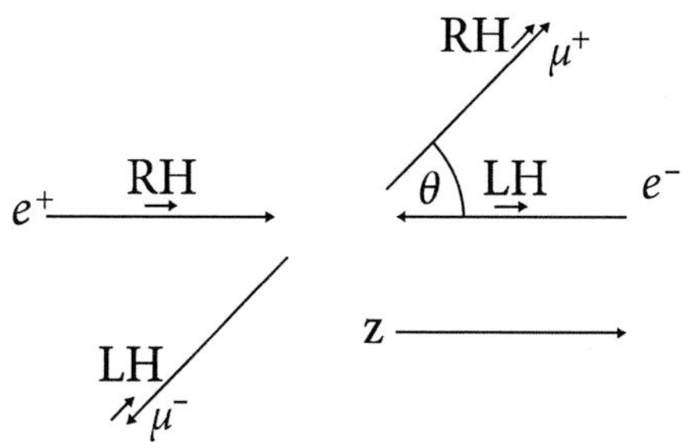

(annihilation process viewed in cms)

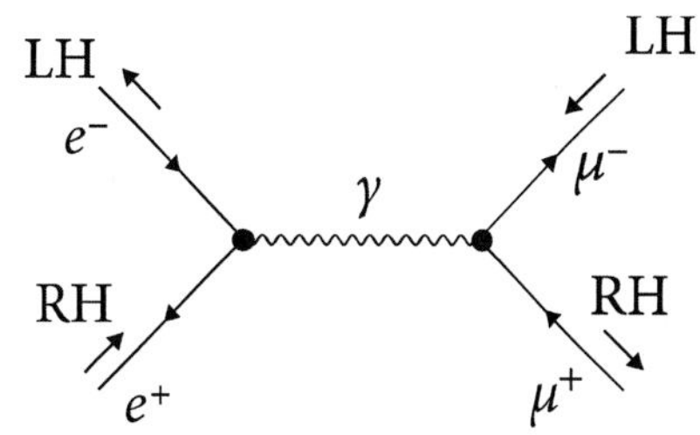

(annihilation process $e^+e^- \to \mu^+\mu^-$)

$\to$ with neglecting effects of spin, $M_{\mathrm{if}} = \dfrac{e^2}{q^2} = \dfrac{4\pi\alpha}{q^2}$ $\left(\because\ \alpha = \dfrac{e^2}{4\pi\hbar c}\right)$

($\because$ squared four-momentum of virtual photon: $q^2 = -m^2 = -E_{cms}^2 = -S$,

In here, we assume that this energy is so large that electron and muon masses can be neglected.)

and

relative velocities in initial and final states are

$v_i = v_f = 2c$

and

$E_{cms}^2 = 4E_1E_2 = 4E_1^2 \Rightarrow E_{cms} = 2E_1 = 2p_fc$, p_f is the cms momentum

$$\Rightarrow p_f = \frac{E_{cms}}{2c} = \frac{\sqrt{S}}{2c}$$

$$\Rightarrow \frac{d\sigma}{d\Omega}(a+b \to c+d) = \frac{1}{4\pi^2\hbar^4}|M_{\mathrm{if}}|^2\frac{p_f^2}{v_iv_f}$$

with natural units

$$\Rightarrow \frac{d\sigma}{d\Omega} = \frac{1}{4\pi^2}\frac{16\pi^2\alpha^2}{S^2}\frac{1}{4}\frac{S}{4} = \frac{\alpha^2}{4S}$$

Now we have to consider effects of spin:

① $e^+e^- \to \mu^+\mu^-$: $\dfrac{d\sigma}{d\Omega}(e^+e^- \to \mu^+\mu^-) = \dfrac{\alpha^2}{4S}(1+\cos^2\theta)$,

where $(1+\cos^2\theta)$ is the spin effect

$$\Rightarrow \sigma(e^+e^- \to \mu^+\mu^-) = \frac{4\pi\alpha^2}{3S} \text{ (piontlike elastic scattering)}$$

② $e^+e^- \to$ hadron: introduced "color" quantum number

$e^+e^- \to$ as a result of two jets experiment (angular distribution)

$\Rightarrow$ spin $\frac{1}{2}$ quark constituents

③ $e^-\mu^+ \to e^-\mu^+$: cross-section is expressed by Mandelstam variables($s,\ t,\ u$)

$$\frac{d\sigma}{d\Omega}(e^+e^- \to \mu^+\mu^-) = \frac{\alpha^2}{4S}(1+\cos^2\theta) = \frac{\alpha^2}{8p^2}\left(\frac{t^2+u^2}{s^2}\right)$$

$\downarrow\ S \to -t$

$$\frac{d\sigma}{d\Omega}(e^-\mu^+ \to e^-\mu^+) = \frac{\alpha^2}{8p^2}\left(\frac{s^2+u^2}{t^2}\right) = \frac{\alpha^2}{8p^2\sin^4\left(\frac{\theta}{2}\right)}\left[1+\cos^4\left(\frac{\theta}{2}\right)\right]$$

↳ in the lab frame,

$$\frac{d\sigma}{dy}(e^-\mu^+ \to e^-\mu^+) = \frac{2\pi\alpha^2 S}{q^4}[1+(1-y)^2]$$

④ $\nu e \to \nu e$ (weak interaction)

$$\frac{d\sigma}{dy}(\nu_e e \to \nu_e e) = \frac{G_F^2 S}{\pi}$$

$$\frac{d\sigma}{dy}(\bar{\nu}_e e \to \bar{\nu}_e e) = \frac{G_F^2 S}{\pi}(1-y)^2$$

⑤ electron − nucleon scattering
⑥ neutrino − nucleon scattering
$\Rightarrow$ valence quark, sea quark, structure functions, ⋮

We know that extreme relativistic particles helicity is conserved in a vector or axial vector interaction.

$\Rightarrow$ in scattering of an electron, $e_L^- \to e_L^-$ and $e_R^- \to e_R^-$

<electron-muon scattering via single-photon exchange>

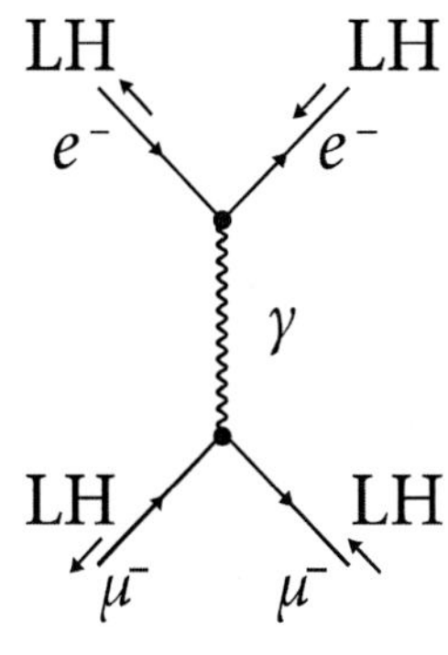

■ For $e^{-}e^{+}$ process

obtained by replacing outgoing electron in the above diagram by an ingoing positron with opposite helicity

<s channel>

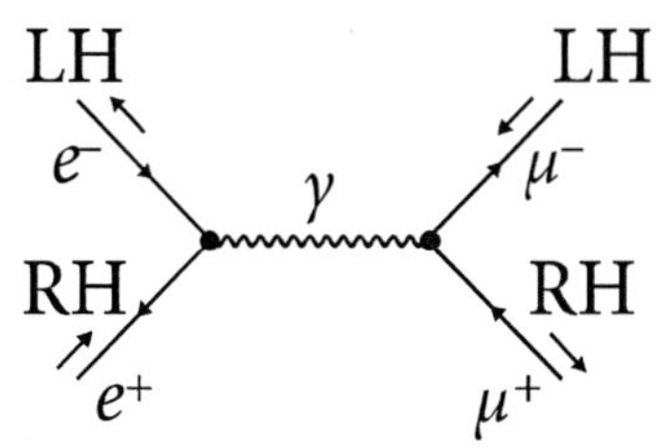

e_L^- and e_R^+ are coupled, but not e_L^- and e_L^+

In the center-of-mass system

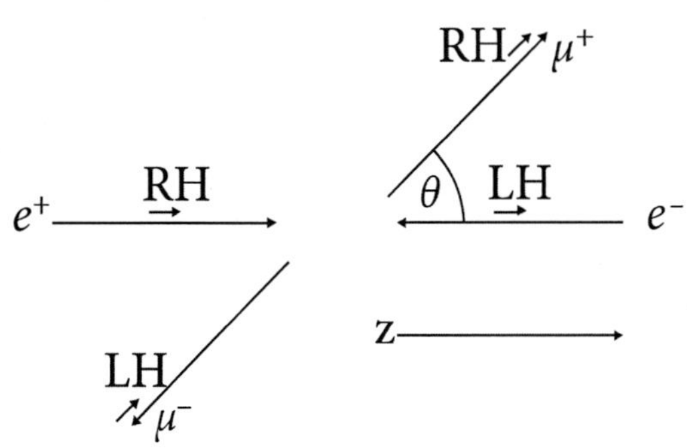

- θ is the emission angle of $\mu^+\mu^-$ pair with respect to $z-$axis
- z is the incident direction

initial state in $J_z = +1$ for RL combination,

or (in here R is positively charged lepton)

$J_z = -1$ for LR combination

so, exchanged photon is transverse:

transverse photon is superposition of $J_z = +1$ and $J_z = -1$ spin substates.

The probability amplitude for this pair to have $J_z = +1$, as in the initial states, is

$$A_{RL\to RL} = d^J_{m,m'}(\theta) = d^1_{1,1}(\theta) = \frac{1+\cos\theta}{2} \qquad (\leftarrow d\text{: d-function})$$

Meanwhile,

$$A_{RL\to LR} = d^1_{1,-1}(\theta) = \frac{1-\cos\theta}{2}$$

$$\Rightarrow A^2_{RL\to RL} + A^2_{RL\to LR} = \frac{1+2\cos\theta+\cos^2\theta}{4} + \frac{1-2\cos\theta+\cos^2\theta}{4} = \frac{1+\cos^2\theta}{2} = P(\theta),$$

$$\left[\because\ M^2 \sim \left(\frac{1+\cos\theta}{2}\right)^2 + \left(\frac{1-\cos\theta}{2}\right)^2\right]$$

where $P(\theta)$ is the angular dependence

to include $LR\to RL$ and $LR\to LR$, we have to double the above equation

so,

$$\frac{d\sigma}{d\Omega}(e^+e^- \to \mu^+\mu^-) = \frac{\alpha^2}{4S}(1+\cos^2\theta),$$ where $(1+\cos^2\theta)$ is the spin effect

$$\Rightarrow \sigma(e^+e^- \to \mu^+\mu^-) = \frac{\alpha^2}{4S}2\pi\frac{8}{3} = \frac{4\pi\alpha^2}{3S}$$

$$[\because \int_0^\pi (1+\cos^2\theta)\sin\theta d\theta = \int_0^\pi \sin\theta d\theta - \int_0^\pi \cos^2\theta d(\cos\theta)$$

$$= [-\cos\theta]_o^\pi + \int_{-1}^1 x^2 dx = 2 + \frac{1}{3}[x^3]_{-1}^1 = \frac{8}{3}$$

and

2π is from azimuthal angle]

$$= 86.8\ \frac{nb}{S}(\mathrm{GeV}^2)$$

$$[\because\ \frac{4\pi\alpha^2}{3}\times(0.1975\times10^{-15})^2\quad(\leftarrow 1\ \mathrm{GeV}^{-1} = 0.1975\ \mathrm{fm})]$$

$\Rightarrow$ This applies to pointlike leptons annihilating via single-photon exchange, at energies where lepton masses can be neglected.

[Supplement-1]

- $e^+ + e^- \rightarrow \mu^+ + \mu^-$: through a virtual photon

  $\Rightarrow$ total angular momentum is $J = 1$

  - take angular momentum quantization axis z along the positron line of flights

    $\rightarrow$ third components of spins of electron and positron can be either both $+1/2$ or both $-1/2$ (not one $+1/2$ and one $-1/2$)

  - In final state, choose as quantization axis z', the line of flight of μ^+

    $\rightarrow$ third component of the orbital momentum is zero

    $\rightarrow$ third component of total angular momentum can be $m' = +1$ or $m' = -1$

    $\rightarrow$ components of final spins must be either both $+1/2$ or both $-1/2$

$\Rightarrow$ In total, there are four cases

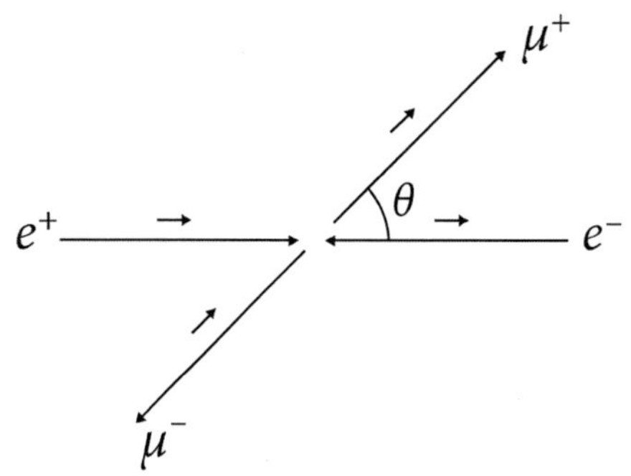

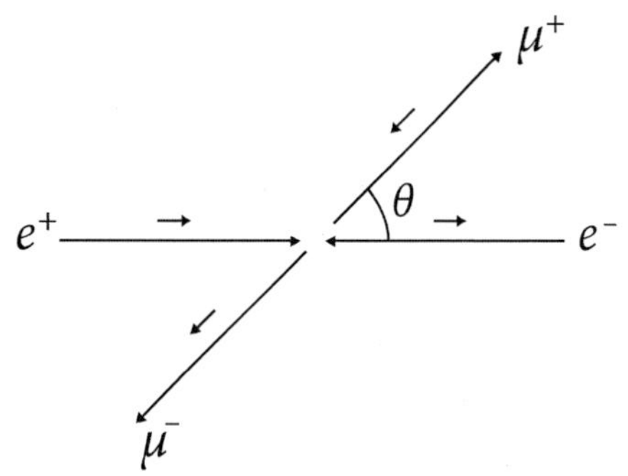

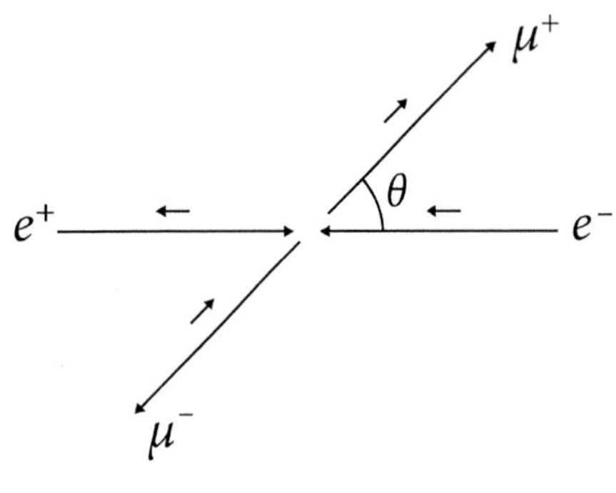

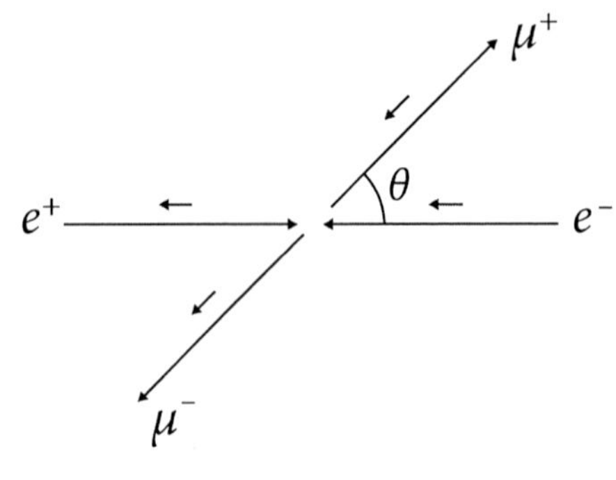

The matrix element for each $J=1,\ m,\ m'$ case is proportional to the rotation matrix from the axis z to the axis z', namely to $d^1_{m,m'}(\theta)$,

※ d-functions (rotation matrices)

i.e. the four contributions are proportional to

$$d^1_{1,1}(\theta) = d^1_{-1,-1}(\theta) = \frac{1}{2}(1+\cos\theta)$$

$$d^1_{1,-1}(\theta) = d^1_{-1,1}(\theta) = \frac{1}{2}(1-\cos\theta)$$

※ The contributions are distinguishable and we must sum their absolute squares.

⇒ We obtain the angular dependence $(1+\cos^2\theta)$

⇒ This result is valid for all the spin $\frac{1}{2}$ particles

[Supplement-2]

- Generator of a finite rotation $\Delta\phi$ about z-axis,

$$R = \lim_{n\to\infty}\left(1+\frac{iJ_z\Delta\phi}{n\hbar}\right)^n = e^{i\frac{J_z\Delta\phi}{\hbar}}$$, in here R is unitary operator, i.e. $R^* = R^{-1}$

⇒ $e^{-i\theta J_y}\phi(j,m) = \sum_{m'} d^j_{m',m}(\theta)\phi(j,m')$, where $d^j_{m',m}$ is called rotation matrices

$m' = -j,\ -j+1,\ \cdots,\ j-1,\ j\ (2j+1 \text{ states})$

for fixed m' and for $j=1$ state,

$$\phi^*(j,m')e^{-i\theta J_y}\phi(j,m) = d^j_{m',m}(\theta),$$

where $e^{-i\theta J_y} = 1 - i\theta J_y - \frac{\theta^2}{2!}J_y^2 + i\frac{\theta^3}{3!}J_y^3 + \cdots$

$$J_{\pm}\phi(j,m) = \sqrt{j(j+1)-m(m\pm1)}\,\phi(j,m\pm1)$$

$$\begin{cases} J_+ = J_x + iJ_y \\ J_- = J_x - iJ_y \end{cases} \Rightarrow J_y = -\frac{i}{2}(J_+ - J_-)$$

so,

$$J_y\phi(1,1) = -\frac{i}{2}(J_+ - J_-)\phi(1,1) = \frac{i}{2}J_-\phi(1,1) = \frac{i}{2}\sqrt{2}\,\phi(1,0) = \frac{i}{\sqrt{2}}\phi(1,0)$$

$$J_y^2\phi(1,1) = J_y[J_y\phi(1,1)] = J_y\left[\frac{i}{\sqrt{2}}\phi(1,0)\right] = -\frac{i}{2}(J_+ - J_-)\left[\frac{i}{\sqrt{2}}\phi(1,0)\right]$$

$$= \frac{1}{2\sqrt{2}}[J_+\phi(1,0) - J_-\phi(1,0)] = \frac{1}{2}[\phi(1,1) - \phi(1,-1)]$$

then,

$$e^{-i\theta J_y}\phi(1,1) = \phi(1,1) - i\theta\frac{i}{\sqrt{2}}\phi(1,0) - \frac{\theta^2}{2!}\frac{1}{2}[\phi(1,1) - \phi(1,-1)] + \cdots$$

$$d_{1,1}^1(\theta) = \phi^*(1,1)e^{-i\theta J_y}\phi(1,1) = \phi^*(1,1)\phi(1,1) - \frac{\theta}{\sqrt{2}}\phi^*(1,1)\phi(1,0)$$

$$-\frac{1}{2}\frac{\theta^2}{2!}[\phi^*(1,1)\phi(1,1) - \phi^*(1,1)\phi(1,-1)] + \cdots$$

$$= 1 - \frac{1}{2}\frac{\theta^2}{2!} + \frac{1}{2}\frac{\theta^4}{4!} + \cdots = \frac{1}{2}(1+\cos\theta)$$

※ show that $\begin{cases} d_{1,0}^1(\theta) = -\frac{1}{\sqrt{2}}\sin\theta \\ d_{0,0}^1(\theta) = \cos\theta \end{cases}$

$$J_y\phi(1,-1) = -\frac{i}{2}(J_+ - J_-)\phi(1,-1) = -\frac{i}{2}\sqrt{2}\,\phi(1,0) = -\frac{i}{\sqrt{2}}\phi(1,0)$$

$$J_y^2\phi(1,-1) = -\frac{i}{\sqrt{2}}J_y\phi(1,0) = -\frac{1}{2\sqrt{2}}(J_+ - J_-)\phi(1,0) = -\frac{1}{2}[\phi(1,1) - \phi(1,-1)]$$

Then,

$$e^{-i\theta J_y}\phi(1,\ -1) = \phi(1,\ -1) + i\theta\frac{i}{\sqrt{2}}\phi(1,\ 0) - \frac{\theta^2}{2!}\frac{1}{2}[\phi(1,\ 1) - \phi(1,\ -1)] + \ \cdots$$

$$d^1_{1,-1}(\theta) = \phi^*(1,\ 1)e^{-i\theta J_y}\phi(1,\ -1) = \frac{1}{2}\frac{\theta^2}{2!} - \frac{1}{2}\frac{\theta^4}{4!} + \cdots = \frac{1}{2}(1 - \cos\theta)$$

$$\leftarrow \ \phi^*(1,\ 1)e^{-i\theta J_y}\phi(1,\ -1) \ = d^1_{1,-1}(\theta)$$

$$\therefore \ d^1_{1,1}(\theta) = \frac{1}{2}(1 + \cos\theta) = d^1_{-1,-1}(\theta)$$

$$d^1_{1,-1}(\theta) = \frac{1}{2}(1 - \cos\theta) = d^1_{-1,1}(\theta)$$

※ from $d^j_{m',m} = (-1)^{m-m'}d^j_{m,m'} = d^j_{-m,-m'}$,

$d^1_{1,1}(\theta) = d^1_{-1,-1}(\theta)$ and

$d^1_{1,-1}(\theta) = (-1)^{-1-1}d^1_{-1,1}(\theta) = (-1)^{-2}d^1_{-1,1}(\theta) = d^1_{-1,1}(\theta)$

※ with the same ways as the above, show that $\begin{cases} d^1_{1,0}(\theta) = -\dfrac{1}{\sqrt{2}}\sin\theta \\ d^1_{0,0}(\theta) = \cos\theta \end{cases}$

<point cross section>

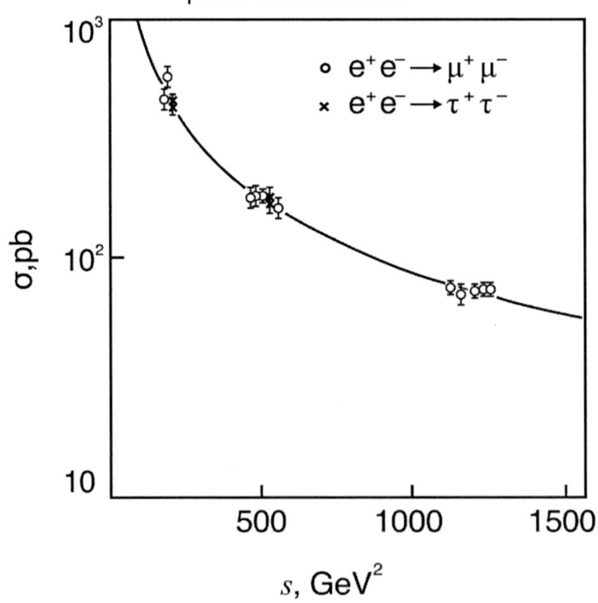

⇒ Experimental data is well agree with the prediction.

<angular distribution: interference of Z^0>

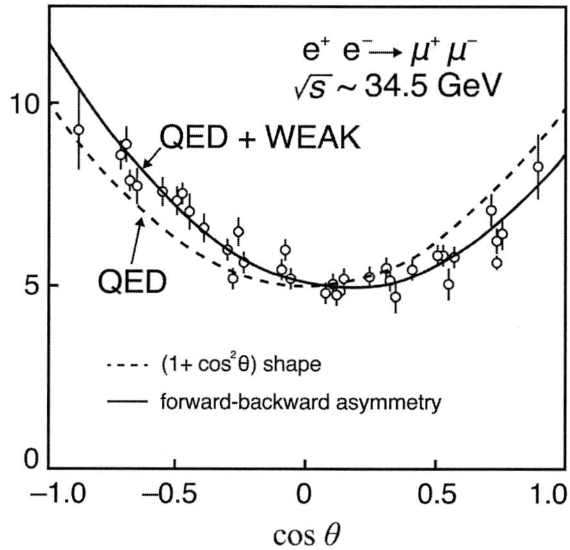

WEAK: there is also parity-violating weak amplitude arising from Z^0 exchange

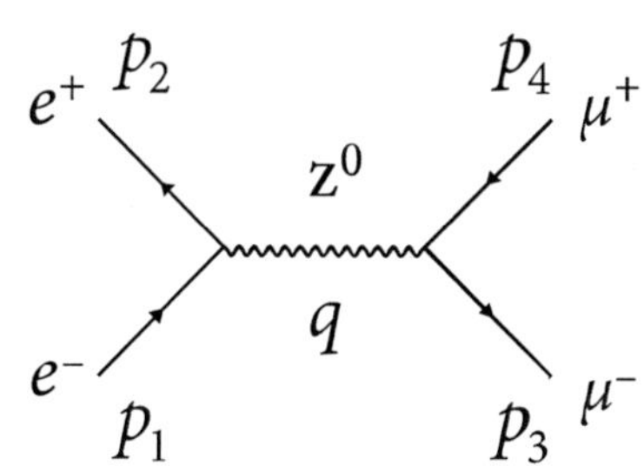

Z^0 boson exchange diagram(Weak Interaction)

$\Rightarrow$ $\gamma - z^0$ interference term is expected to make a relative contribution of magnitude.

$$f = \frac{a_{weak}a_{em}}{a_{em}^2} \sim \frac{G_F^2}{4\pi\alpha} \sim 10^{-4}S, \text{ where } S \text{ is in GeV}^2$$

($\because$ weak matrix element is of order G_F(Fermi constant) $\simeq 10^{-5}$ GeV^{-2})

Asymmetry: $\dfrac{B-F}{B+F} = f \simeq 10\%$ at $S = 1000\ \mathrm{GeV}^2$

Such asymmetries are important tests of electroweak theory (in Chapter 8).

■ e^+e^- annihilation to hadrons

$e^+e^- \rightarrow q\bar{q}$ and $q\bar{q} \rightarrow \mathrm{hadrons}$ ("jet" axis is approximately axis of emission of $q\bar{q}$ pair)

$$R = \frac{\sigma_{(e^+e^- \rightarrow \mathrm{hadrons})}}{\sigma_{(e^+e^- \rightarrow \mu^+\mu^-)}}, \text{ where } \sigma_{e^+e^- \rightarrow \mu^+\mu^-} = \frac{4\pi\alpha^2}{3S}$$

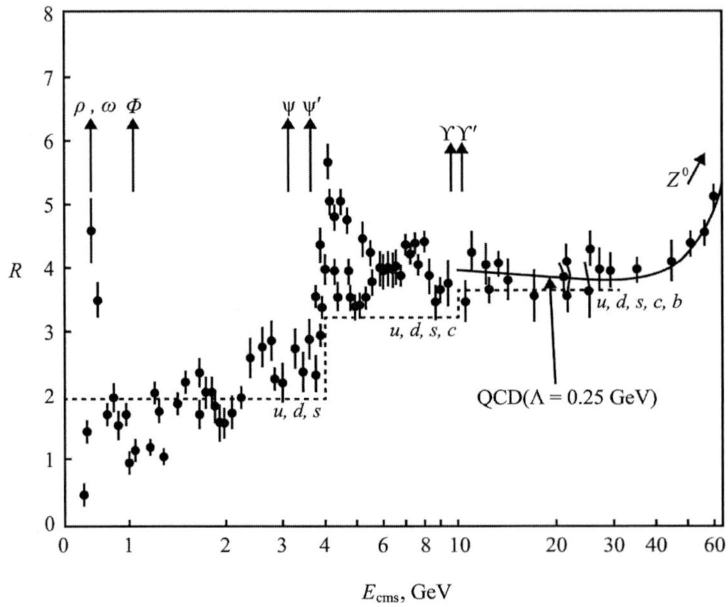

In crossed channel of electron-quark scattering,

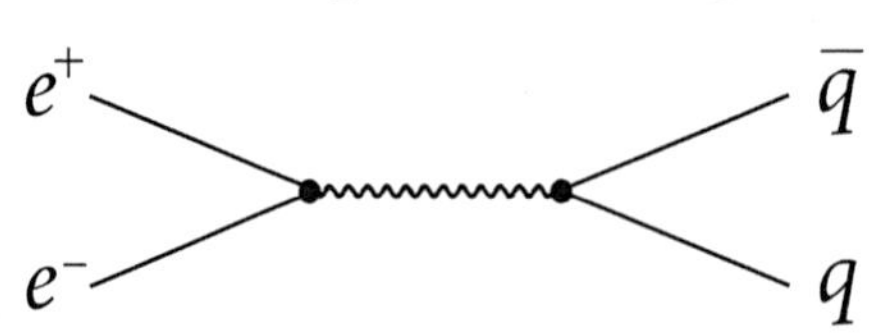

the cross-section is proportional to $\sum Q_i^2$, where i is for all contributing quark flavors.

$$\Rightarrow R = \sum Q_i^2 = \overset{u}{\left(\frac{2}{3}\right)^2} + \overset{d}{\left(-\frac{1}{3}\right)^2} + \overset{s}{\left(-\frac{1}{3}\right)^2} = \frac{2}{3} \text{ below } c\bar{c} \text{ threshold}$$

$$= \overset{u}{\left(\frac{2}{3}\right)^2} + \overset{d}{\left(-\frac{1}{3}\right)^2} + \overset{c}{\left(\frac{2}{3}\right)^2} + \overset{s}{\left(-\frac{1}{3}\right)^2} + \overset{b}{\left(-\frac{1}{3}\right)^2} = \frac{11}{9} \text{ at high } S\text{-values}$$

$\Rightarrow$ does not agree with data in the above Figure

$\Rightarrow$ one of primary reasons to introduce "color" quantum number

(※ the first example is for Δ^{++} and this corresponds to the second example.)

$\Rightarrow$ $q\bar{q}$ flavor combination occurs in three substates $r\bar{r}$, $g\bar{g}$, $b\bar{b}$

$\Rightarrow$ multiply by a factor 3 in R-values

$\Rightarrow$ good agreement with experiment data

<Angular distribution in $e^+e^- \to q\bar{q}$ depends on spin of quark constituents>

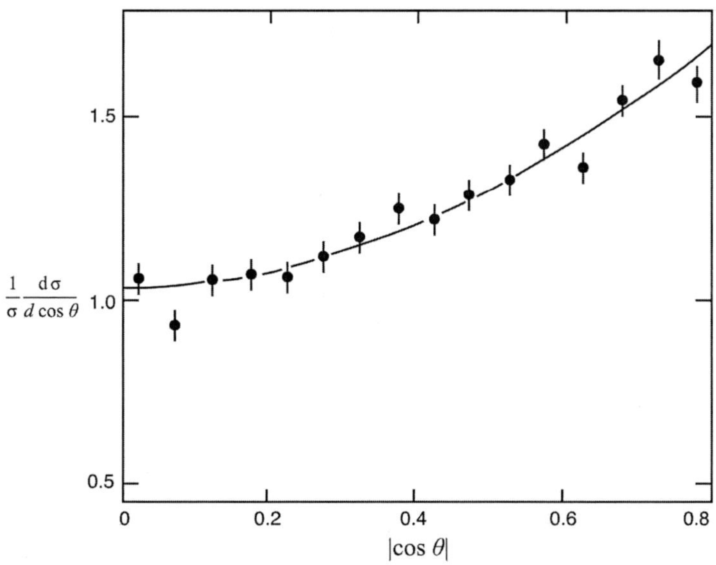

In here, θ is the angle with respect to the beam axis

- consistent with the expected for spin $\frac{1}{2}$ quark constituents $\left[\frac{dN}{d\Omega} \propto (1+\cos^2\theta)\right]$
- $|\cos\theta|$: It is not possible to say which jet corresponds to $\bar{q}$, only half the angular distribution is plotted.

- Summary of $e^+e^- \rightarrow$ hadrons

(i) constancy of R is the evidence for the pointlike(parton) constituent of hadrons.

(ii) The value of R is equal to that expected if the charged partons are quarks, with fractional charges and the color quantum number.

(iii) angular distribution of the two jets is the proof of spin $\frac{1}{2}$ for partons.

■ Electron-muon scattering $e^-\mu^+ \rightarrow e^-\mu^+$ (electromagnetic scattering)

let`s express the cross-section in terms of Mandelstam variables $s,\ t,\ u$ (all positive quantities), which are relativistic invariants, defined as

$a+b \rightarrow c+d,$

(i) $s=-(k_1+k_2)^2=-(k_3+k_4)^2=-2k_1k_2=-2k_3k_4,$

where s is the cms energy squared, and

k_1, k_2, k_3, k_4 are 4-momenta of particles

(ii) $a-c$ four momentum transfer:

$$t=q^2=(k_1-k_3)^2=(k_2-k_4)^2=-2k_1k_3=-2k_2k_4,$$

where t is 4-momentum transfer squared

(iii) $a-d$ four momentum transfer:

$$u=(k_2-k_3)^2=(k_1-k_4)^2=-2k_2k_3=-2k_1k_4$$

where u is the crossed momentum transfer squared

In the center-of-mass system,

$$s=-2\vec{p}_3\cdot\vec{p}_4+2E_3E_4=2\vec{p}^2+2\vec{p}^2=4p^2,$$

where p is the magnitude of 3-momentum

$(\because |\vec{p}_3|=|\vec{p}_4|\equiv p$

$$E_3 = \sqrt{\vec{p}_3^{\,2} + m_3^2} \approx p_3 = p = E_4,$$

in here, lepton rest masses are neglected)

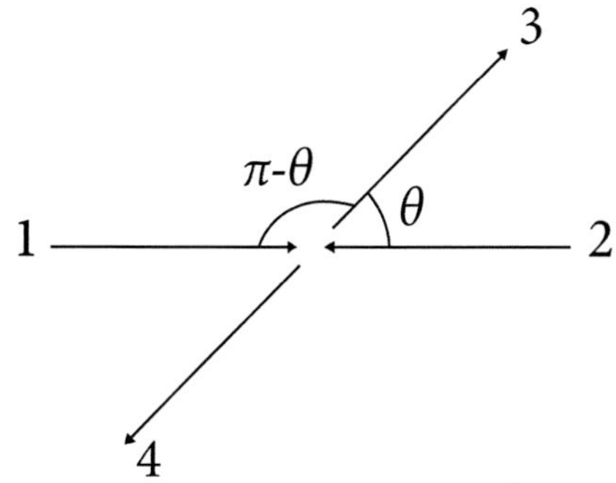

$$t = -2\vec{p}_2 \cdot \vec{p}_4 + 2E_2E_4 = 2p^2(1-\cos\theta) = 4p^2\sin^2\frac{\theta}{2}$$

$$u = -2\vec{p}_1 \cdot \vec{p}_4 + 2E_1E_4 = -2p^2\cos(\pi-\theta) + 2p^2 = 2p^2(1+\cos\theta) = 4p^2\cos^2\frac{\theta}{2}$$

$\Rightarrow$ $s = u + t$: they are not independent variables.

Then, (in terms of Mandelstam variables)

$$\frac{d\sigma}{d\Omega}(e^+e^- \to \mu^+\mu^-) = \frac{\alpha^2}{4s}(1+\cos^2\theta) = \frac{\alpha^2}{8p^2}\left(\frac{t^2+u^2}{s^2}\right)$$

↳∵ $t^2 = 4p^4(1-\cos\theta)^2$, $u^2 = 4p^4(1+\cos\theta)^2$

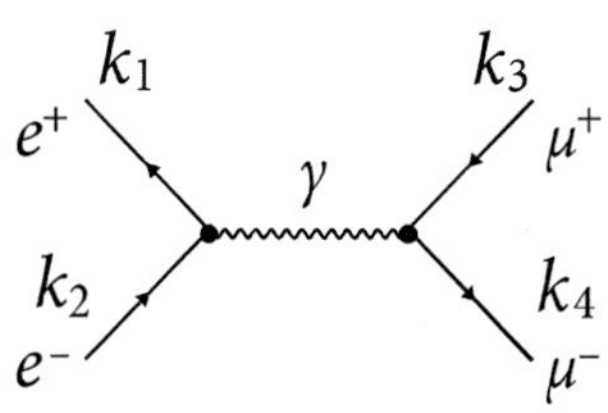

(annihilation process)

$\Rightarrow$ replace incoming e^+ by outgoing e^- and rotate through 90^0
$\Rightarrow$ assign k momentum

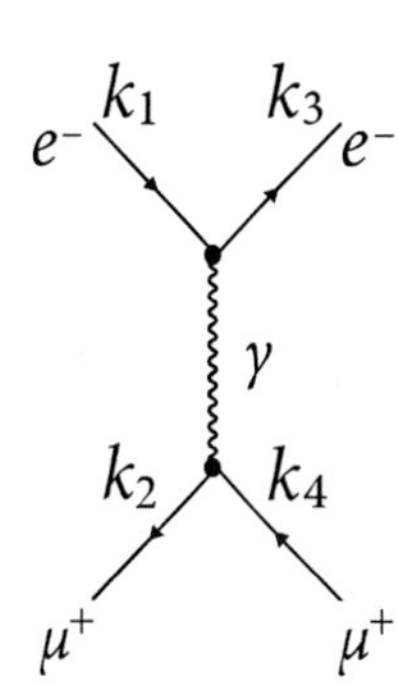

(scattering process)

$q = k_1 + k_2$ in (a) is replaced by $q = k_1 - k_3$ in (b)

$\Rightarrow k_2$ replaces $-k_3$

$\Rightarrow s \rightarrow -t$ (this amounts to the interchanges $s \rightarrow -t$)

$$\Rightarrow \frac{d\sigma}{d\Omega}(e^-\mu^+ \rightarrow e^-\mu^+) = \frac{\alpha^2}{8p^2}\left(\frac{s^2+u^2}{t^2}\right) = \frac{\alpha^2}{8p^2\sin^4\frac{\theta}{2}}\left(1+\cos^4\frac{\theta}{2}\right)$$

※ interference term $\propto \dfrac{2u^2}{st}$

So far, all of expressions are for the center-of-momentum system.

It is also useful to express the cross-section in term of lab. quantities, i.e. those measured in the rest frame of the target muon .

Let`s consider the laboratory system.

Suppose the muon is a target at rest in the laboratory

(Note. this is not practical since lifetime of muon is $2.2\ \mu s$)

γ is the boost from the cms to lab.

p and θ are the projectile parameters in the cms]

E_e is the incident electron energy in the laboratory system.

E_μ is the recoiling muon energy after collision.

$$\begin{cases} \text{scattered muon: } E_\mu[=\gamma(E_\mu - \beta p_x)] = \gamma(p - \beta p\cos\theta) = \gamma p(1-\cos\theta) \\ \text{incident electron: } E_e = 2\gamma p\ [\because E_e = \gamma(p+\beta p)] \end{cases}$$

$$\Rightarrow y = \frac{E_\mu}{E_e} = \frac{1-\cos\theta}{2} \quad (0 < y < 1) \quad \rightarrow \quad dy = \frac{1}{2}d(\cos\theta)$$

then

$$\begin{cases} \cos^2\frac{\theta}{2} = \frac{1}{2}(1+\cos\theta) = 1-y \\ d\Omega = 2\pi d(\cos\theta) = 4\pi dy \end{cases}$$

so,

$$\frac{d\sigma}{d\Omega}(e^-\mu^+ \rightarrow e^-\mu^+) = \frac{\alpha^2}{8p^2\sin^4\frac{\theta}{2}}\left[1+\cos^4\frac{\theta}{2}\right]$$

$d\Omega \rightarrow 4\pi dy$:

$$\Rightarrow \frac{d\sigma}{dy}(e^-\mu^+ \to e^-\mu^+) = 4\pi\frac{\alpha^2 16p^4}{8p^2q^4}[1+(1-y)^2]$$

$$= \frac{2\pi\alpha^2 s}{q^4}[1+(1-y)^2] \quad \cdots\cdots (1)$$

$$\left(\because\ q^2 = 4p^2\sin^2\frac{\theta}{2} \Rightarrow \sin^4\frac{\theta}{2} = \frac{q^4}{16p^4},\quad s = 4p^2 \Rightarrow p^2 = \frac{1}{4}s\right)$$

In Eq. (1),

1 corresponds to scattering between particles of the same helicity.("same helicity scattering").

$(1-y)^2$ corresponds to scattering between particles of the opposite helicity ("opposite helicity scattering").

$[1+(1-y)^2]$ is the typical y distribution for scattering between two spin $\frac{1}{2}$ particles (Mott scattering).

where,

$$\tilde{p}^2 = (\tilde{p}_1 + \tilde{p}_2)^2 \Rightarrow -s = -m_e^2 - m_\mu^2 + 2\vec{p}_e \cdot \vec{p}_\mu - 2E_e E_\mu \text{ (in here } E_\mu = m_\mu\text{) in lab}$$

$$\Rightarrow s = 2m_\mu E_e + m_\mu^2 + m_e^2 \simeq 2m_\mu E_e$$

If muon recoil energy is small, $y \to 0$

$$q^2 = 4p^2\sin^2\frac{\theta}{2} = 4p^2\frac{1}{2}(1-\cos\theta) = 2p^2(1-\cos\theta) = 4p^2 y$$

$$\Rightarrow y = \frac{q^2}{4p^2} \Rightarrow \frac{dy}{dq^2} = \frac{1}{4p^2}$$

then,

$$\frac{d\sigma}{dy}\frac{dy}{dq^2} = \frac{d\sigma}{dq^2} = \frac{2\pi\alpha^2 s}{q^4}[1+1]\frac{dy}{dq^2} = \frac{2\pi\alpha^2}{q^4}4p^2\cdot 2\frac{dy}{dq^2} = \frac{16\pi\alpha^2}{q^4}p^2\frac{dy}{dq^2} = \frac{4\pi\alpha^2}{q^4}$$

$$\left(\because\ \frac{dy}{dq^2} = \frac{1}{4p^2}\right)$$

$$\therefore \ \frac{d\sigma}{dq^2}(y\to 0) = \frac{4\pi\alpha^2}{q^4} : \text{Rutherford formula for spin is neglected}$$

$$\text{c.f. } \frac{d\sigma}{d\Omega} = \left(\frac{Z_1 Z_2 e^2}{8\pi\epsilon_0 m v_0^2}\right)^2 \operatorname{cosec}^4\left(\frac{\theta}{2}\right)$$

[Supplement-3]

moving along x-axis with velocity βc

$$\begin{pmatrix} \gamma & -\beta\gamma & 0 & 0 \\ -\beta\gamma & \gamma & 0 & 0 \\ 0 & 0 & \gamma & 0 \\ 0 & 0 & 0 & \gamma \end{pmatrix} \begin{pmatrix} E \\ p_x \\ p_y \\ p_z \end{pmatrix}_{cms} = \begin{pmatrix} E' \\ p_x' \\ p_y' \\ p_z' \end{pmatrix}_{lab}$$

$$\Rightarrow E' = \gamma E - \beta\gamma p_x = \gamma(E - \beta p_x)$$

■ Neutrino-electron scattering, $\nu_e e \to \nu_e e$ (weak interaction)

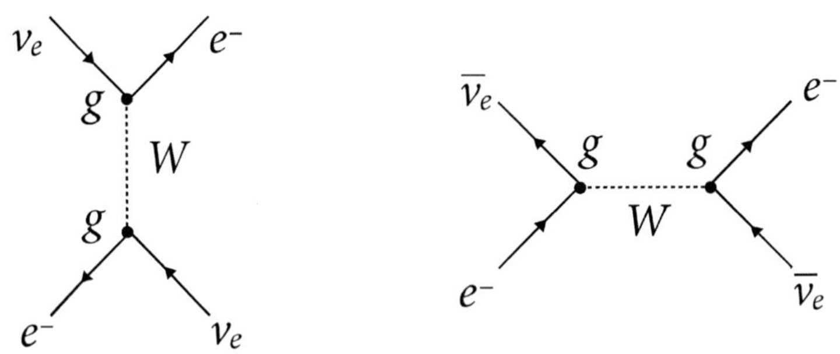

- g is the weak coupling of W boson to the lepton

Consider only the charged-current reaction

(matrix element)

$$M(\nu e \to \nu e) = \frac{(g/\sqrt{2})^2}{q^2 + M_W^2}, \text{ where } \frac{1}{\sqrt{2}} \text{ is a conventional normalization}$$

factor in he electroweak theory.

$$\simeq \frac{g^2}{2M_w^2}$$

($\because$ $M_W = 80$ GeV, $q^2_{\max} \simeq 2mE_\nu \Rightarrow$ for even TeV neutrino energies, $q^2 \ll M_W^2$)

Then, differential cross-section is

$$\frac{d\sigma}{dq^2} = \frac{2\pi}{v} |M|^2 p^2 \frac{dp}{dE_f} \frac{d\Omega}{dq^2} \frac{2}{(2\pi)^3} \quad \cdots\cdots (1),$$

where v is the relative velocity ($v = 2c$)

$$\sqrt{p^2 + m_e^2} + \sqrt{p^2 + m_\nu^2} = E_f \Rightarrow 2p = E_f$$

$$\Rightarrow 2dp = dE_f \Rightarrow \frac{dp}{dE_f} = \frac{1}{2}$$

2 in $\frac{2}{(2\pi)^3}$ comes from that summing over final-state spins and averaging over initial-state spins

and $q^2 = 2p^2(1 - \cos\theta) \Rightarrow dq^2 = 2p^2 d(\cos\theta)$

$$\frac{d\Omega}{dq^2} = \frac{2\pi d(\cos\theta)}{dq^2} = \frac{2\pi}{2p^2} = \frac{\pi}{p^2}$$

Put them all in Eq (1),

$$\frac{d\sigma}{dq^2} = \frac{2}{\pi}\left(\frac{g^2}{8M_W^2}\right)^2, \ c = 1$$

where $\frac{G_F}{\sqrt{2}} \equiv \frac{g^2}{8M_w^2}$: process was treated as a point like i.e. S-wave, interaction of the four fermion states involved, and decay rate was specified by G_F, defined for vector transition.

$$= \frac{G_F^2}{\pi}$$

※ $\sigma = \frac{G_F^2}{\pi} S$, since $q^2_{\max} = S$

$$\left[\because \ \sigma = \frac{G_F^2}{\pi} \int_{q^2=0}^{q^2_{\max}} dq^2 = \frac{G_F^2}{\pi} (q^2)_0^{q^2_{\max}} = \frac{G_F^2}{\pi} q^2_{\max} = \frac{G_F^2}{\pi} S\right]$$

• Let`s consider spin effects: (leptons participating in weak interactions are longitudinally polarized particles)

$$\text{Polarization}(P) \equiv \frac{I_+ - I_-}{I_+ + I_-},$$

where I_+ is the number of leptons with $\vec{\sigma}$ and $\vec{p}$ parallel and

I_- is the number of leptons with $\vec{\sigma}$ and $\vec{p}$ antiparallel

$= \alpha \frac{P}{E}$, ← by $V-A$ theory of weak interactions

where P is the 3-momentum

$= \alpha \frac{v}{c}$,

where $\alpha = -1$ for leptons and

$\alpha = +1$ for antileptons

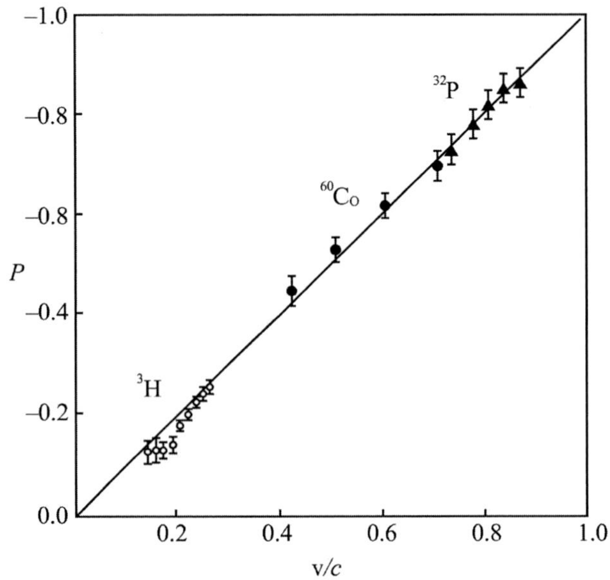

for relativistic leptons, $v \simeq c \Rightarrow P = \pm 1$: they are nearly pure helicity states

for zero mass leptons, $v = c \Rightarrow$ pure helicity eigenstates

$$\Rightarrow \begin{cases} H=-1 \text{ for neutrinos} \\ H=+1 \text{ for antineutrinos} \end{cases}$$

(i) for neutrino

$$J = 0 \text{ (isotropic)}$$

$$\overset{\leftarrow}{\nu \longrightarrow}\overset{\rightarrow}{\longleftarrow e^-} \rightarrow z$$

both leptons are LH ("same helicity")

$\Rightarrow$ total angular momentum $J = 0$

$\Rightarrow$ angular distribution is isotropic

$\Rightarrow$ differential cross-section is given by the end equation in the previous page

$\Rightarrow$ total cross-section is

$$\int d\sigma = \frac{G_F^2}{\pi}\int_{q^2=0}^{q^2_{\max}=S=4p^2} dq^2 \Rightarrow \sigma(\nu e \rightarrow \nu e) = \frac{G_F^2}{\pi}S \text{ , where } S \simeq 2mE_\nu$$

(i) for antineutrino

$$J = 1 \text{ and } JZ = +1 \text{ (opposite helicity)}$$

$$\overset{\rightarrow}{\bar{\nu} \longrightarrow}\overset{\rightarrow}{\longleftarrow e^-} \rightarrow z$$

$\Rightarrow$ angular momentum conservation requires $J_Z = +1$ in the final states

$\Rightarrow$ d-function $d^1_{1,1} = \frac{1}{2}(1+\cos\theta)$ contribution to probability amplitude

so, $$\frac{d\sigma}{dq^2} = \frac{G_F^2}{\pi}\left[\frac{1}{2}(1+\cos\theta)\right]^2$$

$$\Rightarrow \frac{d\sigma}{d(\cos\theta)}(\bar{\nu}_e e \rightarrow \bar{\nu}_e e) = 2p^2\frac{G_F^2}{\pi}\frac{1}{4}(1+\cos\theta)^2 = \frac{G_F^2}{8\pi}S(1+\cos\theta)^2$$

$$[\because q^2 = 2p^2(1+\cos\theta),\ dq^2 = 2p^2 d(\cos\theta)]$$

$$\left(\because S = 4p^2,\ p^2 = \frac{1}{4}S\right)$$

$$\Rightarrow \sigma(\bar{\nu}_e e \rightarrow \bar{\nu}_e e) = \frac{G_F^2}{3\pi}S : \text{ small cross-section}$$

($\because$ $J_Z = +1$ requirement in the final state only one allowed out of 3 possible)

※ from (i) and (ii),

$$\sigma(\nu e \to \nu e) = 3\sigma(\bar{\nu}_e e \to \bar{\nu}_e e)$$

[Supplement-4]

so, $\frac{d\sigma}{dy}(\nu_e e \to \nu_e e) = \frac{G_F^2 S}{\pi}$

$$\frac{d\sigma}{dy}(\bar{\nu}_e e \to \bar{\nu}_e e) = \frac{G_F^2 S}{\pi}(1-y)^2$$

$$\because \frac{d\sigma}{dq^2}\frac{dq^2}{dy} = \frac{G_F^2}{\pi}\frac{1}{4}(1+\cos\theta)^2\frac{dq^2}{dy} = \frac{G_F^2}{4\pi}(1+\cos\theta)^2 4p^2$$

$$= \frac{G_F^2}{\pi}p^2(1+\cos\theta)^2 = \frac{G_F^2}{\pi}p^2 4\cos^4\left(\frac{\theta}{2}\right)$$

$$= \frac{G_F^2}{\pi}p^2 4(1-y)^2 = \frac{G_F^2}{\pi}S(1-y)^2$$

■ Elastic lepton-nucleon scattering

electron (neutrino) beams incident on nucleon targets

$\Rightarrow e+p \to e+p,\ \nu_\mu + n \to \mu^- + p$

These reactions are dominated by "form factors" $F(q^2)$, which determine how the scattering is reduced from its value for a pointlike nucleon.

※ If the nucleon were a pointlike particle, the cross-section follows the $e\mu \to e\mu$ calculation:

$$\frac{d\sigma}{dq^2} = \frac{2\pi\alpha^2}{q^4}\left[1+\left(1-\frac{q^2}{S}\right)^2\right]$$

But this equation must be modified by the addition of form factor $\frac{d\sigma}{dq^2} \to \frac{d\sigma}{dq^2}F(q^2)$, therefore the nucleon has structure.

- form factor $F(q^2)$ is rapidly decreased as q^2

- $|F(q^2)|^2$ measures the probability that nucleon holds together and recoils intact under impact q

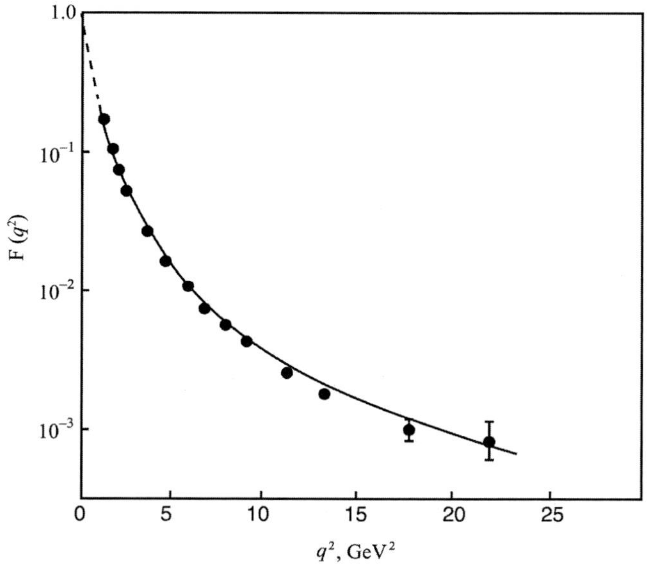

empirically,

$$F(q^2) \simeq \frac{1}{\left(1+\frac{q^2}{M_V^2}\right)^2} : \text{“dipole formula”},$$

where V is the vector coupling and $M_V \simeq 0.9$ GeV

- In neutrino scattering

$$\sigma \sim \frac{G_F^2 S}{\pi}, \quad S \simeq 2mE_\nu$$

$\Rightarrow$ neutrino cross-section is expected to rise with neutrino energy S However, elastic and quasielastic cross-sections are saturated due to strong q^2 dependence of the form factor, therefore this rise is halted by strong q^2 dependence in the form factor

⇒ cross-section tends to a constant value

<$\nu n \rightarrow \mu^- p$: neutrino-nucleon scattering>

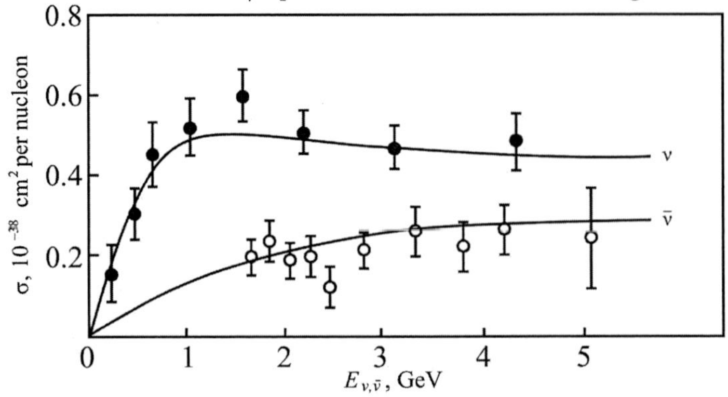

⇒ this appears to represent an exponential charge distribution

■ Deep inelastic scattering partons

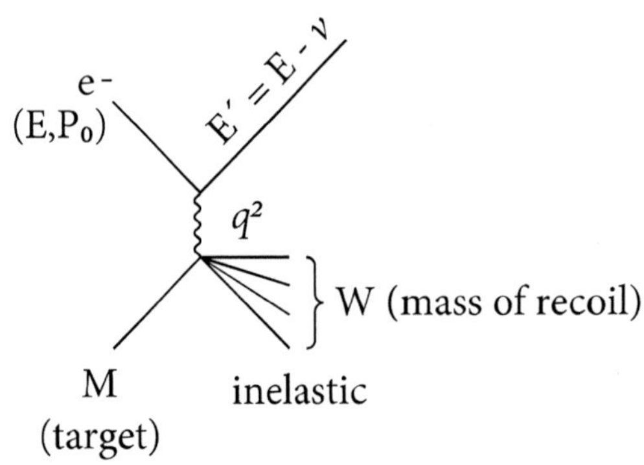

lepton + nucleon → lepton'+ anything: inclusive process

These inclusive reactions are observed to be only weakly q^2 dependent, unlike case for exclusive process.

<electron-parton collision via single-photon exchange>

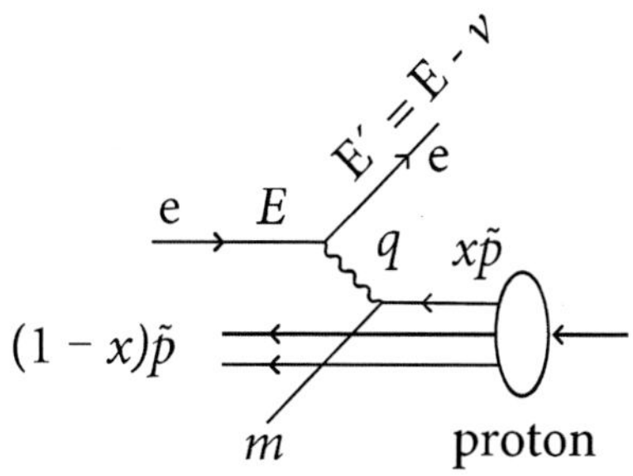

where $\tilde{p} = (p, 0, 0, ip)$, proton mass is neglected in infinite-momentum frame i.e. very large proton momentum and $0 < x < 1$

Suppose that one parton of mass m is scattered elastically by absorbing 4-momentum q of the scattered lepton.

$\Rightarrow (x\tilde{p} + q)^2 = -m^2 \simeq 0$

└> $x^2\tilde{p}^2 + q^2 + 2x\tilde{p}q$

$\Rightarrow$ for $\left| x^2\tilde{p}^2 \right| = x^2M^2 \ll q^2$,

$$q^2 + 2x\tilde{p}q = 0 \Rightarrow x = \frac{-q^2}{2\tilde{p}q} = \frac{q^2}{2M\nu} \quad \cdots\cdots (1)$$

($\because$ in lab system, nucleon is at rest and energy transfer is ν)

This model assumes that the process occurs in two stages:

(i) one parton is scattered: collision time $t_1 \sim \dfrac{\hbar}{\nu}$

※ ν is the energy transfer in the lab. i.e. $\nu = E - E'$

(ii) parton recombine to form final state ($M = W$): time $t_2 \sim \dfrac{\hbar}{W}$, $\quad t_2 \gg t_1$

※ W^2 is the square of mass of hadronic system

If a free particle(parton) of mass m, stationary in lab system, undergoes elastic collision in which it receives energy ν and 3-momentum q' i.e. $\tilde{q} = (q', i\nu)$,

4-momentum transfer by elastic relation is

$$q^2 = q'^2 - \nu^2 = [(\nu+m)^2 - m^2] - \nu^2$$

$$= 2m\nu \quad \cdots\cdots (2)$$

$(\because\ q'^2 = (\text{energy})^2 - \text{m}^2 = (\nu + \text{m})^2 - \text{m}^2,$

in here, $(\nu+m)$ is the sum of the invariant mass energy of a particle m at rest plus the energy ν transferred.)

Substituting Eqs. (2) into (1),

$$x = \frac{2m\nu}{2M\nu} = \frac{m}{M},$$

where x is the fractional mass of the nucleon carried by a parton.

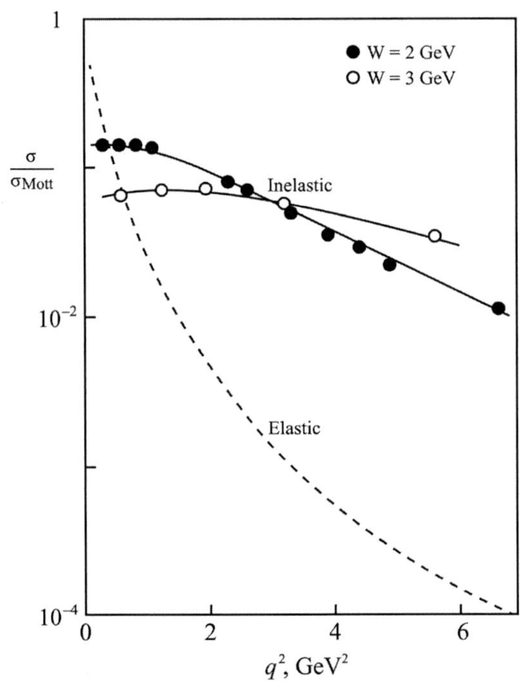

- σ_{Mott} is for the pointlike cross-section.

(i) These inclusive reactions are observed to be only weakly q^2 dependent unlike case for exclusion process.

(ii) This is a sign of elastic scattering from point like constituents with in the nucleon.

[Supplement-5]

Evidence that partons are quarks:

$$\frac{d\sigma}{d\Omega}(e^-\mu^+ \to e^-\mu^+) = \frac{\alpha^2}{8p^2\sin^4\frac{\theta}{2}}\left(1+\cos^4\frac{\theta}{2}\right)$$

angular distribution characterized by two terms:

(i) $\cos^2\frac{\theta}{2}$: electric (Mott)

(ii) $\sin^4\frac{\theta}{2}$: magnetic (spin-flip)

$$1+\cos^4\frac{\theta}{2} = 2\cos^2\frac{\theta}{2}+\sin^4\frac{\theta}{2}$$

Data reveals just this dependence.

Deep inelastic scattering and quarks

(I) electron-nucleon scattering

let $u(x)dx$ and $d(x)dx$ be number of u and d quarks in the proton that carry a fractional momentum in the range $x \sim x+dx$

$$\frac{d\sigma}{dy}(e^-\mu^+ \to e^-\mu^+) = \frac{2\pi\alpha^2 S}{q^4}[1+(1-y)^2]$$

$$\Rightarrow \quad \frac{d^2\sigma(ep)}{dydx} = \frac{4\pi\alpha^2}{q^4}xS\frac{F_2^{ep}(x)}{x}\left[\frac{1+(1-y)^2}{2}\right],$$

where $F_2^{ep}(x)$ is the structure function, involves quark density weighted by squares of quark charges

$$\Rightarrow \begin{cases} \dfrac{F_2^{ep}(x)}{x} = \dfrac{4}{9}[u(x)+\bar{u}(x)]+\dfrac{1}{9}[d(x)+\bar{d}(x)+s(x)+\bar{s}(x)] \\ \dfrac{F_2^{en}(x)}{x} = \dfrac{4}{9}[d(x)+\bar{d}(x)]+\dfrac{1}{9}[u(x)+\bar{u}(x)+s(x)+\bar{s}(x)] \end{cases}$$

($\because$ by isospin invariance, we expect that u, $\bar{u}$ population in the neutron is equal to d, $\bar{d}$ population in the proton)

$\Rightarrow$ For a nucleon target which has an equal admixture of protons and neutrons.

$$\frac{F_2^{en}(x)}{x} = \frac{1}{2} \times \frac{4}{9}[u(x) + \bar{u}(x) + d(x) + \bar{d}(x)]$$

$$+ \frac{1}{2} \times \frac{1}{9}[d(x) + \bar{d}(x) + s(x) + \bar{s}(x) + u(x) + \bar{u}(x) + s(x) + \bar{s}(x)]$$

$$= \frac{5}{18}[u(x) + \bar{u}(x) + d(x) + \bar{d}(x)] + \frac{1}{9}[s(x) + \bar{s}(x)]$$

- Valence and Sea quarks

※ partons = quarks and gluons

- two types of quarks

valence quarks: contribute to quantum numbers of quarks
sea quarks: virtual quark-antiquark($q\bar{q}$) pairs

Parton distribution function $f_i(x)$ =

probability to find parton with $x_i \in [x, x + dx]$

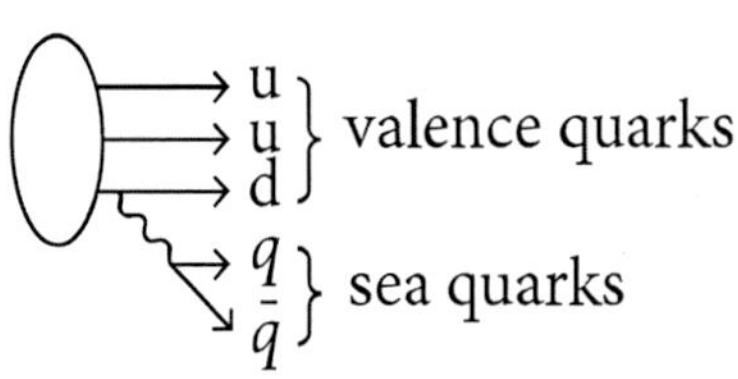

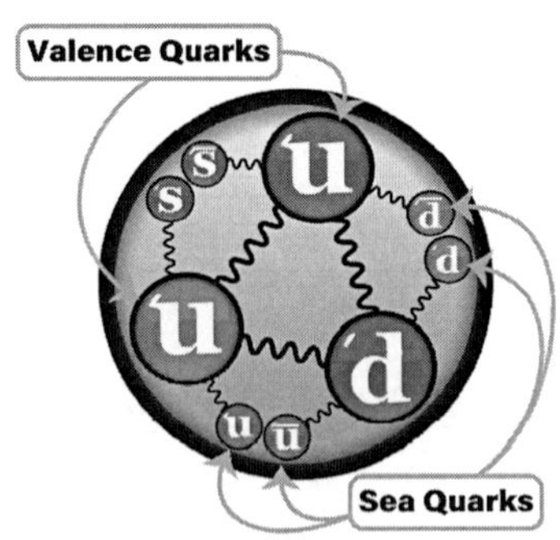

(http://www.jicfus.jp/en/promotion/pr/mj/brian_colquhoun/)

① quark composition of the proton

$u_v + u_v + d_v + (u_s + \bar{u}_s) + (d_s + \bar{d}_s) + (s_s + \bar{s}_s)$, where $u(x)$, ⋯ quark density of u

sea quarks: heavy quark contribution strongly suppressed

electrons do not distinguish quarks from antiquarks:

$$\frac{F_2^{ep}(x)}{x} = \sum_i Q_i^2 f_i(x)$$

$$= \frac{4}{9}[u^p(x) + \bar{u}^p(x)] + \frac{1}{9}[d^p(x) + \bar{d}^p(x)] + \frac{1}{9}[s^p(x) + \bar{s}^p(x)]$$

② quark composition of the neutron

$$= \frac{4}{9}[u^n(x) + \bar{u}^n(x)] + \frac{1}{9}[d^n(x) + \bar{d}^n(x)] + \frac{1}{9}[s^n(x) + \bar{s}^n(x)]$$

similarly,

$$\frac{F_2^{en}(x)}{x} = \frac{4}{9}[d^n(x) + \bar{d}(x)] + \frac{1}{9}[u(x) + \bar{u}(x)] + \frac{1}{9}[s(x) + \bar{s}(x)]$$

⇒ isospin symmetry (invariance) $u(x)$

$$\begin{cases} u^p(x) = d_n(x) & \bar{d}_p(x) = \bar{u}_n(x) \\ d_p(x) = u_n(x) & \bar{u}_p(x) = \bar{d}_n(x), \text{ where } d_p(x) = d(x),\ u^p(x) = u(x) \\ s^n(x) = s^p(x) & \bar{s}_p(x) = \bar{s}_n(x) \end{cases}$$

• Sum rules

$$\begin{cases} q(x) = q_v(x) + q_s(x) \\ \bar{q}(x) = \qquad\quad \bar{q}_s(x) \end{cases}$$

※ naive guess

(a) $u_v(x) = 2d_v(x)$

(b) $s_v(x) = \bar{u}_v(x) = \bar{d}_v(x) = \bar{s}_v(x) = 0$

(c) $u_s(x) = \bar{u}_s(x) = d_s(x) = \bar{d}_s(x) = s_s(x) = \bar{s}_s(x) \equiv k$

$$\text{valence}\begin{cases}\int_0^1 u_v(x)dx = 2\\ \int_0^1 d_v(x)dx = 1\end{cases}$$

$$\int_0^1 [q_s(x) - \bar{q}_s(x)]dx = 0$$

[II] Neutrino-nucleon scattering

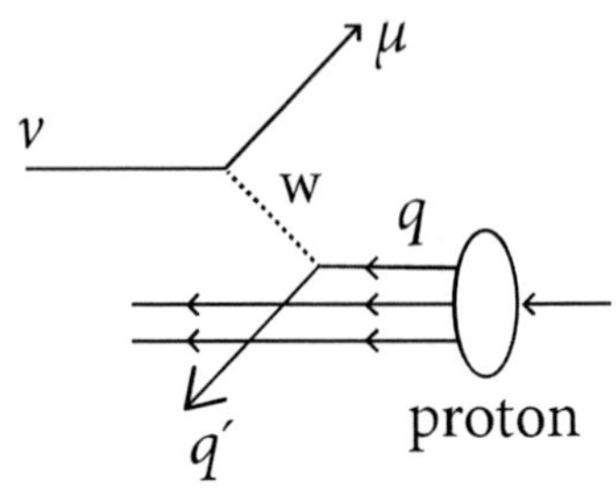

These are very powerful probes because they see different quarks.

※ How to make neutrino beam: proton beams hit the target, select π and K with a specific momentum using dipole magnet from secondaries. Then, focus with quadrupole magnet and sent it to the decay tunnel.

$[\pi(\sim 100\%),\ K(\sim 64\%) \rightarrow \mu + \nu_\mu]$

After that, block μ with steel and earth.

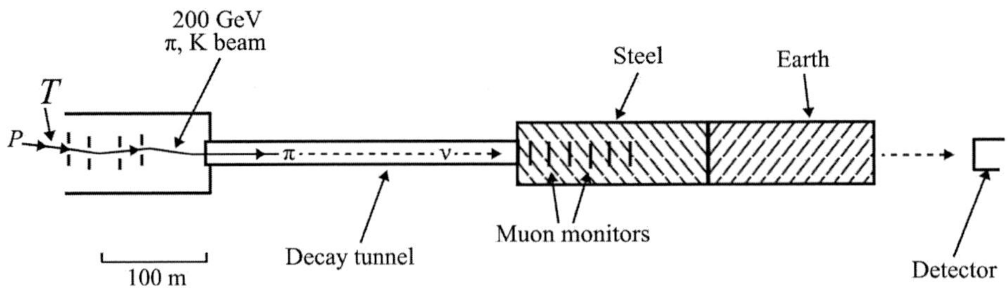

$$\frac{d\sigma}{dy}(\nu_e e \rightarrow \nu_e e) = \frac{G_F^2 S}{\pi} = \frac{d\sigma}{dy}(\overline{\nu e} \rightarrow \overline{\nu e})$$

$$\frac{d\sigma}{dy}(\overline{\nu}_e e \rightarrow \overline{\nu}_e e) = \frac{G_F^2 S}{\pi}(1-y)^2 = \frac{d\sigma}{dy}(\nu \overline{e} \rightarrow \nu \overline{e})$$

Dominant charged current($W^{\pm}$) interactions:

$$\nu_\mu + d \xrightarrow[W^+]{} \mu^- + u, \qquad \nu_\mu + \overline{u} \xrightarrow[W^+]{} \mu^- + \overline{d}$$ (via W exchange)

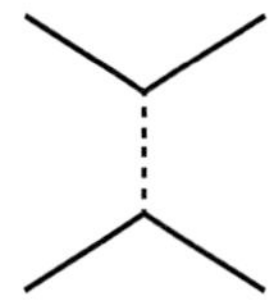

$$\overline{\nu}_\mu + u \xrightarrow[W^-]{} \mu^+ + d, \qquad \overline{\nu}_\mu + \overline{d} \xrightarrow[W^-]{} \mu^+ + \overline{u},$$ are allowed.

※ One might expect to have four independent processes, but only two of them are such, as can be seem. We then consider the proton target only:

$$\nu_\mu + p \rightarrow \mu^+ + X, \quad \overline{\nu}_\mu + p \rightarrow \mu^- + X$$

by measuring the cross-section of the deep inelastic scattering, one extracts the structure functions as Eq. (3) in the below.

※ $\begin{cases} \nu_\mu + u \rightarrow \mu^+ + d \\ \nu_\mu + \overline{d} \rightarrow \mu^+ + \overline{u} \\ \overline{\nu}_\mu + d \rightarrow \mu^- + u \\ \overline{\nu}_\mu + \overline{u} \rightarrow \mu^- + \overline{d} \end{cases}$ violates the lepton number and are forbidden.

then,

$$\begin{cases} \dfrac{d^2\sigma(\nu p)}{dydx} = \dfrac{G_F^2 x S}{\pi}[d(x) + \overline{u}(x)(1-y)^2] \\ \dfrac{d^2\sigma(\nu n)}{dydx} = \dfrac{G_F^2 x S}{\pi}[u(x) + \overline{d}(x)(1-y)^2] \end{cases}$$ ← through isospin invariance

for equal numbers of neutrons and protons (“isoscalar”target: $\#p = \#n$)

$$\frac{d^2\sigma(\nu N)}{dydx}=\frac{G_F^2 xS}{2\pi}\{u(x)+d(x)+[\bar{u}(x)+\bar{d}(x)](1-y)^2\}\ \cdots\cdots\ (1)$$

Similarly, for antineutrinos, by interchanging $(1-y)^2 \leftrightarrow 1$

$$\frac{d^2\sigma(\bar{\nu} N)}{dydx}=\frac{G_F^2 xS}{2\pi}\{[u(x)+d(x)](1-y)^2+[\bar{u}(x)+\bar{d}(x)]\}\ \cdots\cdots\ (2)$$

$$\because \begin{cases} \dfrac{d^2\sigma(\bar{\nu}p)}{dydx}=\dfrac{G_F^2 xS}{\pi}[u(1-y)^2+\bar{d}] \\ \dfrac{d^2\sigma(\bar{\nu}n)}{dydx}=\dfrac{G_F^2 xS}{\pi}[d(1-y)^2+\bar{u}] \end{cases}$$

define neutrino structure functions

$$\begin{cases} \dfrac{F_2^{\nu N}(x)}{x}=u(x)+d(x)+\bar{u}(x)+\bar{d}(x) \\ F_3^{\nu N}(x)=u(x)+d(x)-\bar{u}(x)-\bar{d}(x) \end{cases}\ \cdots\cdots\ (3)$$

※ from Eq. (3) and $F_2^{eN}\approx\frac{5}{18}x[u+\bar{u}+d\bar{d}]$, we obtain $F_2^{eN}\approx\frac{5}{18}F_2^{\nu N}$.

$$\Rightarrow \frac{d^2\sigma^{\nu N,\bar{\nu}N}}{dydx}=\frac{G_F^2 ME}{\pi}\left\{\left[\frac{F_2(x)\pm xF_3(x)}{2}\right]+\left[\frac{F_2(x)\mp xF_3(x)}{2}\right](1-y)^2\right\},$$

where $S\approx 2ME$,

$$(\because\ \tilde{p}=\tilde{p}_\nu+\tilde{p}_N \Rightarrow -S^2=-m_N^2+2\vec{p}_\nu\cdot\vec{p}_N-2E_\nu E_N$$
$$=-2E_\nu E_N=-2E_\nu M\ )$$

(a) Apply $S=2ME$

(b) eliminate x

[Supplement-6]

- Total spin determines inelastic distribution

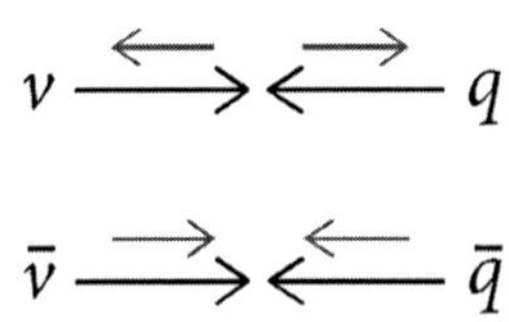

νq or $\overline{\nu}\,\overline{q}$

total spin $= 0$

$\Rightarrow$ flat in y

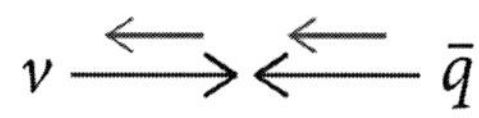

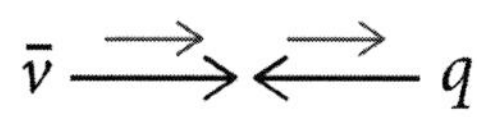

$\nu\overline{q}$ or $\overline{\nu}q$

total spin $= 1$

$$\Rightarrow \frac{1}{4}(1+\cos\theta)^2 = (1-y)^2$$

$$\Rightarrow \int (1-y)^2 dy = \frac{1}{3}$$

- Neutrino/anti-neutrino CC each produce particular Δq in scattering:

$$\begin{cases} \nu d \rightarrow \mu^- u \\ \overline{\nu} u \rightarrow \mu^+ d \end{cases}$$

$$\begin{cases} \dfrac{d^2\sigma(\nu p)}{dydx} = \dfrac{G_F^2 xS}{\pi}[d(x) + \overline{u}(x)(1-y)^2] \\ \dfrac{d^2\sigma(\overline{\nu} p)}{dydx} = \dfrac{G_F^2 xS}{\pi}[\overline{d}(x) + u(x)(1-y)^2] \end{cases}$$

From Eq. (1),

$$\frac{d^2\sigma(\nu N)}{dydx} = \frac{G_F^2 xS}{2\pi}[u + d + (\overline{u} + \overline{d})(1-y)^2]$$

$$\Rightarrow \frac{d\sigma}{dx} = \frac{G_F^2 xS}{2\pi}\left[u + d + \frac{1}{3}(\overline{u} + \overline{d})\right]$$

$$\Rightarrow \sigma = \frac{G_F^2 S}{2\pi}\left[\int x(u+d)dx + \frac{1}{3}\int x(\overline{u} + \overline{d})dx\right]$$

※ total cross-section is proportional to energy

$\Rightarrow$ consequence of pointlike scattering

$$= \frac{G_F^2 ME}{\pi}\left(Q + \frac{1}{3}\overline{Q}\right)$$

$$\therefore \begin{cases} Q = \int x[u(x) + d(x)]dx \\ \overline{Q} = \int x[\overline{u}(x) + \overline{d}(x)]dx \end{cases}, \text{ integer over } x,\ y = 0 \rightarrow 1,$$

Then Eqs. (1) and (2),

$$\sigma(\nu N) = \frac{G_F^2 ME}{\pi}\left(Q + \frac{1}{3}\overline{Q}\right) \qquad \left(\because \int_0^1 (1-y)^2 dy = \left[y - y^2 + \frac{1}{3}y^3\right]_0^1 = \frac{1}{3}\right)$$

$$\sigma(\overline{\nu}N) = \frac{G_F^2 ME}{\pi}\left(\overline{Q} + \frac{1}{3}Q\right)$$

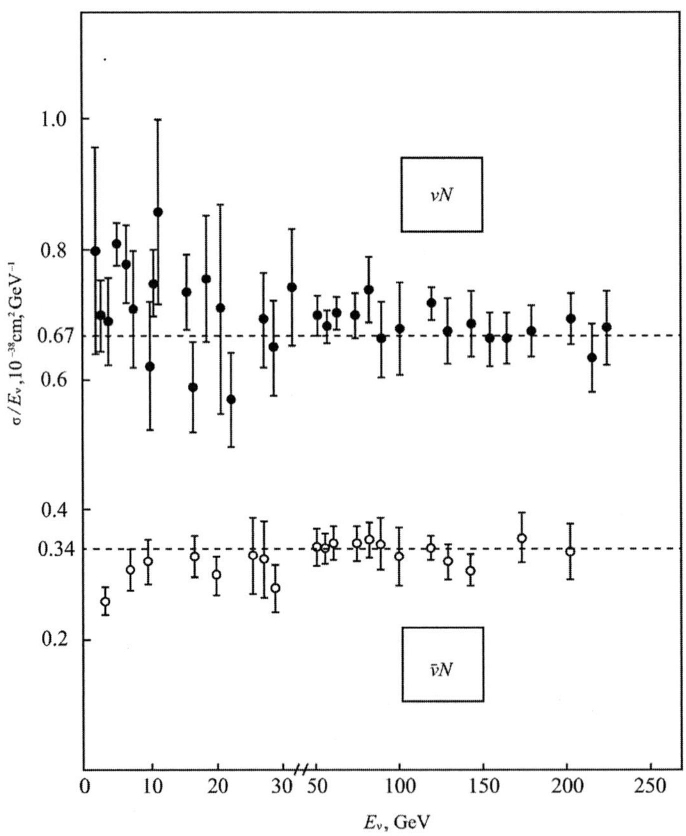

⇒ expected ratio of antineutrino to neutrino cross-section is

$$R = \frac{\sigma(\overline{\nu}N)}{\sigma(\nu N)} = \frac{\overline{Q}/Q + 1/3}{1 + \overline{Q}/Q} = \frac{1 + 3\overline{Q}/Q}{3 + \overline{Q}/Q}$$

※ We measured $R \approx 0.45$ and from this result, we get $\overline{Q}/Q = 0.15$.

⇒ nucleon contains antiquark as well as quark

■ Experimental results on quark distributions in the nucleon:

A number of predictions from experiments

(i) total neutrino-nucleon cross-section is proportional to energy:

$$S = (E_\nu + E_p)^2 - E_\nu^2 = 2E_\nu m_p = 2EM$$

⇒ this is a consequence of pointlike scattering

(ii) $R = \dfrac{\sigma(\overline{\nu}N)}{\sigma(\nu N)} = \dfrac{1 + 3\overline{Q}/Q}{3 + \overline{Q}/Q} \simeq 0.45$

⇒ $\dfrac{\overline{Q}}{Q} \simeq 0.15$ ⇒ nucleon contains antiquark as well as quarks

(iii) $$\frac{F_2^{eN}(x)}{x} = \frac{5}{18}[u(x) + \overline{u}(x) + d(x) + \overline{d}(x)] + \frac{1}{9}[s(x) + \overline{s}(x)]$$

$$\frac{F_2^{\nu N}(x)}{x} = u(x) + \overline{u}(x) + d(x) + \overline{d}(x)$$

※ electron does not see the difference of quarks but neutrino see the difference of quarks.

⇒ comparison establish fractionally charged quarks are real dynamical entities

① if s, $\overline{s}$ quark population is ignored,

$$F_2^{eN}(x) = \frac{5}{18} F_2^{\nu N}(x)$$

→ agreement with data is that nucleon contained fractionally charged quarks as real dynamical entities

② total momentum fraction carried by quarks and antiquarks is

$$\int_0^1 x[u(x)+d(x)+\overline{u}(x)+\overline{d}(x)]dx = \int F_2^{\nu N}(x)dx$$

$$= \frac{18}{5}\int F_2^{eN}(x)dx \approx 0.5 \text{(experimental result)}$$

⇒ 50% of the nucleon momentum is carried by partons without weak or electromagnetic coupling

⇒ In fact, this fraction is carried by the strongly interacting neutral gluons.

(iv) In $x(Q-\overline{Q})$ distribution, peak at 0.15 and quite broad

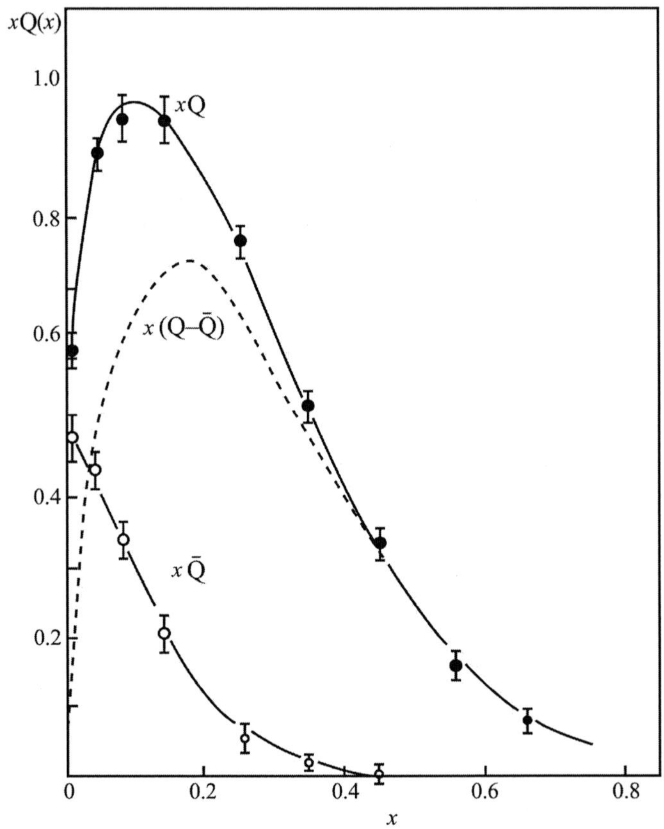

⇒ if valence quarks are quasi-free, peak at $\frac{1}{3}$ is expected

(∵ nucleon had consisted only of three quasi-free valence quarks)

⇒ broad distribution and smaller x-value at the peak is due to nucleon momentum is shared between valence quarks, quark-antiquark pairs and gluons.

※ broad distribution is due to the quark confinement so that distribution of Δp is broad from $\Delta x \Delta p \sim \hbar$ relationship.

⇒ As shown in the above Figure,

(a) both Q and $\overline{Q}$ are finite as $x \rightarrow 0$

(b) valence quark distribution, $Q - \overline{Q}$, tends to zero there

[Supplement-7]

$$F_2^{eN} = \frac{1}{2}[F_2^{ep} + F_2^{en}] = \frac{5}{18}x[u + \overline{u} + d + \overline{d}] + \frac{1}{9}x[s + \overline{s}]$$

$$\approx \frac{5}{18}x[u + \overline{u} + d + \overline{d}] = \frac{5}{18}[\text{sum of all quark momenta}]$$

↳ small s quark contribution is neglected

Naively one expects

$$\frac{18}{5}\int_0^1 F_2^{eN}(x)dx \approx 1$$

Experimental observation,

$$\frac{18}{5}\int_0^1 F_2^{eN}(x)dx \approx 0.5$$

⇒ - probed quarks and antiquarks carry only 50% of nucleon momentum

- remaining momentum carried by gluons

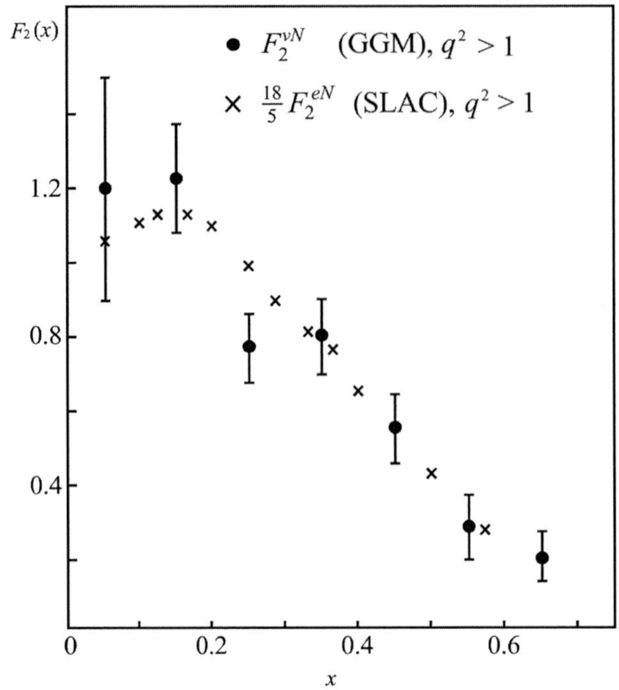

- GGM: at CERN in Gargamelle bubble chamber

Theory predicts

$$F_2^{eN}(x) = \frac{5}{18}F_2^{\nu N}(x)$$

∵ neutrino-nucleon scattering is not sensitive to charge.

Meanwhile, electron-nucleon scattering is sensitive to parton charge.

⇒ agreement with data is that nucleon contained fractionally charged quarks as real dynamical entities

For F_2^{ep} and F_2^{en}, the following relation can be obtained (Nachtmann).

$$\frac{1}{4} < \frac{F_2^{en}(x)}{F_2^{ep}(x)} < \frac{\frac{4}{9}(d+\bar{d}) + \frac{1}{9}(u+\bar{u}+s+\bar{s})}{\frac{4}{9}(u+\bar{u}) + \frac{1}{9}(d+\bar{d}+s+\bar{s})} < 4 \quad (\text{for all } x)$$

$$\because \quad \frac{F_2^{ep}(x)}{x} = \frac{4}{9}[u+\overline{u}] + \frac{1}{9}[d+\overline{d}+s+\overline{s}] = \frac{1}{9}[4u_v + d_v] + \frac{4}{3}s$$

$$\frac{F_2^{en}(x)}{x} = \frac{4}{9}[d+\overline{d}] + \frac{1}{9}[u+\overline{u}+s+\overline{s}] = \frac{1}{9}[4u_v + 4d_v] + \frac{4}{3}s$$

and

$$\frac{1}{x}[F_2^{ep}(x) - F_2^{en}(x)] = \frac{1}{3}[u_v(x) - d_v(x)]$$

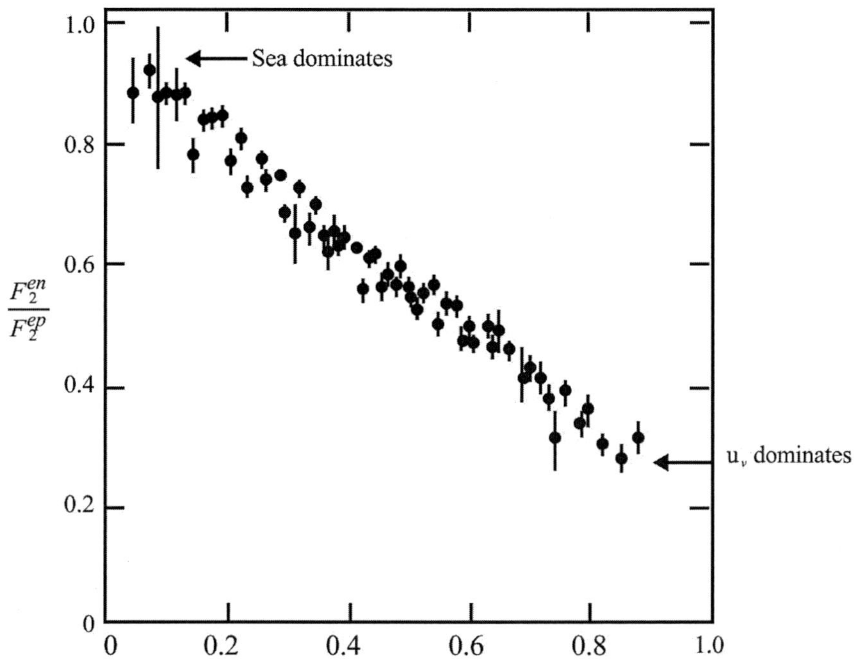

(i) for valence quarks dominate (no momentum left for sea quarks)

$$\frac{F_2^{en}(x)}{F_2^{ep}(x)} \sim \frac{4d_v + u_v}{4u_v + d_v} \geq \frac{1}{4} \quad (\text{for } x \to 1, \quad u_v \gg d_v \text{ for proton})$$

(ii) for sea quarks dominate (created by gluon splitting)

- gluon spectrum $\sim \frac{1}{x}$

$$\frac{F_2^{en}(x)}{F_2^{ep}(x)} \sim 1$$

($\because$ for $u_s \approx d_s \approx s_s \approx \overline{u}_s \approx \overline{d}_s \approx \overline{s}_s \equiv s$

$$\frac{4}{9}[u_s + \bar{u}_s] + \frac{1}{9}[d_s + \bar{d}_s + s_s + \bar{s}_s] = \left(\frac{8}{9} + \frac{4}{9}\right)s = \frac{4}{3}s$$

The ratio $\frac{F_2^{en}}{F_2^{ep}}$ allows to probe the sea and valence quark contributions.

Sum Rules

[I] Gottfried sum rule

$$\frac{F_2^{ep}(x)}{x} = \frac{4}{9}[u(x) + \bar{u}(x)] + \frac{1}{9}[d(x) + \bar{d}(x) + s(x) + \bar{s}(x)]$$

$$\frac{F_2^{en}(x)}{x} = \frac{4}{9}[d(x) + \bar{d}(x)] + \frac{1}{9}[u(x) + \bar{u}(x) + s(x) + \bar{s}(x)]$$

$$\leftarrow \quad u = u_v + u_s, \quad \bar{u} = u_s, \quad \int_0^1 u_v(x)dx = 2$$

$$d = d_v + d_s, \quad \bar{d} = d_s, \quad \int_0^1 d_v(x)dx = 1$$

$$\Rightarrow \int_0^1 [F_2^{ep}(x) - F_2^{en}(x)]\frac{dx}{x} = \frac{1}{3} + \frac{2}{3}\int_0^1 [\bar{u}(x) - \bar{d}(x)]dx,$$

where $\frac{1}{3}$ is due to assumption of the same amount of $\bar{u}$ and $\bar{d}$ quarks,

therefore, $\frac{1}{3}\int (u_v - d_v)dx = \frac{1}{3}(2-1) = \frac{1}{3}$

deviates from quark parton model

⇒- quark-antiquark sea is not flavor symmetric

- in proton, there are more $\bar{d}$ quarks than $\bar{u}$ quarks

⇒ observed value is 0.24±0.03

[II] Gross Llewellyn Smith sum rule

$$F_3^{\nu N}(x) = u(x) + d(x) - \bar{u}(x) - \bar{d}(x)$$

$$\int_0^1 xF_3^{\nu N}(x)\frac{dx}{x} = \int_0^1 [u_v(x) + d_v(x)]dx = 3,$$

where $u_v = u - \bar{u}$ and $d_v = d - \bar{d}$

This prediction is for the naive "free" parton model.

This rule is satisfied to about 10% (the sum rule is verified in practice to an accuracy ~ 10%) → (confirmed within 10%)

[III] Spin structure functions: Bjorken sum rule (spin structure)

• unpolarized structure function

$$F_1(x) \equiv \frac{1}{2}\sum Z^2[q\uparrow(x)+\bar{q}\uparrow(x)+q\downarrow(x)+\bar{q}\downarrow(x)],$$

where ┌ ↑: quark spin parallel to proton spin
└ ↓: quark spin antiparallel to proton spin

• polarized structure function

$$g_1(x) \equiv \frac{1}{2}\sum Z^2[q\uparrow(x)+\bar{q}\uparrow(x)-q\downarrow(x)-\bar{q}\downarrow(x)],$$

where Ze is the quark charge and the sum is over all quark flavors

Then, polarization asymmetry

$$A(x) = \frac{d\sigma^{\uparrow\uparrow}(x)-d\sigma^{\uparrow\downarrow}(x)}{d\sigma^{\uparrow\uparrow}(x)+d\sigma^{\uparrow\downarrow}(x)},$$

where $d\sigma^{\uparrow\uparrow}$ is for parallel lepton and proton spin

$d\sigma^{\uparrow\downarrow}$ is for antiparallel lepton and proton spin

For sea quarks are neglected, expected asymmetry $\dfrac{g_1(x)}{F_1(x)}$ is

$$\Rightarrow \begin{cases} A_p = \dfrac{5}{9} \\ A_n = 0 \end{cases}$$

The most important sum rule for polarized scattering is that due to Bjorken for the difference:

$$I = \int_0^1 (g_1^p - g_1^n)dx = \frac{1}{6}\left(\frac{g_A}{g_V}\right)\left(1-\frac{\alpha_S}{\pi} + \cdots\right),$$

where $\dfrac{g_A}{g_V} = 1.26$ is the ratio of axial to vector coupling in neutron β-decay

$\Rightarrow$ good to about 10% $\left(\leftarrow \dfrac{0.18 - 0.16}{0.18} \right)$

※ recent measurement at $q^2 = 5\ \mathrm{GeV}^2$, $I = 0.16$

Estimation from the above equation is 0.18.

Chapter 6

Quark Interactions and QCD

■ Color quantum number

① Quantum chromodynamics (QCD) is the formal theory of the strong color interactions between quarks.

- interquark interaction mediated by gluons are assumed to be invariant under color interchange
- gluon carries color and anticolor, and postulated to belong to an octet of states: 8 effective gluon states

color octet (possible basis for gluons):

$$r\bar{b},\ r\bar{g},\ b\bar{g},\ b\bar{r},\ g\bar{r},\ g\bar{b},\ \frac{1}{\sqrt{2}}(r\bar{r} - b\bar{b}), \frac{1}{\sqrt{6}}(r\bar{r} + b\bar{b} - 2g\bar{g})$$

(← blue instead of u, red instead of d, green instead of s)

color singlet (=color neutral):

$$\frac{1}{\sqrt{3}}(r\bar{r}+b\bar{b}+g\bar{g}) \rightarrow \text{carries no net color} \rightarrow \text{excluded}$$

※ hadrons are always color neutral (or color singlet) i.e. $Q_r\overline{Q}_r$, $Q_rQ_bQ_g$

② "Color" was invented to explain:

- existence of $\triangle^{++} = uuu$
- R measurement $e^+e^- \rightarrow$ hadrons

Color of a quark has three possible values: blue, red, green

antiquark carries anticolor: antiblue, antired, antigreen

Boson mediating the quark-quark interaction are called gluons

- gluons are to the strong force (what the photon is to the EM force)
- gluons carry a color and an anticolor:

  9 possible combinations of color and anticolor

  $\rightarrow$ $r\bar{r} + b\bar{b} + g\bar{g}$ is color neutral, leaving 8 effective color combinations

③ Other possible combinations, qq or $qq\bar{q}\bar{q}$ are not found, i.e. the interaction is found to be repulsive, so that QCD correctly predicts that only two of all the possible quark combinations should exist in nature.

④ Interquark potential: $U_s = -\frac{4}{3}\frac{\alpha_s}{r} + kr,$

where $-\frac{4}{3}\frac{\alpha_s}{r}$ is the Coulomb type and

kr is for the confine quark in a hadron

$\Rightarrow$ This form successfully accounts for energy levels in charmonium and bottomonium bound states.

⑤ Gluon exchanges are analogous to photon exchanges, but different

<gluons carry color-charge>

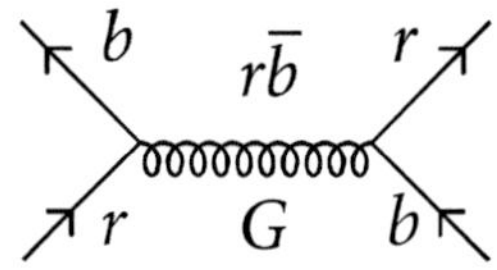

< photons do not carry electric charge>

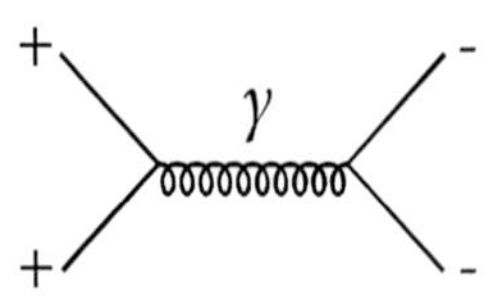

■ The QCD potential at short distance

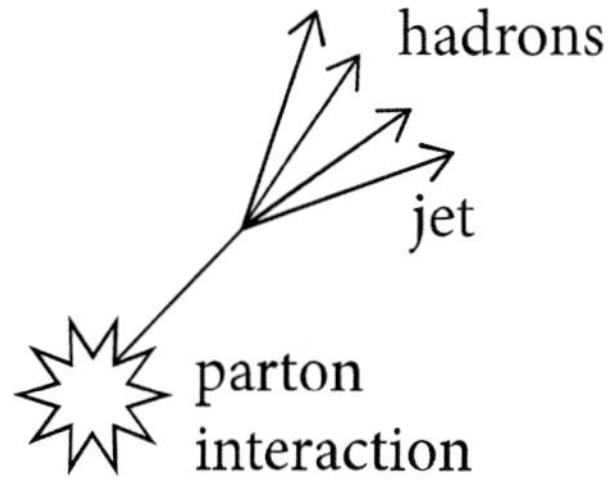

① When a quark or gluons leaves a high energy collision, it generates a narrow cone of particle, known as “jet”.

- Jet produced in high energy collisions are direct support for short range behavior.
- Such events can be interpreted as arising from parton-parton elastic interactions.

② Jets are indirect evidence that a quark or gluon was involved in the hard scattering process.

Interpretation: Two-jet events are the elastic scattering of a parton (quark or gluon) in the proton from one in the antiproton

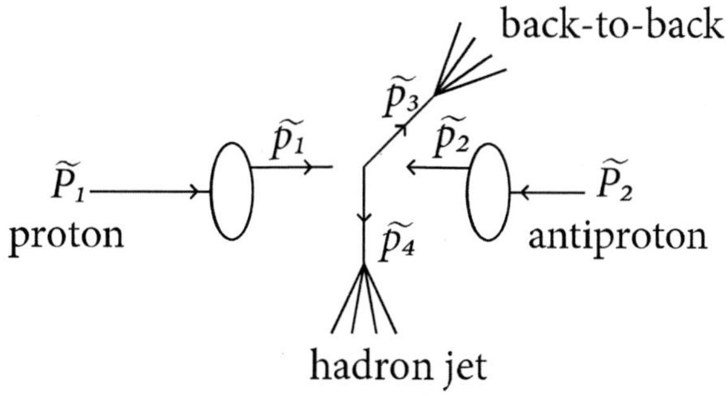

<back-to back in azimuthal angle and z-axis is the beam axis>

where each of interaction parton carry a fraction x of the momentum of hadrons, and

$\tilde{p}_1 = x_1 P_1$ and $\tilde{p}_2 = x_2 P_2$ and $\tilde{p}_3$ & $\tilde{p}_4$ known

E is the energy of each beam, $\tilde{p}_1$ & $\tilde{p}_2$ unknown

Can we express x_1 and x_2 in terms of observables (final state)?

Expected cross-section is

$$\frac{d^3\sigma(p\bar{p} \to 2jets)}{dx_1 dx_2 d(\cos\theta)} = \sum_{i,j}\left[\frac{F_i(x_1)}{x_1}\right]\left[\frac{F_j(x_2)}{x_2}\right]\frac{d\sigma_{ij}}{d(\cos\theta)},$$

where $\left[\frac{F_i(x_1)}{x_1}\right]$ and $\left[\frac{F_j(x_2)}{x_2}\right]$ are the densities of partons of type i and j in the proton and antiproton with fractional momentum x_1 and x_2, and $\frac{d\sigma_{ij}}{d(\cos\theta)}$ is the elementary parton-parton cross-section.

Let`s express x_1 and x_2 in terms of observables (final state)

$\tilde{p}_1 + \tilde{p}_2 = \tilde{p}_3 + \tilde{p}_4$, from energy-momentum conservation

from momentum conservation,

$$p_{1z} + p_{2z} = p_{3z} + p_{4z} \Rightarrow x_1 P_{1z} + x_2 P_{2z} = x_1 E - x_2 E$$

$$\Rightarrow E(x_1 - x_2) = P_{3z} + P_{4z}$$

$$\Rightarrow x_1 - x_2 = \frac{P_{3z} + P_{4z}}{E} \equiv x_F \quad \cdots\cdots (1)$$

$$\Rightarrow (\tilde{p}_3 + \tilde{p}_4)^2 = (x_1\tilde{P}_1 + x_2\tilde{P}_2)^2 = x_1^2\tilde{P}_1^2 + x_2^2\tilde{P}_2^2 + 2x_1x_2\tilde{P}_1\tilde{P}_2$$

(Ignore the mass term in the following expression)

$$= 2x_1x_2\overrightarrow{P_1}\cdot\overrightarrow{P_2} - 2x_1x_2EE = -2x_1x_2P_1P_2 - 2x_1x_2EE$$

$$= -4x_1x_2E^2$$

$$(\because\ P_1^2 = E_1^2 - m_1^2 \Rightarrow P_1 = E_1 = E)$$

$\rightarrow$ so, we can relate initial momentum to the final momenta.

$$\Rightarrow x_1 x_2 = -\frac{(\tilde{p}_3 + \tilde{p}_4)^2}{4E^2} = \tau \cdots\cdots (2),$$ i.e. Eqs. (1) and (2) are known

From Eqs. (1) and (2),

$$x_1 = \frac{x_F \pm \sqrt{x_F^2 + 4\tau}}{2} = \frac{x_F + \sqrt{x_F^2 + 4\tau}}{2}$$: known since x_F and τ are known

$$x_2 = \frac{-x_F \pm \sqrt{x_F^2 + 4\tau}}{2} = \frac{x_F - \sqrt{x_F^2 + 4\tau}}{2}$$: x_2 cannot be negative & known

pull "minus" and take the absolute value

4-momentum transfer is

$$\tilde{q} = \tilde{p}_3 - \tilde{p}_1 = \tilde{p}_2 - \tilde{p}_4 \qquad \cdots\cdots (3)$$

Since x_1 and x_2 are known, (then $\tilde{p}_1$ and $\tilde{p}_2$ can be obtained) momentum transfer $\tilde{q}$ can be also calculated (or determined).

Scattering angle of partons relative to beam direction is

$$\cos\theta = \frac{(\vec{p}_3 - \vec{p}_4) \cdot (\vec{p}_1 - \vec{p}_2)}{|\vec{p}_3 - \vec{p}_4|\ |\vec{p}_1 - \vec{p}_2|},$$

where the scattering angle can be extracted in the cms from the laboratory scattered particles.

- since we don't know which incident parton belongs to which jet, there is ambiguity between θ and $\pi - \theta$

$\Rightarrow$ smaller θ value of two values is for the predicted distribution when comparing with theory

① Scattering proceed through moment transfer in t-channel

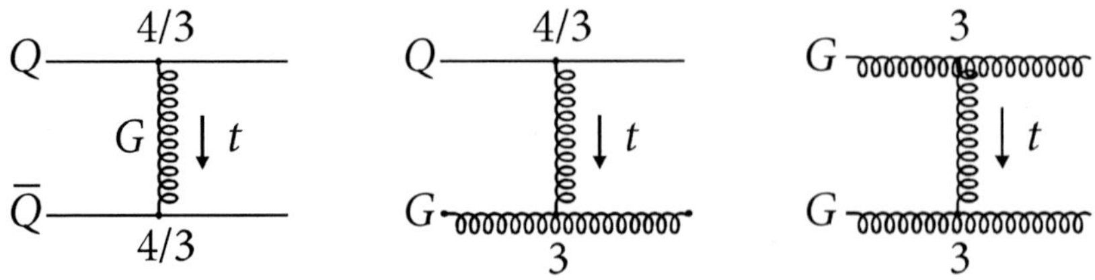

$$t = \tilde{q}^2 = (\tilde{p}_1 - \tilde{p}_3)^2\text{: } t\text{-channel cross-section } \frac{d\sigma}{dt} \simeq \frac{1}{t^2}\left(\text{or } \frac{1}{q^4}\right)$$

② $s = -(\tilde{p}_1 + \tilde{p}_2)^2 = 4x_1x_2E^2$: s-channel cross-section $\dfrac{d\sigma}{ds} \simeq \dfrac{1}{s^2}$

(※ $4x_1x_2E^2$ is the square of cms energy of partons)

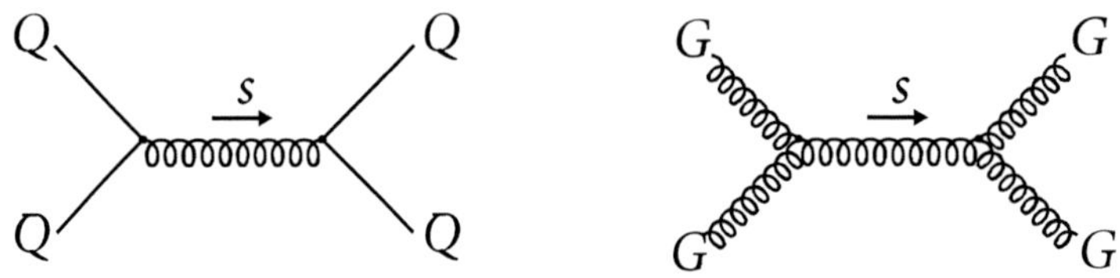

In real experiment $t \ll s$ so that t-channel dominant

① Structure function will include gluons and quarks:

$$F(x) = G(x) + \frac{4}{9}[Q(x) + \overline{Q}(x)],$$

where $Q(x)$, $\overline{Q}(x)$ and $G(x)$ are quark, antiquark and gluon densities weighted by momentum fraction x and

4/9 comes from difference in couplings $\begin{cases} Q\overline{Q}G \rightarrow \frac{4}{3}\alpha_S \\ GGG \rightarrow 3\alpha_s \end{cases}$

($\because$ QQG or $\overline{Q}\,\overline{Q}G$ coupling factor is $\frac{4}{3}\alpha_s$, while for GGG vector is $3\alpha_s$)

<The parton scattering follows the Rutherford scattering for $\frac{1}{r}$ potential with α_s replacing α>

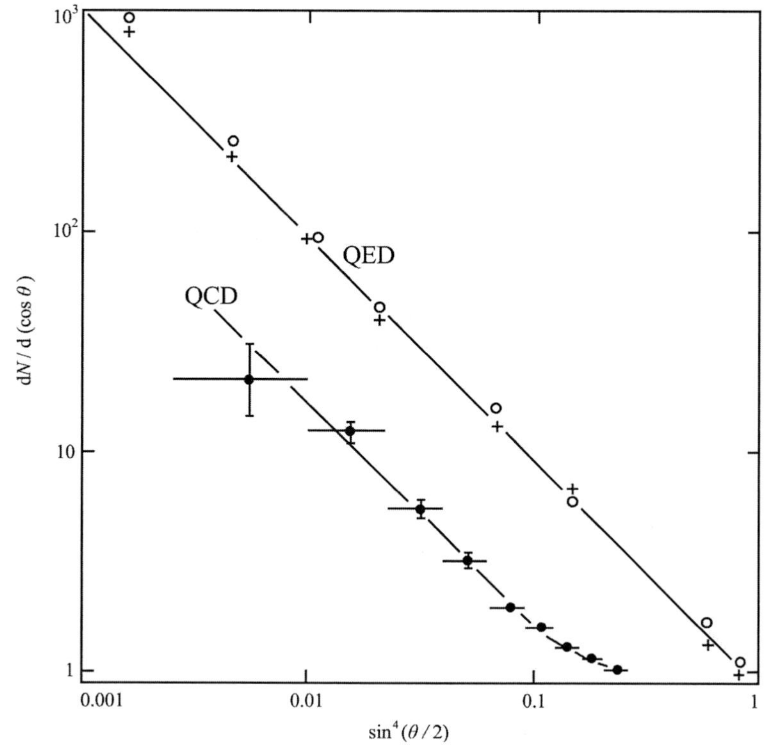

- QED data: Geiger and Marsden (1909), α scattering on Au and Ag
- QCD data: CERN, $p\bar{p} \rightarrow 2\ \text{jets}$ at $q^2 = 2000\ \text{GeV}^2$

② Scattering angle θ is small

$$\frac{d\sigma}{d\Omega} = \frac{\alpha^2}{8p^2\sin^4\left(\frac{\theta}{2}\right)}\left[1+\cos^4\left(\frac{\theta}{2}\right)\right] \xrightarrow[\text{small angle}]{\text{for}} \frac{d\sigma}{d\Omega} \propto \frac{\alpha_s^2}{4p_0^2\sin^4\left(\frac{\theta}{2}\right)},$$

where $p_0 = \sqrt{\frac{s}{2}}$ is the cms momentum of each parton.

■ **The QCD potential at large distances**: the string model

What accounts for the below plots?

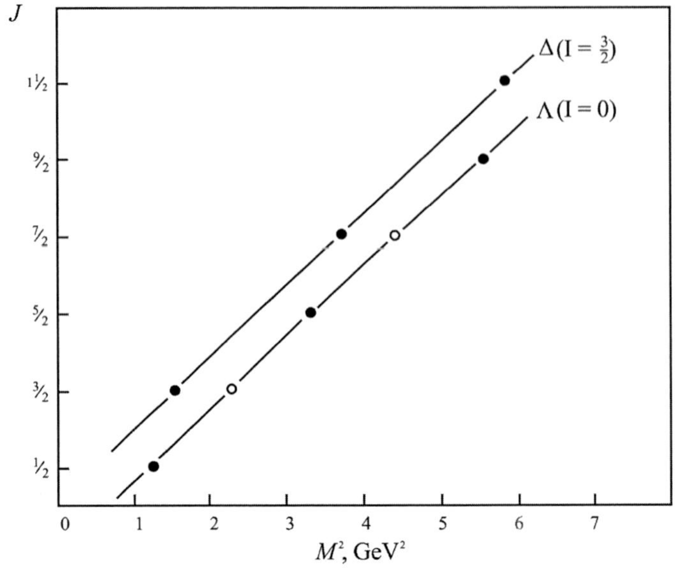

- "Chew-Frautschi" plot: $J = \alpha' E^2 + \text{constant}$

<Figure shows that linear dependence of angular momentum J as a function of mass square: linear dependence of J an M^2>

⇒ called a "Chew-Frautschi" plot

⇒ relevant to the form of QCD potential at large distance

<quarks are held together in form of string or tube by color lines of force>

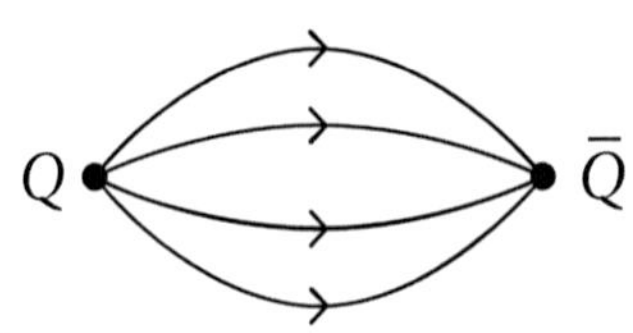

※ self-coupling of gluons pulls the line together by strong self-interaction of gluons, so the system can be approximated by a tube or string.

let`s calculate angular momentum by assuming that ends of tube rotate with velocity c

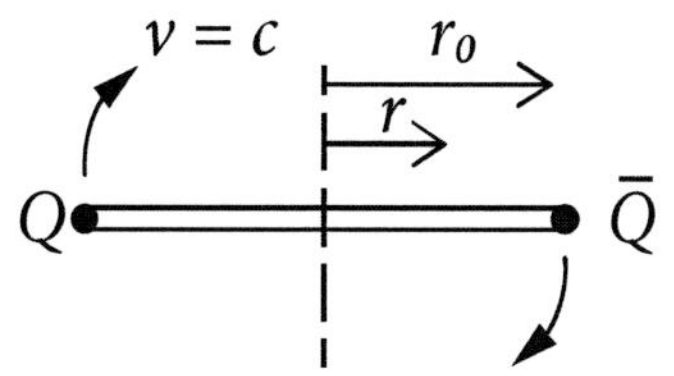

"string model": $v = c$

※ orbital angular momentum of quark pair is equal to angular momentum of gluon tube

$\Rightarrow$ local velocity at radius r is $\frac{v}{c} = \frac{r}{r_0}$ $(\because\ v = \frac{r}{r_0}c)$

$\Rightarrow$ relativistic total mass is

$$E = Mc^2 = 2\int_0^{r_0} \frac{kdr}{\sqrt{1-v^2/c^2}} = kr_0\pi \ \cdots\cdots (1) \quad \because\ E = mc^2/\sqrt{1-v^2/c^2}$$

where k is the energy density per unit length of a string

and

orbital angular momentum is

$$J = \frac{2}{\hbar c^2}\int_0^{r_0} \frac{krvdr}{\sqrt{1-v^2/c^2}} = \frac{kr_0^2\pi}{2\hbar c} \ \cdots\cdots (2)$$

$(\because\ J = rmv = rv\frac{E}{c^2} = rv\frac{kdr}{c^2\sqrt{1-v^2/c^2}}$ and $\hbar$ is introduced for dimension.)

By eliminating r_0 between Eqs. (1) and (2), we expect that

$$J = \alpha' E^2 + \text{constant}, \text{ where } \alpha' = \frac{1}{2\pi k \hbar c}$$

(c.f. for $U = kr^n$, $J \propto E^{(1+1/n)}$)

$\Rightarrow$ This result holds for constant energy density k of string

i.e. for potential of form $U = kr$

- The experimental results in the previous page are evidence that the potential is linear potential.
- Once we get the slope, we obtain the value of k.
- From measurement, $\alpha' = 0.93\ \text{GeV}^{-2} \rightarrow k = 0.87\ \text{GeV fm}^{-1}$

※ Alternate, elementary estimate of k from size of hadrons:

$$M \sim 1\ \text{GeV},\ r \sim 1\ \text{fm} \rightarrow k = \frac{1\ \text{GeV}}{1\ \text{fm}} = 1\ \text{GeV fm}^{-1}$$

■ Gluon jets in e^+e^- annihilation

e^+e^- annihilation to hadrons:

- hadrons appear in the form of two "jets" collimated around $q\bar{q}$-axis

(※ jet: bunch of particle moving in the same direction)

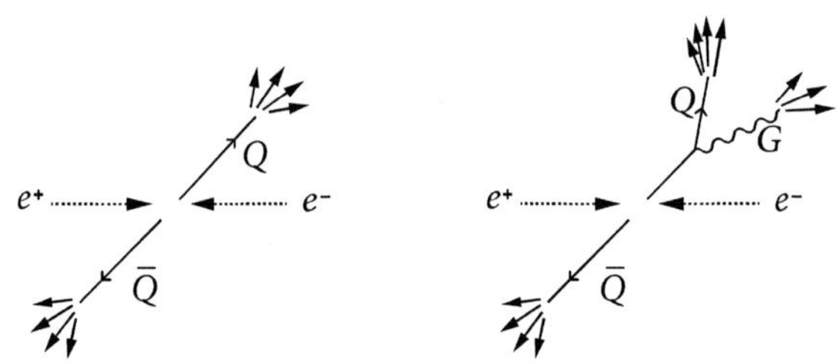

- separate hadronic jets
- quark radiates 'hard' gluon (G), carrying approximately half of quark energy, at large angle

- rate of three-jet events is $f(\alpha_s)$

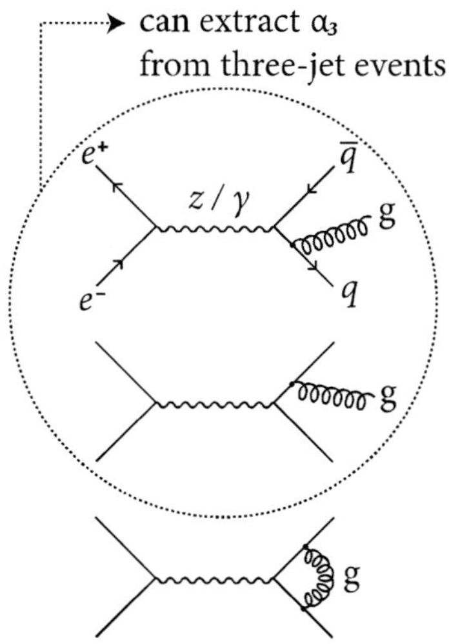

<above two diagrams are the leading order Feynman diagrams proportional to α_s>

- rate of three-jet to two-jet events is determined by α_s
- experiment results $\alpha_s \simeq 0.14$ (Analysis of PETRA, for $\sqrt{s} \sim 30-40\ \mathrm{GeV}$)

※ PETRA: Positron-Electron Tandem Ring Acceleration

- Angular distribution in the three-jet events determines gluon spin.

  steps:

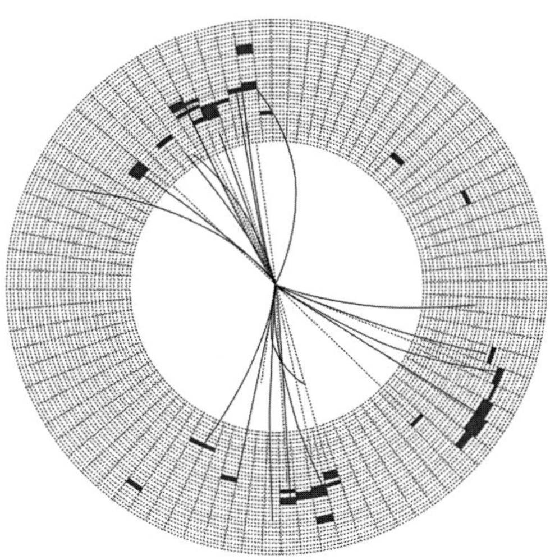

① jets are ordered in energy, $E_1 > E_2 > E_3$

(E_3 is the most likely to be produced by gluon)

② transformation is made to cms frame of jets 2 and 3

③ angle $\tilde{\theta}$ calculated for jet 1 with respect to common line of flight of jets 2 and 3 (in their cms frame)

④ distribution in $\tilde{\theta}$ is sensitive to gluon spin

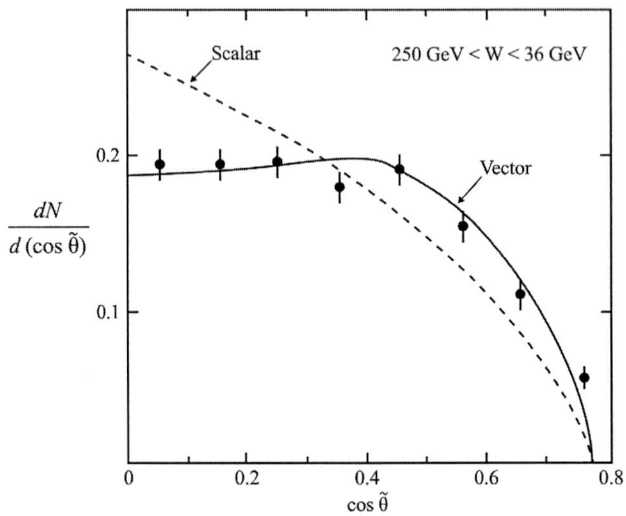

$\Rightarrow$ data favor $J^P = 1^-$ (vector)

■ Running couplings in QED and QCD

charged lepton magnetic moment

(i) lowest order

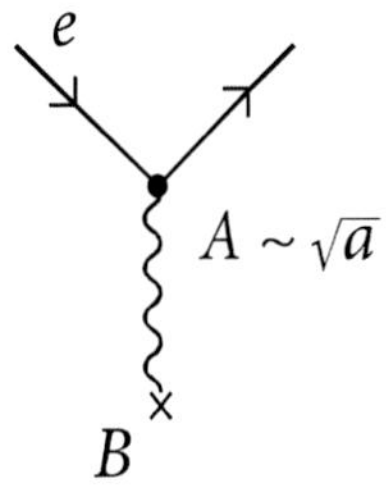

(ii) higher order correction: $\sim e^2 \sim \alpha$

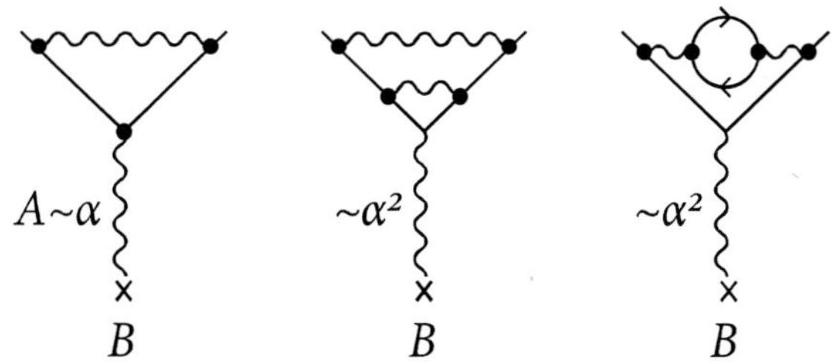

magnetic moment $\mu = g\mu_B s$, where g is the gyromagnetic ratio,

$s = 1/2$ in units of $\hbar$ and

$\mu_B = \dfrac{e\hbar}{2mc}$ is Bohr magneton

$\Rightarrow$ quantity $\dfrac{(g-2)}{2}$ represents the anomaly due to higher order terms

(※ $g = 2$ in the Dirac theory)

$\Rightarrow$ for theory (electron)

$$\frac{g-2}{2} = 0.5\left(\frac{\alpha}{\pi}\right) - 0.32848\left(\frac{\alpha}{\pi}\right)^2 + 1.19\left(\frac{\alpha}{\pi}\right)^3 + \cdots$$

$$= (115965230 \pm 10) \times 10^{-11}$$

[※ experiment $= (11596528.7 \pm 0.4) \times 10^{-11}$]

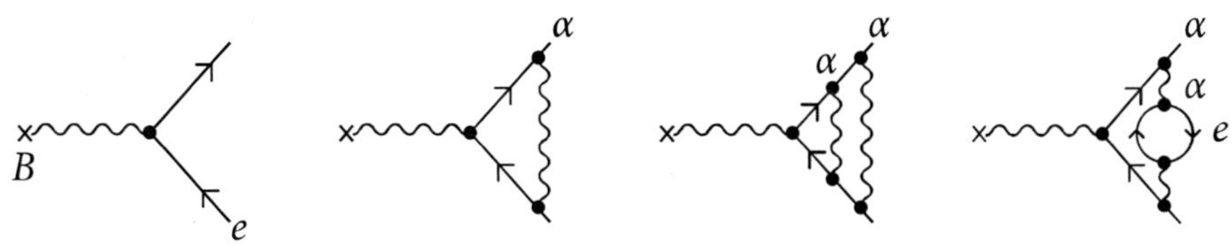

(a) (b) (c)

<leading diagrams contributing to magnetic moment of electron>

or

$\Rightarrow \dfrac{g-2}{2} = \dfrac{0.5}{\pi}\alpha_{eff}$, where α_{eff} depends on the masses on momentum transfers in virtual processes involved)

$\Rightarrow$ this measurement provides very accurate value for the fine structure constant

$\Rightarrow$ α is treated as "running coupling constant"

※ running coupling constant expresses α value at one value of q^2 in terms of that at another value, say $q^2 = \mu^2$

(and thus avoids problem of coupling at $q^2 = \infty$)

without proof:

$$\Rightarrow \frac{1}{\alpha(\mu^2)} = \frac{1}{\alpha(q^2)} + \beta_0 \ln\left(\frac{q^2}{\mu^2}\right) + \cdots \qquad \cdots\cdots (1),$$

where $\beta_0 = \frac{1}{12\pi}(4n_f - 11n_b)$ depends on number of degree of freedom in loop of vacuum polarization (n_f is for fermions degrees of freedom in the vacuum loop and n_b is for bosons degrees of freedom)

: dependence of coupling on momentum transfer follows renormalization group equation

In EM interaction if no loops containing photons and can take $n_f = 3$ for high energy,

$$\beta_0 = \frac{1}{\pi} \quad \cdots\cdots (2)$$

(I) for QED,

In QED, there are no bosons in the loops, so at high energy $n_b = 0$, $n_f = 3$ (where all three families are active)

Substitute Eq. (2) to (1),

$$\alpha_{em}(q^2) = \frac{\alpha(\mu^2)}{\left[1 - \frac{1}{\pi}\alpha(\mu^2)\ln\left(\frac{q^2}{\mu^2}\right)\right]}$$

$\Rightarrow$ if we take $\alpha_{em} = \frac{1}{137}$ at $\mu = 1\ \mathrm{MeV}$,

at Z^0 mass ($q \approx 100\ \mathrm{MeV}$), $\alpha(M_Z) \simeq \frac{1}{129}$

$\Rightarrow$ effective coupling increases very slowly with energy-momentum transfer

※ Reasons to explain why it's increasing

⇒ Running Couplings in QED

Imagine a bare charge, which is screened by polarization charge. At higher energies we probe shorter distance and screening effects are reduced.

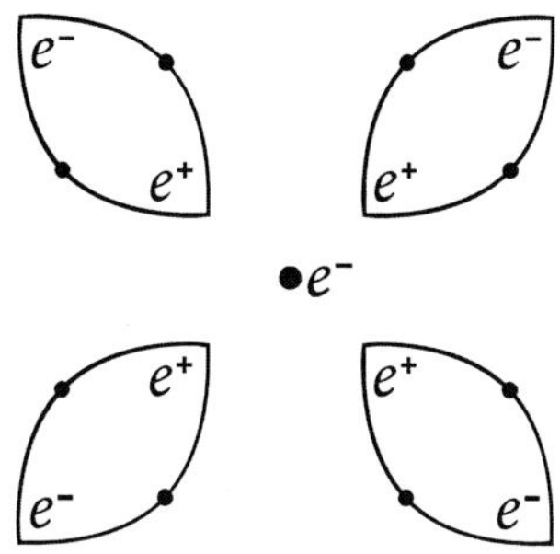

shielding effect is created around a charge, effectively changing the strength of coupling

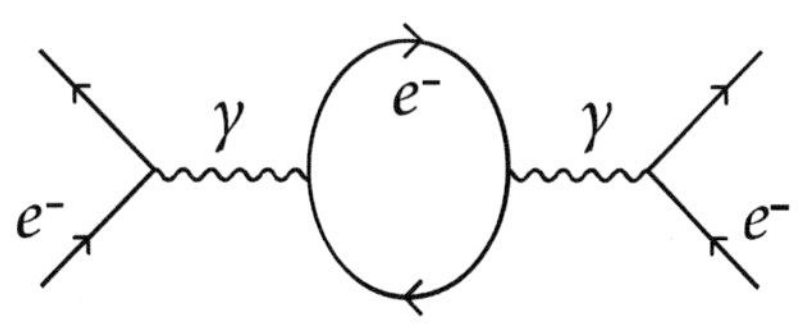

pairs of charge will be created and re-absorbed, producing shielding effect ("vacuum polarization")

(II) for QCD,

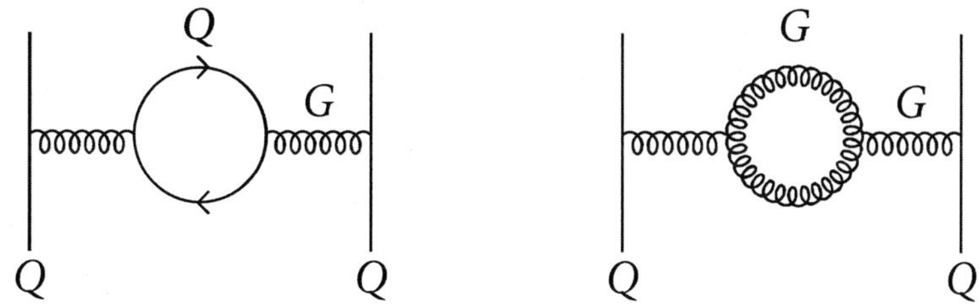

$$n_b = 3 \quad \text{and} \quad n_f = 3 \Rightarrow \beta_0 = -\frac{7}{4\pi} \quad \cdots\cdots (3),$$

where $n_b = 3$ is for three degrees of freedom (3 color)

$n_f = 3$ where all three families are active

Substitute Eq. (2) to (1),

$$\alpha_s(q^2) = \frac{\alpha_s(\mu^2)}{1 + \frac{7}{4\pi}\alpha_s(\mu^2)\ln\left(\frac{q^2}{\mu^2}\right)} \equiv \frac{1}{B\ln\left(\frac{q^2}{\Lambda^2}\right)} \quad \cdots\cdots (4)$$

$$\because \frac{1}{\alpha(\mu^2)} = \frac{1}{\alpha(q^2)} + \beta_0\ln\left(\frac{q^2}{\mu^2}\right) \Rightarrow \frac{1}{\alpha(q^2)} = \frac{1}{\alpha(\mu^2)} - \beta_0\ln\left(\frac{q^2}{\mu^2}\right)$$

$$\Rightarrow \alpha(q^2) = \frac{\alpha(\mu^2)}{1 - \beta_0\alpha(\mu^2)\ln\left(\frac{q^2}{\mu^2}\right)} = \frac{1}{\frac{1}{\alpha(\mu^2)} + B\ln\left(\frac{q^2}{\mu^2}\right)}$$

$\Rightarrow$ α_s decreases with increasing q^2

<Running Coupling in QCD>

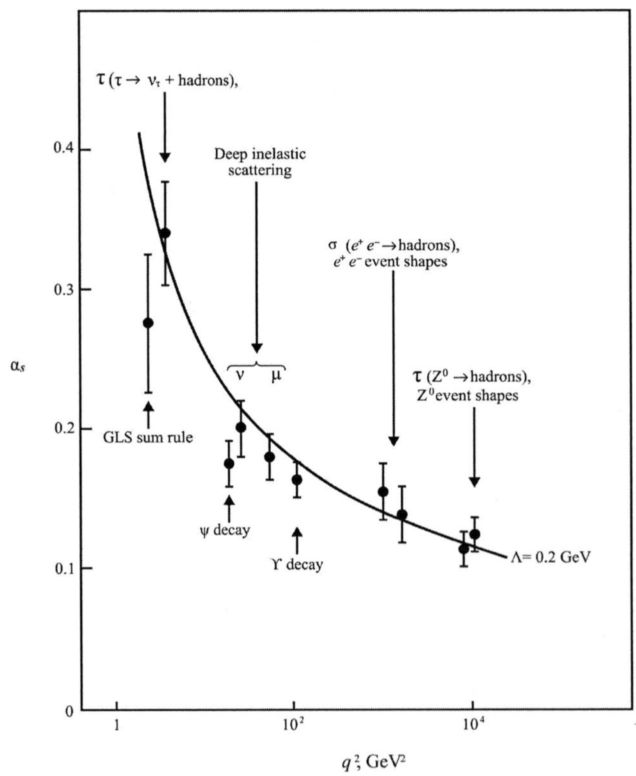

- this is opposite dependence from QED
- it is typical of non-Abelian field, where the field particles carry charge and are self-coupling
- longitudinal gluons have an antishielding effect, spread out color charge.
  └ weakening interaction

In Eq. (4), $B = -\beta_0$ is the positive quantity and $\Lambda^2 = \mu^2 e^{-\frac{1}{B\alpha_s(\mu^2)}}$

┌ at large q^2, $\alpha_s(q^2) \rightarrow 0$: quarks behave as if free, a phenomenon known as asymptotic freedom" and precisely what expected in the naive, free parton, model

└ at low q^2 such that $q \sim \Lambda$,

$\alpha_s(q^2)$ becomes very large and Eq. (1) becomes meaningless.

→ such large interquark coupling is presumably connected with confinement of quarks at large distance

$\xrightarrow[\text{these}]{\text{from}}$ QCD has form of perturbation theory and applies only for $q^2 \gg \Lambda^2$

• Many observables are sensitive to value of α_s

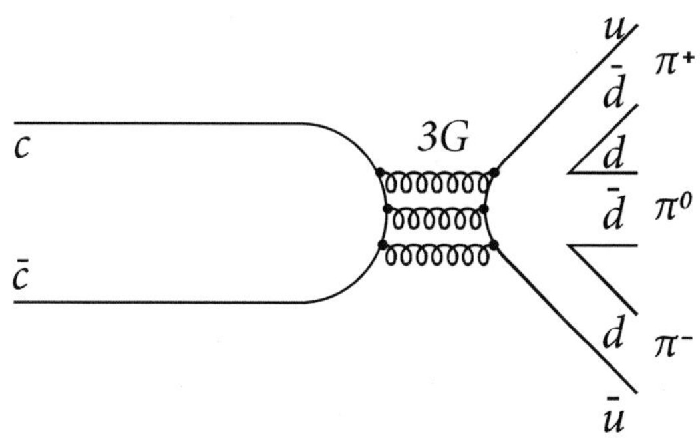

① widths of bounds state of $c\bar{c}$ and $b\bar{b}$

② $R = \dfrac{\sigma(e^+e^- \rightarrow \text{hadrons})}{\sigma(\text{point})}$ requires factor $\left(1 + \dfrac{\alpha_s}{\pi}\right)$

③ event shapes depend on fraction of 3 jets

④ scaling deviations in deep inelastic lepton-nucleon scattering

⑤ hadronic width of Z

<Scaling deviations in deep inelastic lepton-nucleon scattering>

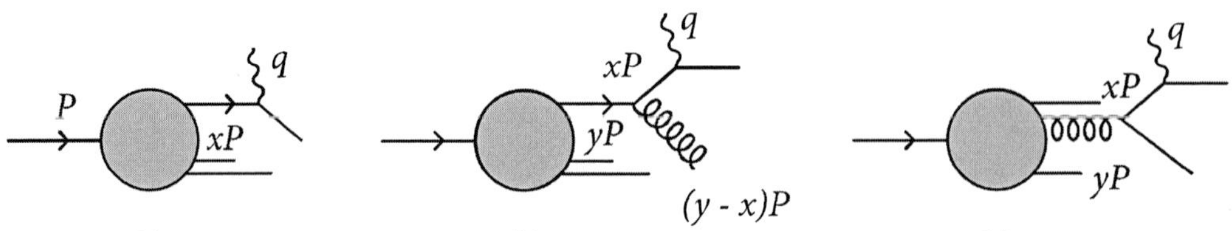

(a) (b) (c)

diagrams illustrating absorption of momentum transfer q in deep inelastic scattering by

(a) a quark with momentum fraction x

(b) a quark with momentum fraction x which has radiated a gluon and which had originally a momentum fraction y

(c) a quark of momentum fraction x created by a gluon with momentum fraction greater than x

■ Evolution of the structure functions in the deep inelastic scattering

$u(x, q^2)dx$: density of valence u quark with momentum fraction $x \sim x+dx$ at a particular value of q^2

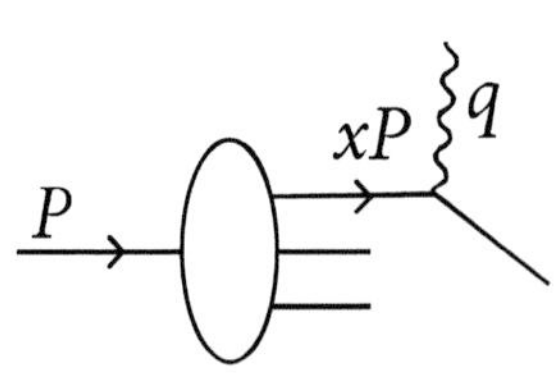

$$\frac{du(x, q^2)}{d(\ln q^2)}dx = \frac{\alpha_s(q^2)}{2\pi}\left[\int_{y=x}^{y=1} u(y, q^2) P_{QQ}\left(\frac{x}{y}\right)\frac{dy}{y}\right]dx$$: **Altarelli-Parisi** equation

where P_{QQ} is the splitting function describing radiation of a vector particle by a fermion

$$P_{QQ}(Z) = \frac{4}{3}\frac{(1+Z^2)}{(1-Z)}, \text{ (by \textbf{Williams-Weizsacker})}$$

where $Z = \frac{x}{y}$ is the probability that quark retains a fraction Z of its momentum and $\frac{4}{3}$ is for the color factor

⇒ increase du in u is proportional to α_s and to the integrated number of quarks with $y > x$ that can radiate a gluon in such a way that they then fall into the interval $x \sim x + dx$

※ When sea quarks are included, an additional term in the equation is needed.

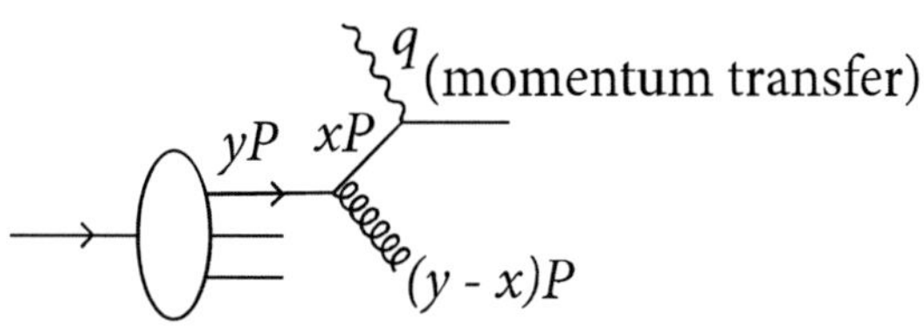

Because of these two terms, the quark distribution shrinks to smaller x as q^2 increases.

⇒ As q^2 increases, the quark distribution shrinks to smaller x
i.e. the density at large x $(x > 0.2)$ progressively decreases while that at small x $(x < 0.2)$ is enhanced.

⇒ The valence quark distribution extends to the largest values of x, with an approximate from $x^{1/2}(1-x)^3$. The sea quarks and gluons are more prominent at small x, the gluon distribution dominating for $x < 10^{-2}$.

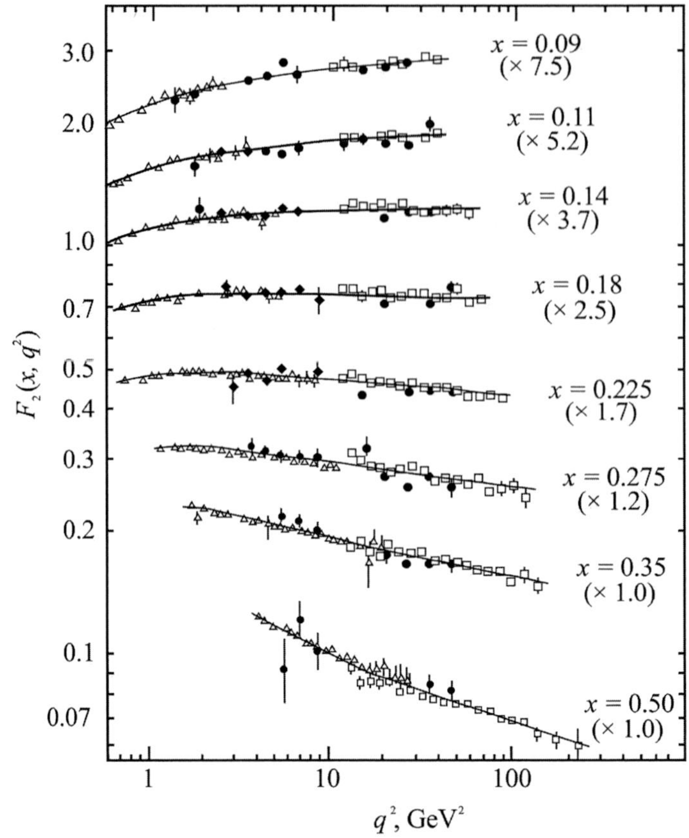

- where $F_2(x, q^2)$ measures the sum of Q and $\overline{Q}$ momentum fractions
- fitted curve is the theoretical predictions such as Altarelli-Parisi (DGLAP theoretical prediction)
- the curves have been calculated with α_s left as free parameter
- the scaling law violation depends on coupling α_s (and on its Q^2 dependence)
- the scaling law is satisfied at small x value: structure function increases with increasing q^2, namely, when we look into proton with increasing resolving power.

■ Gluonium

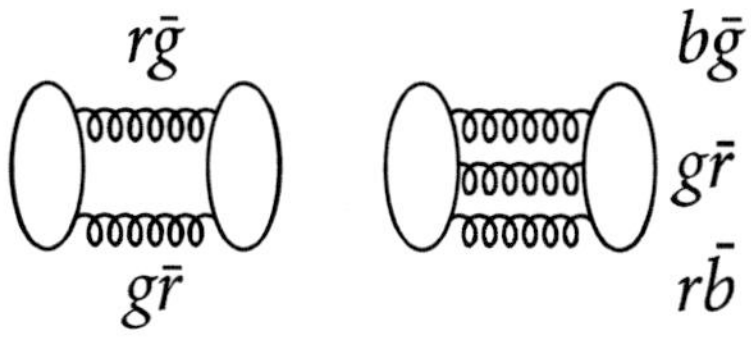

- GG and GGG systems can be formed as color-singlet states, and therefore can occur

  in here, GG: $8 \otimes 8 = (1 \oplus 8 \oplus 27) \oplus (8 \oplus 10 \oplus \overline{10})$

- These state, glueballs, are difficult to identify, but evidence for them does exist
- what state are allowed:

  combine two gluons into color singlet

  gluons are vector particles, members of a color octet

  the allowed states are:

L	J^{PC}
0	0^{++}, 2^{++}
1	0^{-+}, 1^{-+}, 2^{-+}
2	2^{++}, 0^{++}, 4^{++}
3	2^{-+}, 3^{-+}, 4^{-+}

① Two spin-one gluons in an S-wave ($L=0$) must be symmetric and will form states of $J^{PC} = 0^{++}$ or 2^{++}.

② $L=1$ case must have the antisymmetric $S=1$ case and will yield $J^{PC} = 0^{-+}$, 1^{-+}, 2^{-+} (※ $L+S < J < |L-S|$)

Several of these J^{PC} values can not be formed by quark/antiquark.

→ such states are called exotics

→ finding mesons with these quantum number will suggest "glueballs"

ex 1^{-+}, 3^{-+} , $[P = (-1)^{L+1},\ C = (-1)^{L+S}]$

Properties to look for to find glueballs are

- · exotic quantum number
- · flavor symmetry in decay
- · no radiative decays and no two photon decays
- · mass $1-2$ GeV (scalar glueball)
- · decays into $\pi\pi$, $K_S K_S$, K^+K^- and $\eta\eta$ (scalar glueball)
- · width of order $\sim 10's$ of MeV

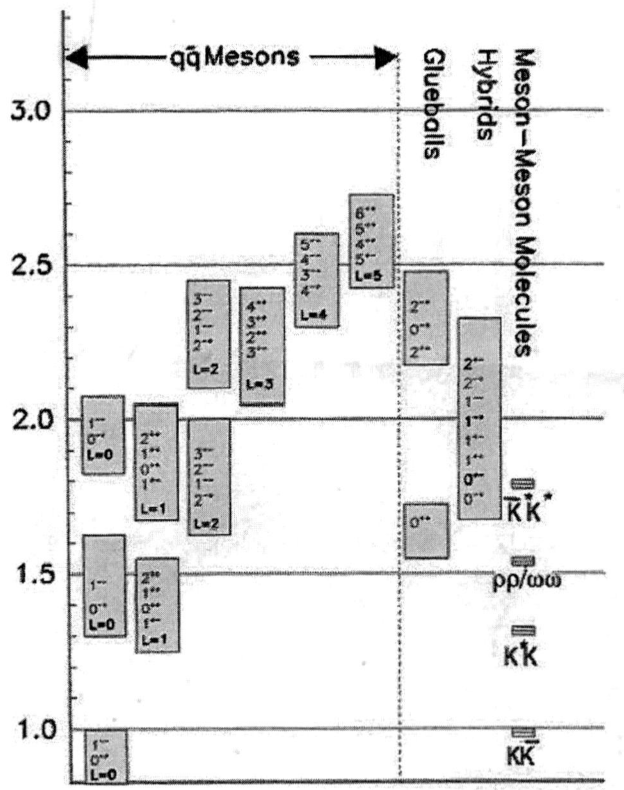

<Mixing effects with $q\bar{q}$ mesons obscure interpretations>

Chapter 7

Weak Interactions

Three vector bosons ($W^{\pm}$, Z^0) mediates the weak interactions.

$\leftarrow$ $W^{\pm}$: weak charge-current interaction
Z^0: weak neutral-current interaction

■ Classification

① weak charged-current(CC) interaction between lepton and quark

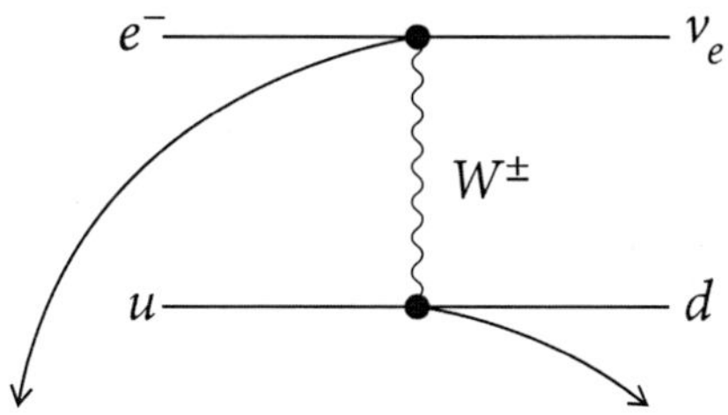

does not change lepton flavor   change quark flavor

- semi-leptonic process: both hadrons and leptons are present

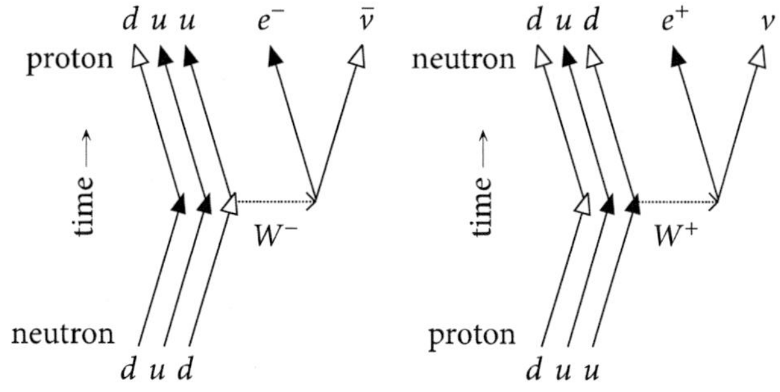

Beta decay (electron emission)   Inverse beta decay (positron emission)

<semi-leptonic processes: beta decay of nucleons>

② weak neutral-current(NC) interaction

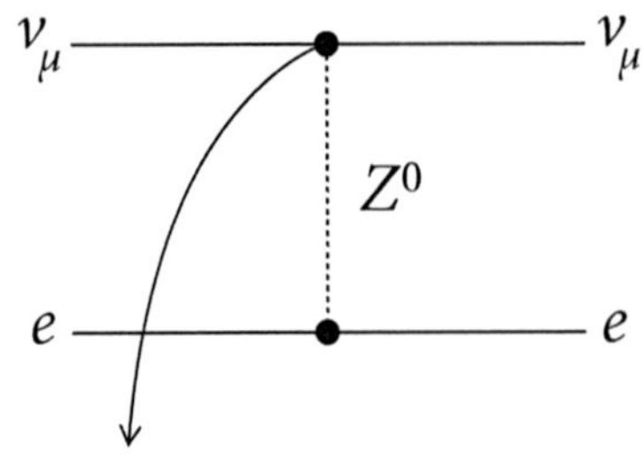

g: weak charge $\leftrightarrow$ e (electric charge)

- leptonic weak interactions are distinguished by the fact that they obey pure $V-A$ theory
- leptonic process: only leptons are present in the initial and final states

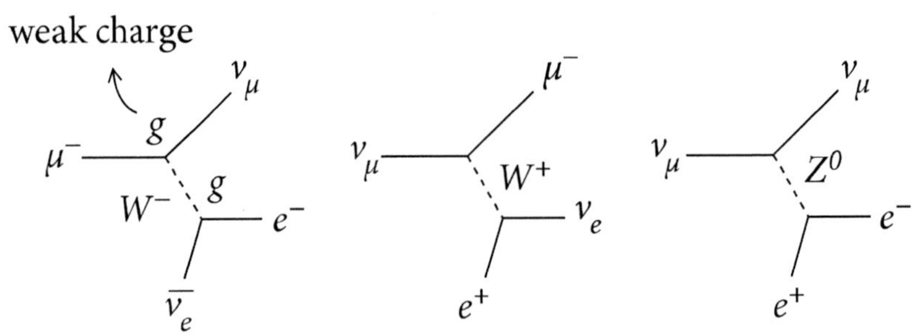

<leptonic processes>

③ non-leptonic weak interaction

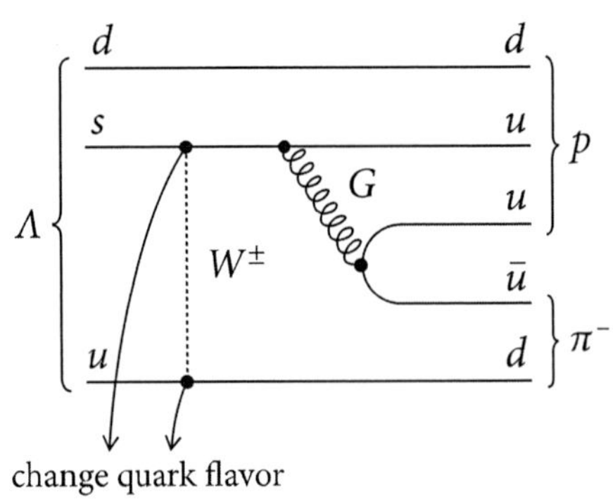

- non-leptonic process: only hadrons are present in initial and final states

<non-leptonic processes>

$$\Lambda^0 \to p + \pi^- \quad (s \to u + \bar{u} + d)$$

always involve a change of strangeness of hadron, $|\Delta s = 1|$

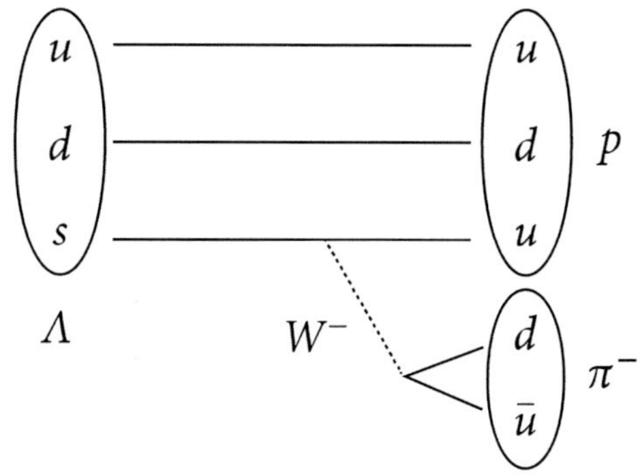

Only in those cases where a strong interaction is forbidden (when the strangeness changes) can nonleptonic weak interactions be observed experimentally.

■ Lepton Universality

As seen in Chapter 2, for $q^2 \ll M_W^2$ (at low energy)

$G_F \sim \dfrac{g^2}{M_W^2}$, where g is the weak charge(=coupling amplitude to $W^{\pm}$)

$[g]$: dimensionless and its magnitude is comparable to em coupling $\sqrt{\alpha}$

$\Rightarrow$ for $q^2 \ll M_W^2$, the decay amplitude (probability amplitude) is proportional to G_F

therefore, $\propto g^2$

$\Rightarrow$ the decay rate is proportional to G_F^2

for muon decay, (μ^+ beta decay: purely weak process $\mu^+ \to e^+ \bar{\nu}_\mu \nu_e$)

$$\Gamma_{\mu\to e\nu_e\bar{\nu}_\mu} = \frac{1}{\tau} \text{ (in } \hbar = c = 1 \text{ units)}$$

$$\propto G_F^2 m_\mu^5 \ (\because \text{ from dimensional analysis,}$$

since $[G_F] = \mathrm{GeV}^{-2}$, five times of invariant mass energy are needed.)

$$= \frac{G_F^2 m_\mu^5}{192\pi^3} \cdots\cdots (1)$$, muon mass determines the decay rate Γ in fifth power

where $192\pi^3$ comes from full calculation using $V-A$ theory of weak interactions

Eq. (1) → determination of G_F requires accurate measurement of muon lifetime and precise measurement of its mass

Since $\Gamma_\mu = \frac{1}{\tau_\mu} = G_F^2 m_\mu^5 \Rightarrow G_F^2 = \frac{1}{m_\mu^5 \tau_\mu} \cdots\cdots (2)$, in here $G_F^2 = g_\mu^4$

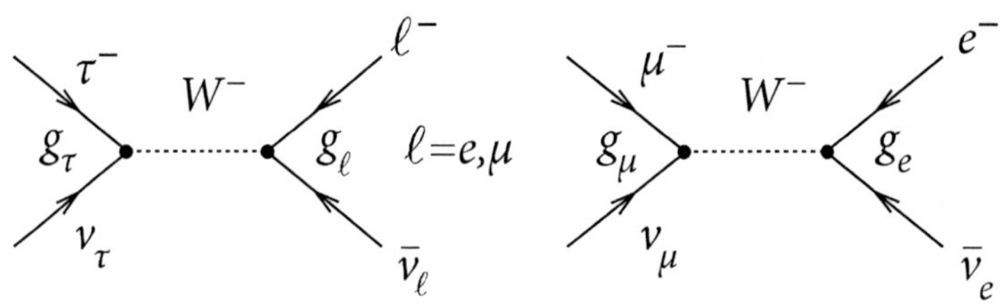

$$\frac{\Gamma_{(\mu\to e\nu_e\bar{\nu}_\mu)}}{\Gamma_{(\tau\to e\nu_e\bar{\nu}_\tau)}} = \frac{1}{\tau_\mu}\frac{\tau_\tau}{BR(\tau\to e\nu_e\nu_\tau)} \quad (\because \text{ muon decays 100\%})$$

$$\Rightarrow \frac{g_\mu^4 m_\mu^5}{g_\tau^4 m_\tau^5} = \frac{1}{\tau_\mu}\frac{\tau_\tau}{BR(\tau\to e\nu_e\nu_\tau)} \quad \text{by Eq. (2)}$$

$$\Rightarrow \left(\frac{g_\tau}{g_\mu}\right)^4 = B(\tau\to e\nu_e\bar{\nu}_\tau)\left(\frac{m_\mu}{m_\tau}\right)^5\left(\frac{\tau_\mu}{\tau_\tau}\right)$$: We can test universality of couplings g_μ and g_τ to $W^\pm$

with $\tau_\mu = 2.197\times 10^{-6}$ sec, $\tau_\tau = (291.0\pm 1.5)\times 10^{-15}$

$m_\mu = 105.658$ MeV, $m_\tau = 1777.0$ MeV

$$B(\tau \rightarrow e\nu_e\bar{\nu}_\tau) = 17.84\%$$

$$\Rightarrow \frac{g_\tau}{g_\mu} = 0.999 \pm 0.003 \quad (\mu - \tau \text{ universality})$$

Meanwhile, for $e-\mu$ universality,

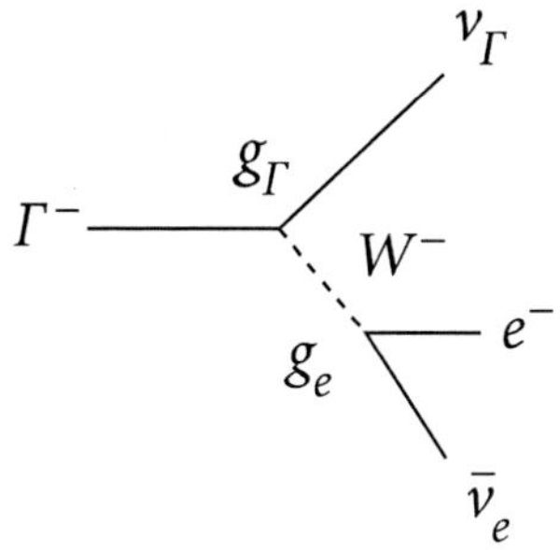

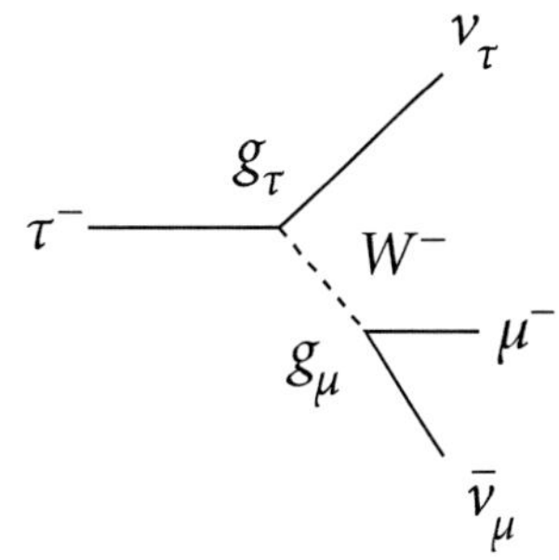

$$\frac{\Gamma_{(\tau^- \rightarrow \mu^- \bar{\nu}_\mu \nu_\tau)}}{\Gamma_{(\tau^- \rightarrow e^- \bar{\nu}_e \nu_\tau)}} = \frac{BR(\tau^- \rightarrow \mu^- \bar{\nu}_\mu \nu_\tau)}{BR(\tau^- \rightarrow e^- \bar{\nu}_e \nu_\tau)}$$

$$= \frac{(17.36 \pm 0.05)\%}{(17.84 \pm 0.05)\%} = 0.974 \pm 0.004$$

$$= \frac{g_\mu^2}{g_e^2}\frac{\rho_\mu}{\rho_e},$$ in here ρ_μ / ρ_e is the ratio of the phase space volume, which can be calculated precisely.

$$\Rightarrow \therefore \quad \frac{g_\mu}{g_e} = 1.001 \pm 0.004 \text{ from } \begin{cases} \tau \rightarrow e\nu_e\bar{\nu}_\tau \\ \tau \rightarrow \mu\nu_\mu\bar{\nu}_\tau \end{cases}$$

$\xrightarrow{\text{conclusion}}$ different flavors of lepton have identical coupling to $W^\pm$

and

different flavors of lepton have identical coupling to Z^0

But, quarks to weak bosons depend on quark flavors involved.

$$\Gamma = \frac{\hbar}{\tau} = \frac{\hbar c}{c\tau} \Rightarrow \Gamma_\mu = 2.996 \times 10^{-19}\ \text{GeV},$$ where τ is the muon lifetime

then, $\dfrac{G_F}{(\hbar c)^3} = 1.1664 \times 10^{-5}\ \text{GeV}^{-2} \quad \left[G_F = 1.1664 \times 10^{-5}\ \text{GeV}^{-2}(\hbar c)^3\right]$

■ Nuclear β-decay

Nuclear β-decay is the prototype weak interaction.

The first weak interaction process to be observed was nuclear β-decay.

$\Rightarrow$ conservation of energy, momentum and angular momentum requires the existence of a light, uncharged particle of spin $\frac{1}{2}$, called neutrino (Pauli, 1933).

$$\langle n \rightarrow p + e^- + \bar{\nu}_e \rangle \text{ (mean lifetime } \sim 10 \text{ min)}$$

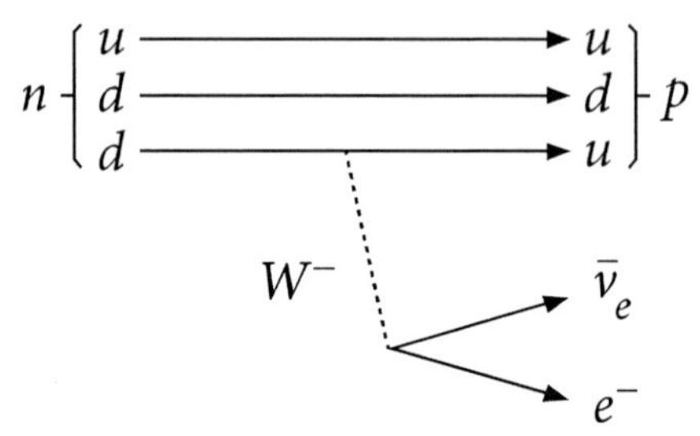

Fermi theory: Non-zero neutrino mass can be detected by observing a distortion in the electron energy spectrum

In Fermi theory, transition probability (decay rate per unit time) can be calculated from first order perturbation theory:

$W = \frac{2\pi}{\hbar} G_F^2 |M|^2 \frac{dN}{dE_0}$ ...... (1), G_F^2: taking G_F out of the matrix element

where $G_F = \frac{g^2}{M_W^2}$ for $q^2 \ll M_W^2$ case (low momentum transfer),

E_0 is the available energy of the final state,

phase space factor $\frac{dN}{dE_0}$ is the density of final states per unit energy interval, and

$|M|$ is the dimensionless and square of matrix element for transition

① "allowed" β-transitions are these in which the electron and neutrino carry off zero orbital angular momentum ($\ell = 0$)

② Then, their total angular momentum is 1 (spins parallel) or 0.

③ If total angular momentum J (leptons) $= 0$, $|M|^2 = 1$: Fermi transition

J (leptons)$= 1$, $|M|^2 = 3$: Gamow-Teller transition

($\because$ spin multiplicity factor $\Leftarrow 2J+1$)

$\Rightarrow$ The above nuclear decay is a combination of both type of transitions.

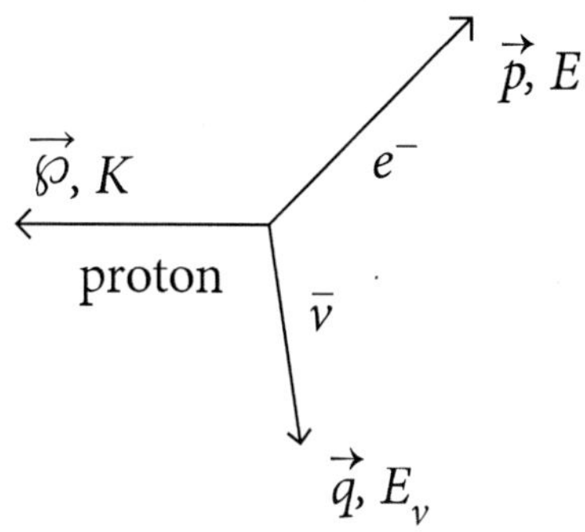

$\vec{p}$, $\vec{\wp}$, $\vec{q}$ are the momenta of e^-, p, $\bar{\nu}_e$, respectively

E, K, E_ν are the kinetic energies of e^-, p, $\bar{\nu}_e$, respectively

In the rest frame of initial states,

$$\vec{\wp} + \vec{q} + \vec{p} = 0$$

$K + E_\nu + E = E_0$, in here E_0 is the final states available energy

where $E_\nu = \sqrt{\vec{q}^{\,2}c^2 + m_\nu^2 c^4} = qc \quad (\because\ m_\nu^2 = 0)$

(※ remember q is the magnitude of 3-momentum vector)

$$K = \frac{\vec{\wp}^{\,2}}{2M} = \frac{(\wp c)^2}{2Mc^2} \approx 10^{-3}\ \text{MeV} \text{ can be neglegible.}$$

($\because$ in order of magnitude $E_0 \simeq 1\ \text{MeV}$, so that $\wp c \simeq 1\ \text{MeV}$)

(in here, M is the recoiling nucleon mass.)

$\Rightarrow\ E_\nu + E = E_0 \ \Rightarrow\ qc + E = E_0$

so, $qc = E_0 - E$, energy E_0 is shared between electron and neutrino

$\Rightarrow q = \dfrac{E_0 - E}{c}$ ······ (2) neutrino momentum is fixed for given values of p and E

A number of states in the phase space available to electron confined to volume V with momentum $\vec{p} \sim \vec{p} + d\vec{p}$ defined inside $d\Omega$ is

$$\frac{Vd\Omega}{(2\pi)^3\hbar^3}p^2dp \Rightarrow \frac{4\pi p^2dp}{(2\pi)^3\hbar^3}$$

(※ by integrating over space angles and normalizing wave function to unit volume)

Similarly, phase space available to neutrino

$$\Rightarrow \frac{4\pi q^2dq}{(2\pi)^3\hbar^3}$$

[※ there is no phase space factor for proton, since its momentum is fixed by $\vec{\wp} = -(\vec{p} + \vec{q})$]

$\Rightarrow$ Number of final states is $d^2N = \dfrac{(4\pi)^2}{(2\pi)^6\hbar^6}p^2q^2dpdq$ ······ (3)

Since $dq = \dfrac{dE_0}{c}$ from Eq. (2), Eq (3) is then

the energy density of final states is $\dfrac{dN}{dE_0} = \dfrac{1}{4\pi^4\hbar^6c^3}p^2(E_0 - E)^2dp$

If $|M|^2$ is regarded as a constant, then Eq. (1) gives the electron spectrum

$$N(p)dp \propto p^2(E_0 - E)^2dp$$

If we plot $\left[\dfrac{N(p)}{p^2}\right]^{\frac{1}{2}}$ against E, a straight line cutting the energy axis at $E = E_0$ should result: This is called a "**Kurie** plot"

[$^3H \rightarrow {}^3He + e^- + \bar{\nu}_e$ (tritium β-decay transition):

very small Q value: $Q = m_H - m_{He} = 18.6$ KeV]

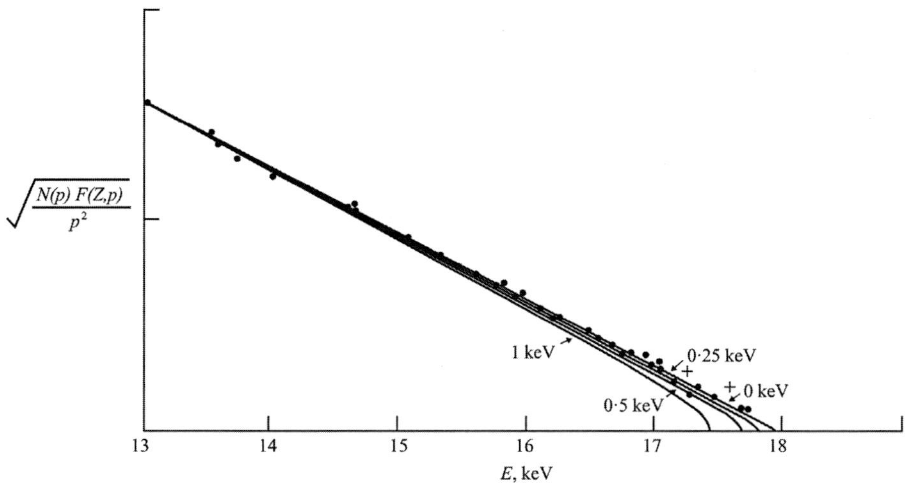

in $\sqrt{\dfrac{N(p)F(z,\rho)}{\rho^2}}$, $F(z,\rho)$ is the correction factor ($\sim$ 1 in the case) to account for the Coulomb effect on the electron wave function.

$1 = \dfrac{m_\nu c^2}{E_0 - E} \Rightarrow E = E_0 - m_\nu c^2$에서 cuts the axis for nonzero neutrino mass

※ This provides a method of determinacy neutrino mass.

By MAINZ experiment in Germany, $m_{\nu_e} < 2.3$ eV

For non-zero neutrino mass

$$N(p)dp \propto p^2(E_0 - E)^2\sqrt{1 - \left(\frac{m_\nu c^2}{E_0 - E}\right)^2}\,dp$$

$$\because\ E_\nu = \sqrt{m_\nu^2 c^4 + q^2 c^2} \rightarrow m_\nu^2 c^4 + q^2 c^2 = (E_0 - E)^2$$

$$\rightarrow q^2 = \frac{1}{c^2}\left[(E_0 - E)^2 - m_\nu^2 c^4\right] \rightarrow 2qdq = \frac{2(E_0 - E)}{c^2}dE_0$$

$$\rightarrow q^2\frac{dq}{dE_0} = q^2\frac{(E_0 - E)}{qc^2} = q\frac{(E_0 - E)}{c^2}$$

$$= \frac{(E_0 - E)}{c^2}\sqrt{\frac{(E_0 - E)^2 - m_\nu^2 c^4}{c^2}}$$

$$= \frac{(E_0 - E)}{c^2}\sqrt{(E_0 - E)^2 - m_\nu^2 c^4}$$

$$= \frac{1}{c^3}(E_0 - E)^2\sqrt{1 - \left(\frac{m_\nu c^2}{E_0 - E}\right)^2}$$

For decay rate,

electrons are relativistic $\Rightarrow E \approx pc$

$\Rightarrow N \simeq \int_0^{E_0} E^2(E_0 - E)^2 dE = \frac{E_0^5}{30}$: then the disintegration constant is proportional to the fifth power of disintegration energy = "**Sargent rule**"

To be more specific,

for $E_0 \gg m_e c^2$

$$W = \frac{2\pi}{\hbar} G_F^2 |M|^2 \frac{dN}{dE_0} \quad \& \quad \frac{dN}{dE_0} = \frac{1}{4\pi^4 \hbar^6 c^3} p^2 (E_0 - E)^2 dp$$

$$\Rightarrow \begin{cases} W = \dfrac{G_F^2 |M|^2 E_0^5}{60\pi^3 (\hbar c)^6 \hbar} \\ \dfrac{G_F}{(\hbar c)^3} = 1.1664 \times 10^{-5}\ \mathrm{GeV}^{-2} \end{cases} \Rightarrow W = \frac{1.11}{10^4} E_0^5 |M|^2, \text{ where } E_0 \text{ in MeV}$$

for muon decay, $E_0 \simeq 100$ MeV

and assuming $|M|^2 = 1$

$\Rightarrow \tau_\mu = \frac{1}{W} \simeq 10^{-6}$ sec (※ actual muon lifetime is 2.2 μs)

■ Inverse β-decay ($\bar{\nu}_e + p \rightarrow n + e^+$): neutrino interactions

$$\frac{d\sigma}{d\Omega}(a + b \rightarrow c + d) = \frac{1}{4\pi^2\hbar^2} |M_{if}|^2 \frac{p_f^2}{v_i v_f}$$

$\xrightarrow{\hbar = c = 1} \sigma_{(\bar{\nu}_e + p \rightarrow n + e^+)} = \frac{G_F^2}{\pi} |M|^2 \frac{p^2}{v_i v_f}$, here, p^2 is the unit of energy^2

※ $[\sigma] = E^{-2}$ ($\because\ v_i \approx v_f \approx c = 1$) or [length]^2

(remember $1\ \mathrm{m} = 5.07\times10^{15}\ \mathrm{GeV}^{-1}$)

for a mixed transition of Fermi transition ($|M|^2 \approx 1$) and Gamow-Teller transition ($|M|^2 \approx 3$),

with $1\ \mathrm{GeV}^{-1} = 1.975\times10^{-14}\ \mathrm{cm}$, $\quad |M|^2 \approx 4$

Then,

$\sigma_{(\bar{\nu}_e + p \rightarrow e^+ + n)} \simeq 10^{-43}\ E^2\ \mathrm{cm}^2$, where $E^2 \approx pc$

(in here, $E^2 \leftarrow p^2$)

→ exceedingly small cross-section

→ corresponding mean free path for antineutrino in water for $E = 1\ \mathrm{MeV}$ is about $10^{20}\ \mathrm{cm}$ or 50 light years.

For a $1000\ \mathrm{MW}$ reactor(∵ uranium fission fragments are neutron-rich and neutrons undergo β-decay), useful flux at a few meters from the core is of order $10^{13}\ \mathrm{cm}^{-2}\mathrm{s}^{-1}$

⇒ low-cross section can be compensated by very high flux

※ 신월성 100만 kW, (4호기)월성~70만 kW, 신 한울 140만 kW,
한울 100만 kW(6호기)

■ Parity nonconservation in β-decay

Lee and Yang came to the conclusion that weak interactions were not invariant under spatial inversion

⇒ weak interaction did not conserve parity

Cosmic ray experiments at 1953:

$$\begin{cases} \theta^+ \rightarrow \pi^+\pi^0 \\ \tau^+ \rightarrow \pi^+\pi^+\pi^- \end{cases}$$

- existence of two apparently identical particles (similar masses and lifetimes) but different only in their decay mode. It became clear that different decay modes of the same particle were being observed (Davies et al. 1955)

$\rightarrow$ now called K meson

from

$K^{\pm} \rightarrow \pi^{\pm}\pi^{+}\pi^{-}$, 3π states has $I = 1$ and $J^{P} = 0^{-}$ from data distribution (Dalitz plot)

However,

$K^{\pm} \rightarrow \pi^{+}\pi^{0}$, parity must be positive if $J = 0$

$\Rightarrow$ The puzzle was solved by T.D. Lee and C. N. Yang in "Rochester Conference" 1956, who made hypothesis that the weak interactions violate parity conservation and by the experiments that immediately confirmed this hypothesis.

Consider the beta decay of a nucleus $N \rightarrow N' + e + \nu$ in cms frame.

- kinematic quantities are $\vec{p}_{N'}$, $\vec{p}_{e}$, $\vec{p}_{\nu}$, we can build, with them,

$$\Rightarrow \begin{cases} \vec{p}_{N'} \cdot \vec{p}_{e} & \text{(being scalar they do not violate parity)} \\ \vec{p}_{N'} \cdot \vec{p}_{e} \times \vec{p}_{\nu} & \text{(it is a psedoscalar and violate parity)} \end{cases}$$

(but it is zero because the three vectors are coplanar.)

$\Rightarrow$ Lee and Yang concluded that parity conservation could be tested only using an axial vector. Such axial vector is provided by polarization.

$\Rightarrow$ One must polarize a sample of nuclei, inducing a non-zero expectation value of the intrinsic angular momentum ($\vec{J}$), and measure an observable proportional to the pseudoscalar $< \vec{J} > \cdot \vec{p}_{e}$.

$\Rightarrow$ The experiment was done by C.S. Wu and collaborators with ^{60}Co.

(※ We will cover it in more detail later.)

- polarization is achieved by orienting the nuclear magnetic moments $\vec{\mu}$, which are parallel to the spins, in a magnetic field $\vec{B}$

- probability of a certain direction of the magnetic moment relative to the field is given by the Boltzmann factor, $e^{-\frac{\vec{\mu}\cdot\vec{B}}{k_B T}}$

- **Polarized nuclei beta decay :**

$$^{60}\mathrm{Co}\,(J^P = 5^+)_\uparrow \rightarrow {}^{60}\mathrm{Ni}^{**}(J^P = 4^+)_\uparrow + e^- + \bar{\nu}_e$$

$$\quad \hookrightarrow {}^{60}\mathrm{Ni}^{*}(J^P = 2^+)_\uparrow + \gamma(1.173\ \mathrm{MeV})$$

$$\quad\quad \hookrightarrow {}^{60}\mathrm{Ni}(J^P = 0^+)_\uparrow + \gamma(1.332\ \mathrm{MeV})$$

- two EM decays are not isotropic, the gamma emission probability is a function of the angle θ with the field. Therefore, we can monitor the polarization of the sample by measuring this anisotropy.

<Experiment by Wu et al. in 1957>

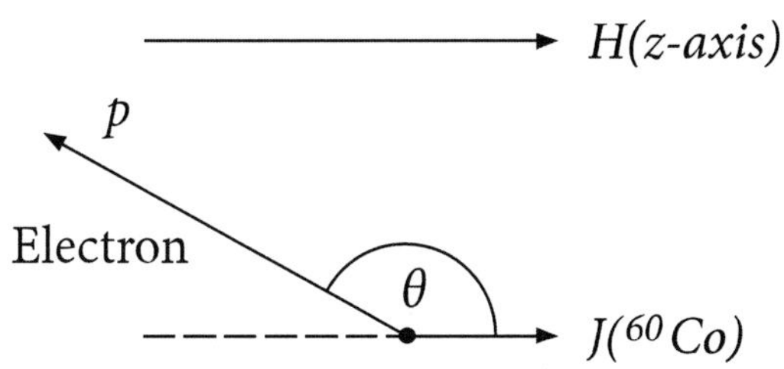

At low temperature, a high proportion of $^{60}\mathrm{Co}$ nuclei (spin $J = 5$) are aligned with the field.

- the daughter nucleus keeps the polarization of parent nucleus
- degree of $^{60}\mathrm{Co}$ alignment is determined from angular distribution of γ-rays from $^{60}\mathrm{Ni}^*$ [$\because$ $^{60}\mathrm{Co}$ undergoes β-decay to $^{60}\mathrm{Ni}^{**}(J = 4)$]

The relative electron intensities along and against field direction were measured and found to be

$$I(\theta) = 1 + \alpha\frac{\vec{\sigma}\cdot\vec{p}}{E} = 1 + \alpha\frac{v}{c}cos\theta,$$

where $\vec{\alpha}$ is a unit spin vector in direction of $\vec{J}$,

$\vec{p}$ and E are the momentum and energy of electron, and

θ is the angle of emitted electron w.r.t. $\vec{J}$

(※ If the beta decay violates the parity, angular distribution of the emitted electrons is asymmetric under $\theta \leftrightarrow \pi - \theta$.

Therefore, the counting rate is expected to depend on angle.)

- If parity is conserved, $\alpha = 0$ and if it is maximally violated, $\alpha = \pm 1$, corresponds to $V \pm A$ structure of interaction. If we assume the interaction to be $V + xA$, the result is compatible with $x = -1$. Wu experiment choose the combination of $V - A$.

⇒ fore-aft asymmetry implies that interaction as a whole violates parity conservation

- $\vec{\sigma}$ being an axial vector does not change sign under reflection
- $\vec{p}$ changes sign under reflection

→ $\vec{\sigma} \cdot \vec{p}$ changes sign under reflection

Therefore, if parity is conserved, $\alpha = 0$ because the sign of $\vec{\alpha} \cdot \vec{p}$ changes under reflection.

Conservation of z-component of angular momentum in the above transition implies that electron spin must also point in the direction of $\vec{J}$ so that if $\vec{\sigma}$ denotes the electron spin vector,

$$I = 1 + \alpha \frac{\vec{\alpha} \cdot \vec{p}}{E},$$

- I^{+}: intensity for $\vec{\sigma}$ parallel to $\vec{p}$
- I^{-}: intensity for $\vec{\sigma}$ antiparallel to $\vec{p}$

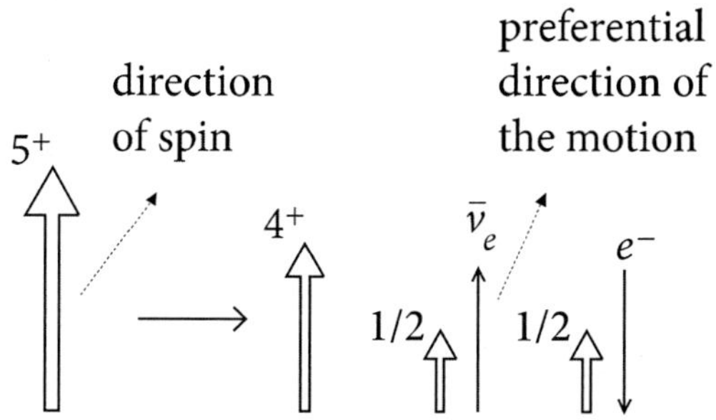

<lengths are such as to satisfy the conservation of the third component of angular momentum>

- Thus, preferential motion of the electron is opposite to the field, and consequently of its spin.
- and antineutrino spin is in the preferential direction of its velocity

⇒ $V-A$ structure of the charged-current weak interaction implies that this behavior is common to all fermions and to all antifermions, respectively.

• **Intrinsic parity**

From Gauge Field Theory, the gauge bosons (γ, gluon, $W^{\pm}$, Z) have $P=-1$.

From Dirac equation, spin $1/2$ particles have opposite parity to spin $1/2$ antiparticle.

<conventional choice> spin $\frac{1}{2}$ particles (e^-, μ^-, τ^-, ν, q) have $P=+1$

and antiparticles have $P=-1$

For Dirac spinors, parity operator is $\begin{pmatrix} 1 & 0 & 0 & 0 \\ 0 & 1 & 0 & 0 \\ 0 & 0 & -1 & 0 \\ 0 & 0 & 0 & -1 \end{pmatrix}$

⇒ net longitudinal polarization is

$$P = \frac{\Gamma^+ - \Gamma^-}{\Gamma^+ + \Gamma^-} = \alpha\frac{v}{c}$$

$$(\because\ \Gamma^+ = 1+\alpha\frac{v}{c},\quad \Gamma^- = 1-\alpha\frac{v}{c}$$

$$\rightarrow \frac{\Gamma^{+}-\Gamma^{-}}{\Gamma^{+}+\Gamma^{-}} = \frac{\left(1+\alpha\frac{v}{c}\right)-\left(1-\alpha\frac{v}{c}\right)}{\left(1+\alpha\frac{v}{c}\right)+\left(1-\alpha\frac{v}{c}\right)} = \alpha\frac{v}{c}\ )$$

$\Rightarrow$ in experiment

$$\alpha = \begin{cases} +1 \text{ for } e^{+}, \text{ thus } P = +\frac{v}{c} \\ -1 \text{ for } e^{-}, \text{ thus } P = -\frac{v}{c} \end{cases}$$

<polarization of emitted electrons in the nuclear β-decay>

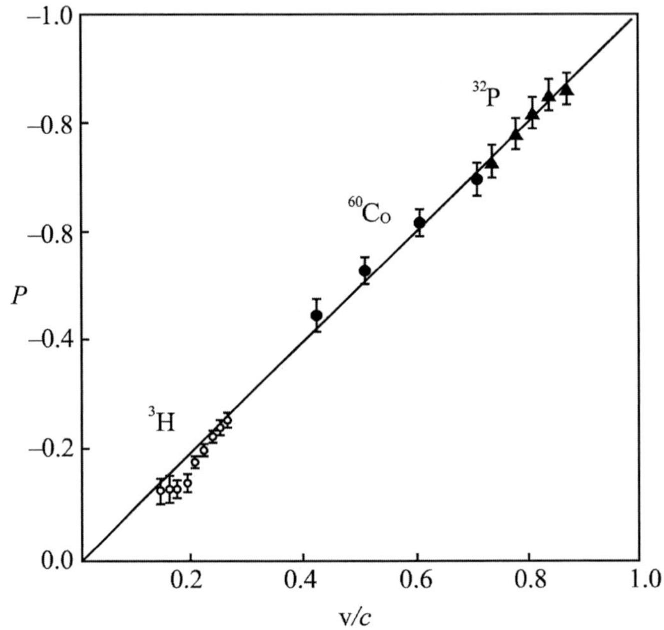

■ Helicity of the neutrino

Neutrino helicity was determined in a classic and beautiful experiment by Goldhaber, Grodzins and Sunyar in 1985. Let`s see how helicity is transferred from the neutrino to the photon.

There are three steps.

① $^{152}\mathrm{Eu}$ (Europium) decays by K-capture of an S-wave electron (to an excited state of $^{152}\mathrm{Sm}$ (Samarium):

$$^{152}\mathrm{Eu}(J=0)+e^{-} \rightarrow {}^{152}\mathrm{Sm}^{*}(J=1)+\nu_e \quad \cdots\cdots (1)$$

Let neutrino direction be quantization axis z. The Sm^{*} direction is $-z$ to conserve angular momentum.

<helicity of $^{152}\mathrm{Sm}^{*}$ is the same as that of the neutrino>

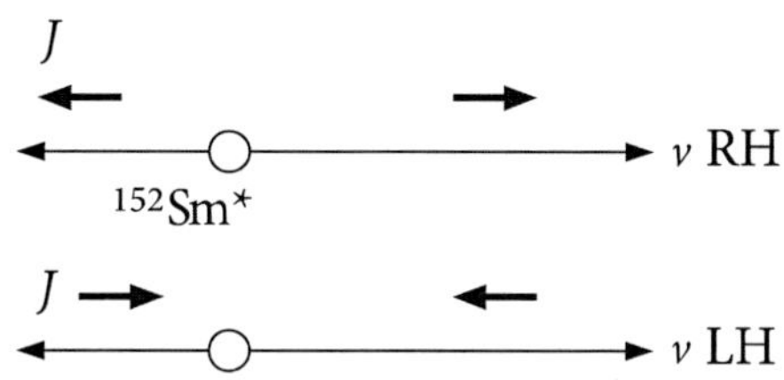

② Select the cases in which Sm^{*} decays emitting a gamma in the forward direction, namely $-z$, by use of the resonant emission-absorption process. The emission process is

$$^{152}\mathrm{Sm}^{*}(J=1) \rightarrow {}^{152}\mathrm{Sm}(J=0)+\gamma(960\ \mathrm{KeV}),$$

which has a lifetime of $10\ \mathrm{fs}$, short enough to satisfy the third condition.

The first three columns of the below Table give all combinations of third components of spins that satisfy angular momentum conservation in the reaction ①.

The fourth column gives the corresponding neutrino helicity.

$S_z(e)$	$S_z(Sm^*)$	$S_z(\nu)$	H_ν	$S_z(\gamma)$	H_γ
$+\frac{1}{2}$	1	$-\frac{1}{2}$	$-$	1	$-$
$+\frac{1}{2}$	0	$+\frac{1}{2}$	$+$	0	$\times$
$-\frac{1}{2}$	-1	$+\frac{1}{2}$	$+$	-1	$+$
$-\frac{1}{2}$	0	$-\frac{1}{2}$	$-$	0	$\times$

Taking into account that the projection of gamma angular momentum on its velocity can`t be zero, only two cases remain.

<gammas emitted forward have the same polarization as the neutrino>

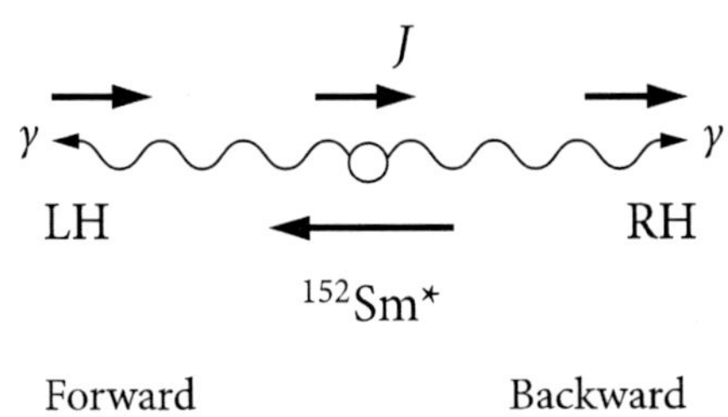

We observe that in both gamma and neutrino have the same helicity

③ Measurement of gamma circular polarization, namely of its helicity.

(※ Remember that spins of electrons responsible of ferromagnetism are oriented opposite to $\vec{B}$. These electrons easily absorb the photons, by flipping their spin, if photon spin has the direction of $\vec{B}$. On the contrary, they cannot do so for photons with spin in the direction opposite to $\vec{B}$. Therefore, the iron slab absorbs the former substantially more than the latter.)

$$\gamma_1 + {}^{152}\mathrm{Sm} \rightarrow {}^{152}\mathrm{Sm}^* \rightarrow \gamma_2 + {}^{152}\mathrm{Sm}$$

If the resonance process takes place, a photon, call it γ_1, emitted by Sm^* is absorbed by a Sm nucleus in the ring. The latter immediately de-excites emitting a photon γ_2, which reaches the detector (in a fraction of cases). The process can happen only of Sm^* was travelling in the right forward direction.

The measured quantity is asymmetry R between counting rates with field oriented in one direction and the other, I_+ and I_-,

$$R = \frac{I_+ - I_-}{I_+ + I_-}$$

From this measurement, the longitudinal polarization of gamma is easily extracted.

The final result is that helicity of the neutrino is negative and compatible with -1. This proves the $V-A$ structure of CC weak interaction.

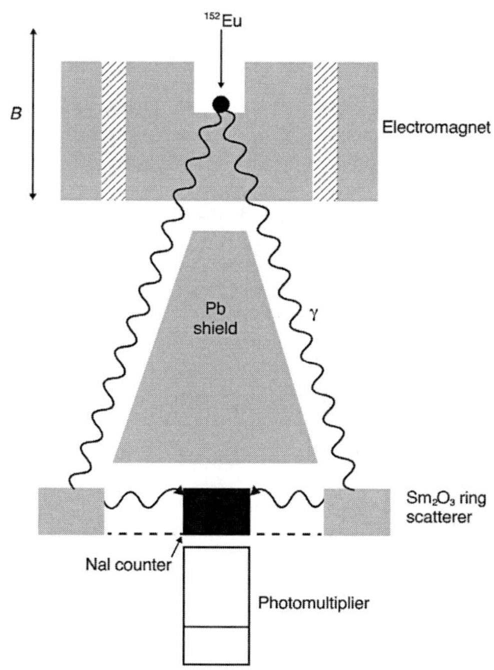

With determination that the neutrino is left-handed, the polarization of the leptons in β-decay are

particle	e^+	e^-	ν	$\bar{\nu}$
polarization	$+\frac{v}{c}$	$-\frac{v}{c}$	-1	+1

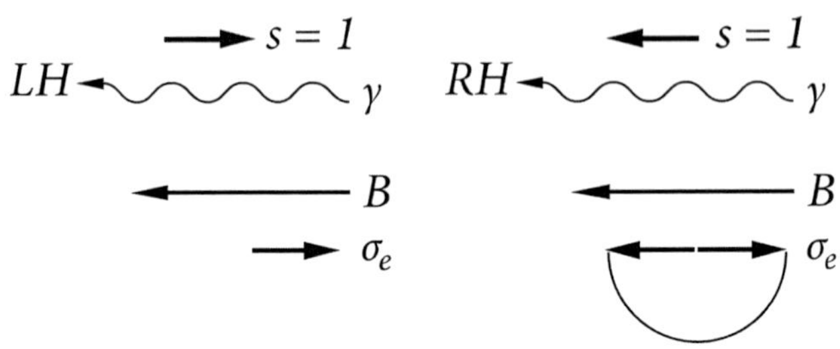

No spin-flip　　Spin-flip

Polarized gammas pass through magnetized iron before impinging on ^{152}Sm absorber.

An electron in the iron can absorb a unit of angular momentum by spin-flip.

$\Rightarrow$ transmission of iron is greater for LH than for RH γ-rays

(※ gammas with spin aligned with B field in electromagnet are more absorbed since electrons are aligned against B)

■ V-A Interaction

- Introduction (history)

Dirac wrote fundamental relativistic wave equation of the electron in 1928.

$\Rightarrow$ equation predicts all the electron properties known from atomic physics, in particular g (gyromagnetic ratio)$=2$ c.f. $\mu = g\mu_B S$

$\Rightarrow$ equation has apparently non-physical negative energy solutions

⇒ in 1930, H. Weyl introduced the mathematical operator C, particle-antiparticle conjugation, finding that antiparticle and particle must have the same mass.

⇒ in 1931, Dirac concluded that yet undiscovered particle (positive and with the same mass as electron) must exist → positron

⇒ Anderson discovered the positron in 1933.

⇒ wave functions of all the spin $\frac{1}{2}$ elementary particles obey the Dirac equation.

- Dirac equation: fundamental relativistic wave equation of the electron

$$\left(i\gamma^{\mu}\partial_{\mu} - m\right)\Psi(x) = 0,$$

where four-component Dirac bi-spinor describes a fermion and its antiparticle $\Psi(x) = \begin{pmatrix}\Psi_1\\ \Psi_2\\ \Psi_3\\ \Psi_4\end{pmatrix} = \begin{pmatrix}\phi\\ \chi\end{pmatrix}$,

two corresponding two-component spinors $\phi = \begin{pmatrix}\phi_1\\ \phi_2\end{pmatrix}$ and $\chi = \begin{pmatrix}\chi_1\\ \chi_2\end{pmatrix}$,

$\gamma_0 = \begin{pmatrix}1 & 0\\ 0 & -1\end{pmatrix}$ and $\gamma_i = \begin{pmatrix}0 & \sigma_i\\ -\sigma_i & 0\end{pmatrix}$

[※ In γ_0 and γ_i, each element stands for a 2×2 matrix, and Pauli matrices: $\sigma_1 = \begin{pmatrix}0 & 1\\ 1 & 0\end{pmatrix}$, $\sigma_2 = \begin{pmatrix}0 & -i\\ i & 0\end{pmatrix}$, $\sigma_3 = \begin{pmatrix}1 & 0\\ 0 & -1\end{pmatrix}$]

[※ (i) spinors represent the particle and antiparticle

(ii) two components of each spinor represent the two states of the third component of spin $S_z = +\frac{1}{2}$ and $S_z = -\frac{1}{2}$]

There are three possibilities (polarization, helicity, chirality) in order to give a physical meaning to ϕ and χ.

(※ polarization: two states have defined third component of spin on an axis, which typically defined by direction of magnetic or electric field.)

ex $\phi^+ = \begin{pmatrix}1\\0\end{pmatrix}$, $\phi^- = \begin{pmatrix}0\\1\end{pmatrix}$

$$\frac{1}{2}\sigma_z\begin{pmatrix}1\\0\end{pmatrix} = +\frac{1}{2}\begin{pmatrix}1\\0\end{pmatrix}, \quad \frac{1}{2}\sigma_z\begin{pmatrix}0\\1\end{pmatrix} = -\frac{1}{2}\begin{pmatrix}0\\1\end{pmatrix}$$

- Whilst polarization and helicity(longitudinal polarization) are properties of a fermion, formally of a two-component spinors.

  (※ helicity: even in the absence of an external field, velocity of the particle defines a direction in any reference frame different from rest frame.

  helicity operator is $\frac{\vec{p}\cdot\vec{\sigma}}{2p}$)

- Chirality is a property of the four-component spinors.

  (※ Two chiral states are eigenstates of γ_5, with two possible eigenstates $+1$ and -1.

  chiral states are called right(R) and left(L), respectively.

  projectors of R and L chiral states are respectively.

  $$\Psi_L = \frac{1}{2}(1-\gamma_5)\Psi, \quad \Psi_R = \frac{1}{2}(1+\gamma_5)\Psi$$

  → Chirality is important because only the left bi-spinor is source and receptor of the charged-current weak interaction.

  → conjugated states: $\overline{\Psi}_L = \overline{\Psi}\frac{1}{2}(1+\gamma_5)$, $\overline{\Psi}_R = \overline{\Psi}\frac{1}{2}(1-\gamma_5)$

  → where $\gamma_5 = \begin{pmatrix}0 & 1\\1 & 0\end{pmatrix} = i\gamma_1\gamma_2\gamma_3\gamma_0$

  with $\gamma_5^2 = 1$, $\gamma_5\gamma_\mu + \gamma_\mu\gamma_5 = 0$, $\mu = 1,\cdots,4$

  γ_5 does not commute with the free Hamiltonian.

  However γ_5 commutes with the Hamiltonian for massless particles (∵ mass of fermion is negligible at high energy).

  [※ projectors: in the ultra-relativistic limit, chiral states correspond to helicity states. applying one of them twice gives the same result as applying it once and that applying both of them results in the null states]

Now, examine fermion and antifermion components of the chiral states:

For free particle, Dirac equation is

$$\left(\gamma_\mu p^\mu - m\right)\Psi = (E\gamma_0 - \vec{p}\cdot\vec{\gamma} - m)\Psi = 0, \quad \text{where } \Psi = ue^{-ip^\mu \chi_\mu}$$

$$\Rightarrow \begin{pmatrix} E-m & -\vec{p}\cdot\vec{\sigma} \\ \vec{p}\cdot\vec{\sigma} & -(E+m) \end{pmatrix}\begin{pmatrix} \phi \\ \chi \end{pmatrix} = 0 \quad \Rightarrow \begin{cases} \phi = \dfrac{\vec{p}\cdot\vec{\sigma}}{E-m}\chi \\ \\ \chi = \dfrac{\vec{p}\cdot\vec{\sigma}}{E+m}\phi \end{cases}$$

for Ψ_L,

$$\Psi_L = \frac{1}{2}(1-\gamma_5)\Psi = \frac{1}{2}\begin{pmatrix} 1 & -1 \\ -1 & 1 \end{pmatrix}\begin{pmatrix} \phi \\ \chi \end{pmatrix} = \frac{1}{2}\begin{pmatrix} \phi-\chi \\ \chi-\phi \end{pmatrix},$$

where $\frac{1}{2}(\phi-\chi)$: fermion, as opposed to the antifermion

$\frac{1}{2}(\chi-\phi)$: antifermion

Consider upper component,

$$\frac{1}{2}(\phi-\chi) = \frac{1}{2}\left(1-\frac{\chi}{\phi}\right)\phi = \frac{1}{2}\left(1-\frac{\vec{p}\cdot\vec{\sigma}}{E+m}\right)\phi$$

(let`s take z-axis in direction of motion and helicity eigenstates ϕ^+ and ϕ^-)

$$= \frac{1}{2}\left(1-\frac{p_z}{E+m}\right)\phi^+ + \frac{1}{2}\left(1+\frac{p_z}{E+m}\right)\phi^- \quad \cdots\cdots (1)$$

→ upper component is not eigenstate of helicity

if particle is massless, $p_z = E$

$\frac{1}{2}(\phi-\chi) = \phi^-$: negative helicity eigenstate

Consider lower component (antifermion): antiparticle is considered as negative energy solution of Dirac equation

$$\frac{1}{2}(\chi-\phi) = \frac{1}{2}\left(1-\frac{\vec{p}\cdot\vec{\sigma}}{E-m}\right)\chi = \frac{1}{2}\left(1-\frac{p_z}{E-m}\right)\chi^+ + \frac{1}{2}\left(1+\frac{p_z}{E-m}\right)\chi^-$$

if particle is massless, $E = -p_z$

$\frac{1}{2}(\chi-\phi) = \chi^+$: massless antiparticle of negative chirality bi-spinor is helicity eigenstate with positive eigenvalue

Chirality can`t be directly determined. We measure the expectation value of the helicity. Considering, for example, the particle spinor of the left bi-spinor, let π_+ and π_-, be the probabilities of finding it in each of two helicity states.

$\Rightarrow$ Considering particle spinor of left bi-spinor

let π_+ and π_- be probability of finding it, spin in the direction of motion or opposite to it

probabilities are squares of two amplitudes Eq. (1).

Then, helicity expectation (or simple helicity) value is

$$h = \frac{\pi_+ - \pi_-}{\pi_+ + \pi_-} = \frac{(E+m-p)^2 - (E+m+p)^2}{(E+m-p)^2 + (E+m+p)^2} = -\frac{p}{E} = -\beta$$

$\rightarrow$ helicity of fermion(upper component) of negative chirality(Ψ_L) is opposite of the ratio between its speed and speed of light

$\rightarrow$ Similarly, helicity of positive chirality fermion is $+\beta$

• $V-A$ interaction

For massless,

operator $P_{R,L} = \frac{1}{2}\left(1 + \frac{\vec{\sigma} \cdot \vec{p}}{E}\right)$ acting on two-component spinors project out particular helicity state from arbitrary superposition of positive and negative helicity states.

$$\Rightarrow \begin{cases} P_R\Psi = \frac{1}{2}\left(1 + \frac{\vec{\sigma} \cdot \vec{p}}{E}\right)(\Psi_L + \Psi_R) = \Psi_R \\ P_L\Psi = \frac{1}{2}\left(1 - \frac{\vec{\sigma} \cdot \vec{p}}{E}\right)(\Psi_L + \Psi_R) = \Psi_L \end{cases},$$

$$\text{where } \Psi_R = \frac{1}{2}(1+\gamma_5)\Psi\ ,\ \ \Psi_L = \frac{1}{2}(1-\gamma_5)\Psi,\ \gamma_5 = \begin{pmatrix} 0 & 1 \\ 1 & 0 \end{pmatrix}$$

Two component spinors χ and ϕ of a solution Ψ of Dirac equation:

$$\chi = \begin{pmatrix} \chi_1 \\ \chi_2 \end{pmatrix},\ \ \phi = \begin{pmatrix} \phi_1 \\ \phi_2 \end{pmatrix}$$

For massive fermions,

analogous operators are $\frac{1}{2}(1 \pm \gamma_5)$

$\Rightarrow$ operator $(1 \pm \gamma_5)$ on fermion wave functions is to project out a state of polarization

$$P = \pm \frac{v}{c} = \pm \beta \quad \left[\text{i.e. } \frac{1}{2}(1 \pm \gamma_5)\Psi = \pm \frac{v}{c}\Psi\right]$$

In Dirac theory,

$e+p \rightarrow e+p$ can be described as interaction of two currents, with matrix element $M \propto \frac{e^2}{q^2} J_{\text{baryon}} J_{\text{lepton}}$, $J \propto \Psi^* \Psi$, Ψ^* denotes incoming amplitude, Ψ denotes outgoing amplitude.

$$J_{\text{lepton}} = \Psi_e^* \gamma_4 \gamma_\mu \Psi_e \equiv \overline{\Psi}_e \gamma_\mu \Psi_e$$

$$J_{\text{baryon}} = \overline{\Psi}_p \gamma_\mu \Psi_p,$$

where $O_{em} = \gamma_4 \gamma_\mu$: em currents are described in Dirac theory by the 4-vector operator

$$\overline{\Psi}_e \equiv \Psi_e^* \gamma_4$$

By analogy, Fermi took for the weak process, for $\nu_e + n \rightarrow p + e^-$

$$M = G_F J_{\text{baryon}}^{\text{weak}} J_{\text{lepton}}^{\text{weak}} = G_F (\overline{\Psi}_p O \Psi_n) \underbrace{(\overline{\Psi}_e O \Psi_\nu)}_{\text{lepton bracket}}$$

(※ Fermi assumed that operator O is vector operator as in electromagnetism)
If weak interaction is parity-conserving, we require that M be scalar quantity under space inversion.

$\rightarrow$ experimentally, not conserving

$\rightarrow$ M should contain both scalar and pseudoscalar parts

$\rightarrow$ We can achieve this by including a factor γ_5 in one of the brackets, lepton bracket

Fine different operator in the matrix element are allowed by relativistic invariance: V(vector), A(axial vector), S(scalar), P(pseudoscalar), T(tensor)

→ S, P and T interactions favor the same helicities for fermion and antifermion

→ early experiment results showed that leptons and antileptons involved in the weak decays have opposite helicity.

→ this narrowed the choice of operators down to V and A operators, (※ massless fermions helicity is conserved in V or A interactions)

→ general combination of V and A amplitudes correspond to an operator of form

$\frac{1}{2}(C_V+\gamma_5 C_A)$, where C_A and C_V are constant coefficients

→ neutrino produced in a pure helicity eigenstate $H=-1$ requires that

$C_A=-C_V$ giving $\frac{1}{2}(1-\gamma_5)$ operator

↳ axial-vector and vector couplings are equal in magnitude but opposite in sign, we obtain the famous $V-A$ interaction

(※ This only holds for leptonic weak interactions but not weak interactions involving hadrons)

therefore, V and A amplitudes, which have odd and even parities under space inversion are equal in magnitude but opposite sign

→ "$V-A$ theory"(principal of "maximal parity violation") postulated by Feynman and Gell-Mann in 1958

vector part(V) of the weak charge(g) is conserved but axial part(A) is not conserved

→ conserved vector current(CVC),

partially conserved axial-vector current($PCAC$)

→ measured ratio $\frac{C_A}{C_V} \neq -1$

ex in neutron β-decay, $\frac{C_A}{C_V} = -1.26$

in Λ β-decay, $\frac{C_A}{C_V} = -0.72$

in $\Sigma^- \rightarrow n + e^- + \bar{\nu}_e$, $\frac{C_A}{C_V} = +0.34$

deep inelastic scattering of neutrinos by quark constituents(quasi-free pointlike particles) of hadrons is described exactly by the $V-A$ theory, with $C_A = -C_V$

	relativistic invariance allows five types of operators
$\bar{\Psi}\Psi$: scalar	V vector γ_μ
$\bar{\Psi}\gamma_\mu\Psi$: four-vector	A axial vector $\gamma_5\gamma_\mu$
$i\bar{\Psi}\gamma_\mu\gamma_\nu\Psi$: six-vector or tensor	S scalar 1
$i\bar{\Psi}\gamma_5\gamma_\mu\Psi$: axial four-vector	P pseudoscalar γ_5
$\bar{\Psi}\gamma_5\Psi$: pseudoscalar	T tensor $\gamma_\mu\gamma_\nu$

- **The weak boson** and Fermi **couplings**

$$G_F \rightarrow \frac{1}{\sqrt{2}} G_F,$$

where G_F has been slightly redefined (replaced by $\frac{G_F}{\sqrt{2}}$) from original Fermi theory, since original theory only had vector interaction, and modern theory has $V-A$.

Since we have both V and A amplitudes,

$1-\gamma_5 \rightarrow \frac{1}{2}(1-\gamma_5)$, where $(1-\gamma_5)$ is helicity operator at the time of $V-A$ theory

⇩ apply at both vertices

① gives an overall factor of 4 (by which g^2 should be divided)

charged-current interactions involve physical charged boson,

$$W^{\pm} = \frac{1}{\sqrt{2}}(W^{(1)} \pm i W^{(2)}),$$

where $W^{(1)}$, $W^{(2)}$, $W^{(3)}$ are the "weak isospin" triplet of fields

② and $\frac{1}{\sqrt{2}}$ is the normalization factor.

From ① and ②, coupling g of a lepton to the $W^{\pm}$ boson is

$$\frac{g^2}{8M_W^2} \equiv \frac{G_F}{\sqrt{2}} \quad \left(\text{c.f. } G_F \sim \frac{g^2}{M_W^2}\right)$$

■ pion and muon decay: $\pi^+ \rightarrow \mu^+ + \nu_\mu (\sim 100\%)$

$\quad\quad\quad\quad\quad\quad\quad\quad\quad\quad\quad\quad\quad\quad\quad\quad\quad$ ↳ $e^+ + \nu_e + \overline{\nu}_\mu$

[※ on the other hand, $\pi^+ \rightarrow e^+ + \nu_e$ (1.23×10^{-4})]

As Wu *et al.* were observing lepton helicities in Co β-decay, others observed effects in decay of pions and muons.

Since pion has spin zero, neutrino and muon must have antiparallel spin vectors.

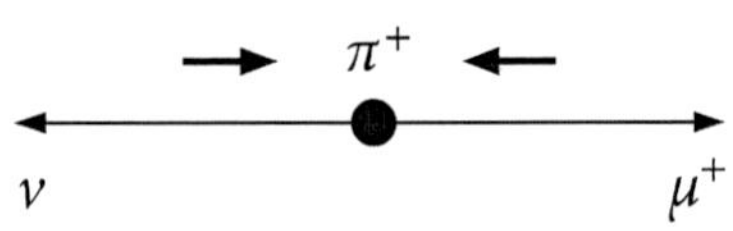

Pion rest frame

(※ In $V-A$ theory, π decay proceeds through A coupling)

→ If neutrino has $H = -1$, μ^+ also has $H = -1$.

In the subsequent muon decays ($\mu^+ \rightarrow e^+ + \nu_e + \overline{\nu}_\mu$),

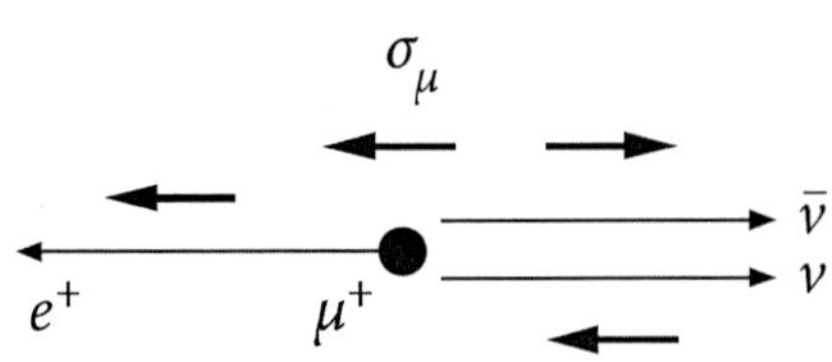

Muon rest frame

<"favored configuration">

→ positron spectrum is peaked in region of maximum energy

→ positron has positive helicity

In the experiments, μ^+ were stopped in a carbon absorber, and angle θ of e^+ momentum vector relative to that of original muon momentum $\vec{p}_\mu$ was measured.

$$\frac{dN}{d\Omega} = 1 - \frac{\alpha}{3}\cos\theta \Rightarrow \text{found that } \alpha = 1$$

⇒ This is exactly the form predicted by $V-A$ theory.

pion decay is possible for two possible modes $\pi^+ \rightarrow \mu^+ + \nu_\mu$ and $\pi^+ \rightarrow e^+ + \nu_e$ and the ratio of the two modes provided a very stringent test of the theory. $V-A$ theory predicts, the net longitudinal polarization of the charged antilepton

$(e^+$ or $\mu^+)$,

$$p = \frac{N_R - N_L}{N_R + N_L} = +\frac{v}{c}$$

The probability that this lepton emerges with the same helicity as neutrino(LH helicity) is

$$\frac{N_L}{N_R + N_L} = \frac{1}{2}\left(1 - \frac{v}{c}\right)$$

However, pseudo(P) coupling favors the same helicity for lepton(ν_e or ν_μ) and antilepton(e^+ or μ^+), and probability is

$$\frac{N_L}{N_R + N_L} = \frac{1}{2}\left(1 + \frac{v}{c}\right)$$

⇒ zero spin of pion and conservation of angular momentum together result in a factor in decay rate of $\left(1 - \frac{v}{c}\right)$ for A coupling and $\left(1 + \frac{v}{c}\right)$ for P coupling.

Other factor determining the rate is the phase-space factor.

in cms (in natural units)

$E_0 = m_\pi = p + \sqrt{p^2 + m^2}$, where m is the charged lepton mass and p is the momentum is cms.

$$\begin{cases} (m_\pi - p)^2 = p^2 + m^2 \Rightarrow p = \dfrac{m_\pi^2 - m^2}{2m_\pi} \\ \sqrt{p^2 + m^2} = \sqrt{\dfrac{(m_\pi^2 - m^2)^2}{4m_\pi^2} + m^2} = \dfrac{(m_\pi^2 + m^2)}{2m_\pi} \\ \dfrac{dp}{dE_0} = \dfrac{1}{1 + \dfrac{p}{\sqrt{p^2 + m^2}}} = \dfrac{m_\pi^2 + m^2}{2m_\pi^2} \end{cases}$$

So, the phase-space factor is

$$p^2 \frac{dp}{dE_0} = \frac{(m_\pi^2 + m^2)(m_\pi^2 - m^2)^2}{4m_\pi^4}$$; phase-space factor for $\pi^+ \to e^+ + \nu_e$ is larger than that for $\pi^+ \to \mu^+ + \nu_\mu$

※ for polarization

$$\begin{cases} 1 + \dfrac{v}{c} = 1 + \beta = 1 + \dfrac{p_{\text{lepton}}}{E_{\text{lepton}}} = 1 + \dfrac{p}{\sqrt{p^2 + m^2}} = 1 + \dfrac{2m_\pi}{m_\pi^2 + m^2} \dfrac{m_\pi^2 - m^2}{2m_\pi} = \dfrac{2m_\pi^2}{m_\pi^2 + m^2} \\ 1 - \dfrac{v}{c} = 1 - \beta = 1 - \dfrac{p_{\text{lepton}}}{E_{\text{lepton}}} = 1 - \dfrac{p}{\sqrt{p^2 + m^2}} = 1 - \dfrac{2m_\pi}{m_\pi^2 + m^2} \dfrac{m_\pi^2 - m^2}{2m_\pi} = \dfrac{2m^2}{m_\pi^2 + m^2} \end{cases}$$

→ [※ factor $\left(1 - \frac{v}{c}\right)$ strongly inhibits decay to the lighter lepton]

Combining the phase-space factor with the polarization

for A coupling decay rate is proportional to

$$p^2 \frac{dp}{dE_0}\left(1 - \frac{v}{c}\right) = \frac{1}{2} m^2 \left(1 - \frac{m^2}{m_\pi^2}\right)^2$$

for P coupling,

$$p^2 \frac{dp}{dE_0}\left(1 + \frac{v}{c}\right) = \frac{1}{2} m_\pi^2 \left(1 - \frac{m^2}{m_\pi^2}\right)^2$$

Then, predicted branching ratios with assumption of lepton universality are for A coupling, $R = \dfrac{\pi \to e+\nu}{\pi \to \mu+\nu} = \dfrac{m_e^2}{m_\mu^2}\dfrac{1}{\left(1-m_\mu^2/m_\pi^2\right)^2} \simeq 1.28\times 10^{-4}$

$$(\because\ m_e^2/m_\pi^2 \ll 1 \to \left(1-m_e^2/m_\pi^2\right)^2 \simeq 1)$$

$\to$ rate is suppressed by requirement that antilepton be left-handed

for P coupling, $R = \dfrac{\pi \to e+\nu}{\pi \to \mu+\nu} = \dfrac{1}{\left(1-m_\mu^2/m_\pi^2\right)^2} \simeq 5.5$

$\Rightarrow$ big difference: angular momentum conservation compels that the electron or muon to have "wrong" helicity for A coupling.

$\Rightarrow$ ratio of two modes provides a very stringent test of the $V-A$ theory.

$\Rightarrow$ $R_{\text{exp}} = 1.23\times 10^{-4}$; agree with prediction of A coupling

$\Rightarrow$ major triumph for $V-A$ theory

<Positron spectrum observed from positive pions stopped by an absorber>

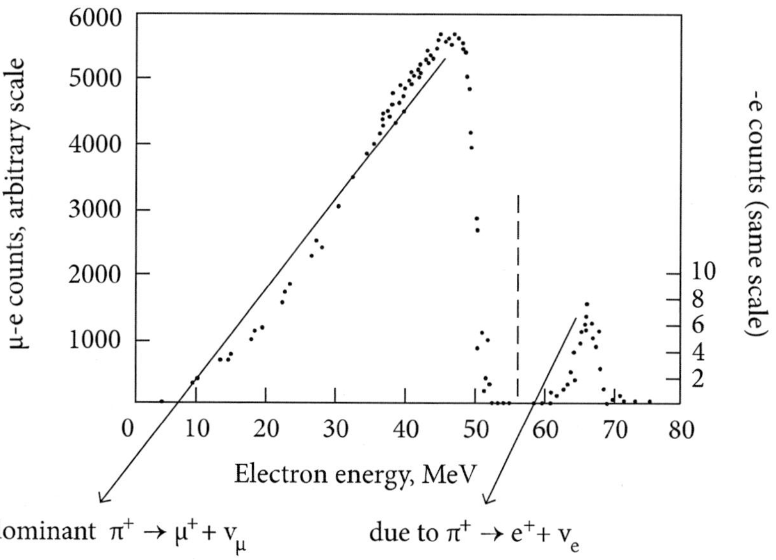

- cuts for rejection of electrons from muon decays
  - (i) momentum
  - (ii) lifetime ($\because$ pion lifetime is 25 ns, muon lifetime is 2200 ns)

(iii) muon pulse in counters

⇒ branching ratio also provides universality that of $W^{\pm}$ coupling to e and μ: $\frac{g_\mu}{g_e}$

because universality is assumed when we derive R (in previous page)

■ Neutral weak currents

Accelerator neutrino beams gave evidence in 1962 for lepton flavors

<charged-current interaction>

$$\left.\begin{array}{l} \nu_\mu + N \to \mu^- + X \\ \nu_e + N \to e^- + X \end{array}\right], \text{ where } X \text{ is the hadronic state}$$

⇒ evidence of neutrino types are different

In 1973, neutrino interactions without a charge lepton were discovered:

<neutral-current interaction>

$$\text{semileptonic} \begin{cases} \nu_\mu + N \to \nu_\mu + X \\ \bar{\nu}_\mu + N \to \bar{\nu}_\mu + X \end{cases}$$

→ These events occur at a fraction of the rate of charged-current events.

<neutral-current interaction>

$$\text{leptonic} \begin{cases} \nu_\mu + e^- \to e^- + \nu_\mu \\ \bar{\nu}_\mu + e^- \to e^- + \bar{\nu}_\mu \end{cases}$$

⇒ discovery of neutral-current events provides favor of the unified model of weak and electromagnetic interactions

⇒ Estimates of the masses of the $W^{\pm}$ and Z^0 were possible.

From the standard Glashow-Salam-Weinberg model, the mass of the W boson is predicted from the coupling constant(g) which can be assumed to be the same in the electromagnetic and weak interactions ($g = e$).

$$\Rightarrow \frac{e^2}{m_W^2} \approx G_F \Rightarrow m_W \approx \frac{e}{\sqrt{G_F}} \approx (90 \sim 100)\ \mathrm{GeV},$$

$$\text{where } e^2 = \frac{4\pi}{137} \left(\leftarrow \alpha = \frac{e^2}{4\pi}\right) \text{ and } G_F \sim 10^{-5}\ \mathrm{GeV}^{-2}$$

Later, the improved model based on the Higgs Mechanism,

$$\frac{G_F}{\sqrt{2}} = \frac{g^2}{8m_W^2} \Rightarrow m_W^2 = \frac{\sqrt{2}\,g^2}{8G_F} = \frac{\sqrt{2}\,e^2}{8G_F \sin^2\theta_w},$$

where θ_w is the weak mixing angle (or Weinberg angle)

$\sin^2\theta_w \approx 0.3 - 0.4$ from Gargamelle

※ It is angle by which spontaneous symmetry breaking rotates the original W^0 and B^0 vector boson plane, producing as a result the Z^0 boson and the photon:

$$\begin{pmatrix} \gamma \\ Z^0 \end{pmatrix} = \begin{pmatrix} \cos\theta_w & \sin\theta_w \\ -\sin\theta_w & \cos\theta_w \end{pmatrix} \begin{pmatrix} B^0 \\ W^0 \end{pmatrix}$$

It also gives the relationship:

$$m_Z = \frac{m_W}{\cos\theta_w}$$

⇒ predicts, $m_W \approx 80\ \mathrm{GeV}$

⇒ Once the masses were known, experiments could be planned to search for the intermediate bosons.

⇒ This led to experiments designed to detect $W^{\pm}$ and Z^0.

■ Observation of $W^{\pm}$ and Z^0 bosons in $p\bar{p}$ collisions(at CERN in 1983)

$W^{\pm}$, Z^0 masses and decay characteristics are predicted in EW theory as described in the previous page.

⇒ required 80 GeV in $q\bar{q}$ collision

- corresponds to 400~600 GeV in pp
- CERN proposed $\sqrt{s} = 540\ \mathrm{GeV}$

$$u + \bar{d} \rightarrow W^+ \rightarrow e^+ + \nu_e, \quad \mu^+ + \nu_\mu \quad \cdots\cdots (1)$$

$$\bar{u} + d \rightarrow W^- \rightarrow e^- + \bar{\nu}_e, \quad \mu^- + \bar{\nu}_\mu \quad \cdots\cdots (2)$$

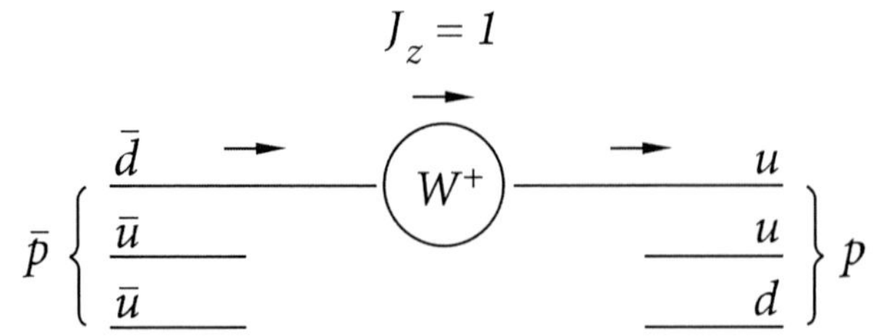

$$\begin{cases} u + \bar{u} \\ d + \bar{d} \end{cases} \rightarrow Z^0 \rightarrow e^+e^-, \ \mu^+\mu^- \quad \cdots\cdots (3),$$

→ this cross-section involves the weak mixing parameter $\sin^2\theta_w$ and turns out be a factor of 10 smaller.

Cross-section for the process (1) is given by Breit-Wigner resonance formula,

$$\sigma_{(u\bar{d} \rightarrow W^+ \rightarrow e^+\nu_e)} = \frac{1}{N_c} \frac{4\pi\lambda\!\!\!^{-}\,\Gamma_{u\bar{d}}\Gamma_{e\nu}/4}{(2S_d+1)(2S_u+1)[(E-M_W)^2+\Gamma^2/4]} \left(\frac{2J+1}{3}\right),$$

where $\frac{1}{N_c}$ is the probability of color matching,

$\Gamma_{u\bar{d}}$ and $\Gamma_{e\nu}$ are the partial widths, Γ is the total width,

S_d and S_u are the quark spin $= 1/2$,

$\frac{2J+1}{3}$ is the spin multiplicity factor for W,

$\lambda\!\!\!^{-} = \frac{\hbar}{p} = \frac{1}{p} = \frac{1}{E/2} = \frac{2}{E}$ is the de Broglie wavelength

(※ In the $V-A$ theory, only LH(RH) helicity states of $u(\bar{d})$ are involved, so the spin-multiplicity factor for W is $\frac{1}{3}(2J+1) = 1$)

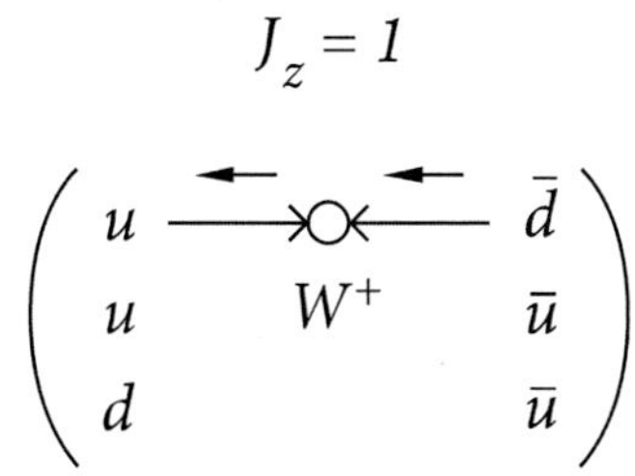

$$\xrightarrow[\text{cross}-\text{section}]{\text{peak}} \sigma_{\max}(u\bar{d} \to W^+ \to e^+\nu_e) = \frac{1}{3}\frac{1}{4}4\pi\lambdabar^2 \frac{\Gamma_{u\bar{d}}\Gamma_{e\nu}}{4}\frac{4}{\Gamma^2}1 = \frac{\pi}{3}\lambdabar^2 \frac{\Gamma_{u\bar{d}}\Gamma_{e\nu}}{\Gamma^2}$$

(※ $\sigma_{\max}$ is σ at $E = M_W$, 1/4 is due to quark spin)

$$= \frac{1}{3}\pi\frac{4}{E^2}\frac{\Gamma_{u\bar{d}}\Gamma_{e\nu}}{\Gamma^2} = \frac{4\pi}{3M_W^2}\frac{\Gamma_{u\bar{d}}\Gamma_{e\nu}}{\Gamma^2}, \text{ where}$$

$$M_W \approx 80\ \text{GeV}$$

$$\frac{\Gamma_{u\bar{d}}}{\Gamma} = \frac{1}{3} \text{ and } \frac{\Gamma_{e\nu}}{\Gamma} = \frac{1}{9},\ \Gamma_{u\bar{d}} = 3\Gamma_{e\nu}$$

(※ BR`s: $BR(ud) = \frac{1}{3}$, $BR(cs) = \frac{1}{3}$, $BR(\ell\nu) = \frac{1}{9}$)

$$= 9.2\ \text{nb}$$

(※ 1 m = 5.07×10^{15} GeV^{-1}, 1 barn = 10^{-28} m^2)

The cross-section for $p\bar{p} \to W^+ \to e^+\nu$ and $p\bar{p} \to Z^0 \to e^+e^-$ result from integrating the elementary quark cross-section over the boson width and the momentum distributions of the quarks inside the nucleon.

$\Rightarrow$ $\sigma_{(p\bar{p} \to W^+ \to e^+\nu)} \sim 1$ nb for $E_p = E_{\bar{p}} = 270$ GeV (540 E_{cms})

$\sigma_{(p\bar{p} \to Z^0 \to e^+e^-)} \sim 0.1$ nb: Z production is 10 times smaller

In the $V-A$ theory, W^+ are polarized

$\Rightarrow$ We expect a parity-violating asymmetry in the decay.

- e^+ from W^+ decay points preferentially in the same direction as the incident antiproton

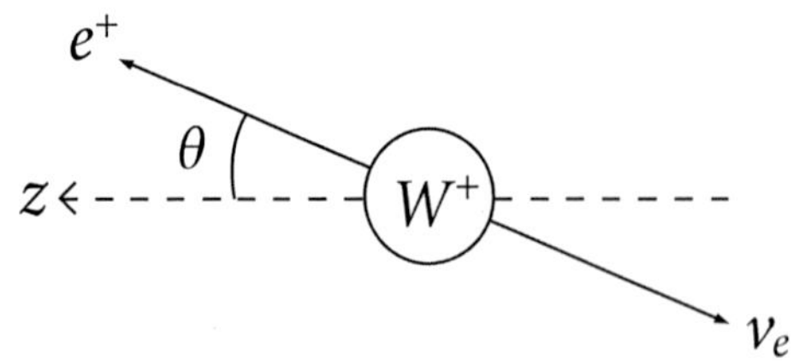

assuming that antiproton beam defines z-axis

└> W^+ is produced in $J_z = +1$ only

Many event observed with systematic balance of P_T of the electron and missing P_T.

Probability that e^+ and ν_e, with helicities $+1$ and -1, respectively, emerge at angle θ and $\pi - \theta$ and with $J_z = +1$ is clearly given by square of appropriate d-function;

$$\frac{dN}{d(\cos\theta)} = \left|d^1_{1,1}\right|^2 = (1+\cos\theta)^2$$

<this angular distribution is well verified experimentally>

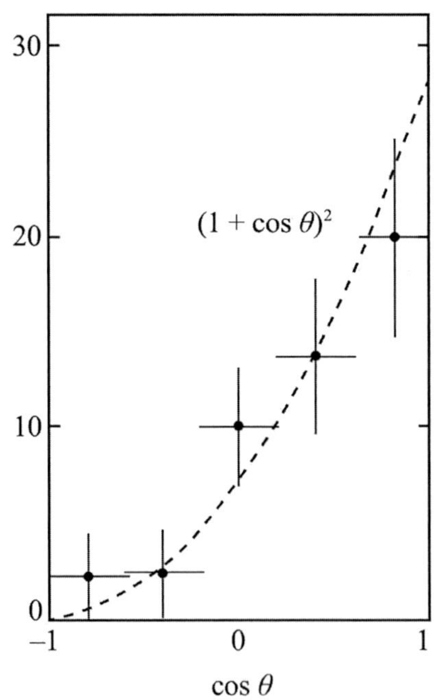

*W*s are polarized in production in $V-A$ interaction

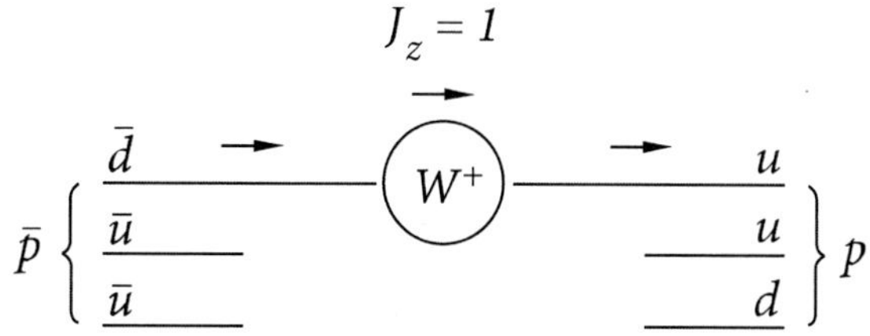

W decay asymmetric by $V-A$ interaction

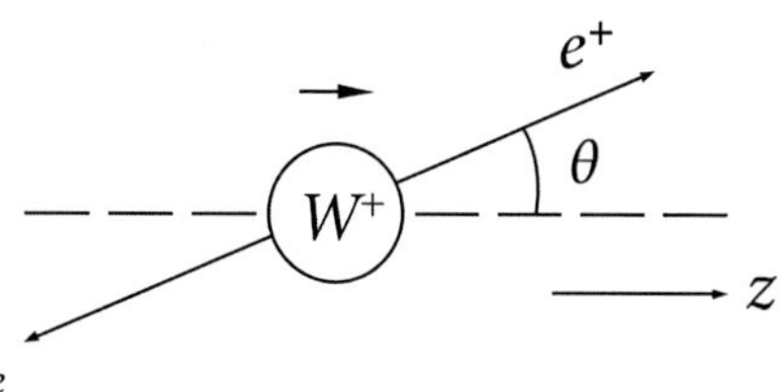

<many events observed with systematic balance of P_T of electron and missing P_T>

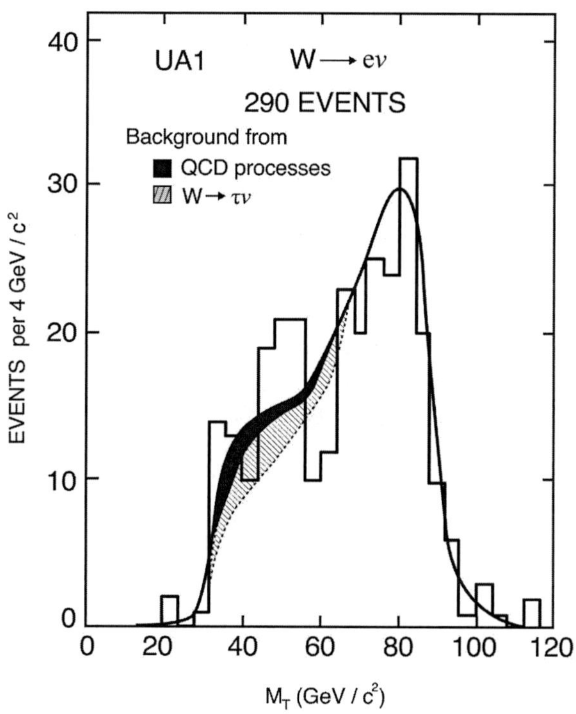

Transverse mass distribution for all W→eν events recorded by UA1 between 1982 and 1985

■ Z^0 production at e^+e^- colliders

Following the discovery of the W and Z bosons at CERN in 1983, LEP at CERN and SLC at SLAC were built and began in 1989 to produce and study the Z with high statistics.

For $e^+e^- \rightarrow Z^0 \rightarrow$ anything,

$$\sigma = \frac{4\pi\lambda\!\!\!^{-}(2J+1)}{(2S+1)^2}\frac{\Gamma_e\Gamma/4}{[(E-E_0)^2+\Gamma^2/4]},$$

where $\lambda\!\!\!^{-} = 2/E \xrightarrow[\text{resonance}]{\text{at}} 2/M_Z$, $S=1/2$ for electron, $J=1$ for Z^0,

E is the cms collision energy, $E_0 = M_Z$, and

Γ_e is the partial width for $Z^0 \rightarrow e^+e^-$

At resonance

$$\sigma_{\max} = \frac{4\pi}{4}\frac{4}{M_Z^2}3\frac{\Gamma_e\Gamma}{4}\frac{4}{\Gamma^2} = \frac{12\pi}{M_Z^2}\left(\frac{\Gamma_e}{\Gamma}\right) \approx 60 \text{ nb}$$

- result is the same if we use relativistic Breit-Wigner formula

Meanwhile, cross-section for pointlike, non-resonant reaction $e^+e^- \rightarrow \mu^+\mu^-$ via single-photon exchange, is

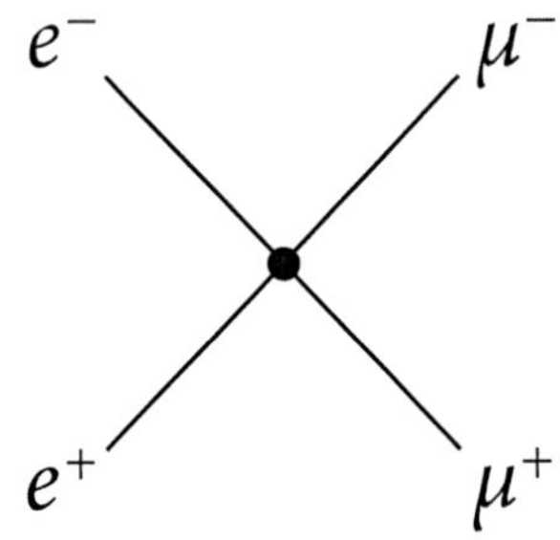

$$\sigma_{(e^+e^-\rightarrow\mu^+\mu^-)} = \sigma_{(\text{point})} = \frac{4\pi\alpha^2}{3E^2} = 87 \text{ nb/S (GeV}^2)$$

At $E = M_Z$ (i.e. Z^0 peak)

$$R = \frac{\sigma_{max}(e^+e^- \to Z^0)}{\sigma_{(point)}} = \frac{12\pi}{M_Z^2}\left(\frac{\Gamma_e}{\Gamma}\right)\frac{3E^2}{4\pi\alpha^2} = \frac{9}{\alpha^2}\left(\frac{\Gamma_e}{\Gamma}\right) \simeq 5700,$$

$$\text{where } \frac{\Gamma_e}{\Gamma} \approx 0.034$$

- this ideal value of $R \approx 5700$ is reduced to $R \approx 3200$ by radiative corrections (initial state radiation).

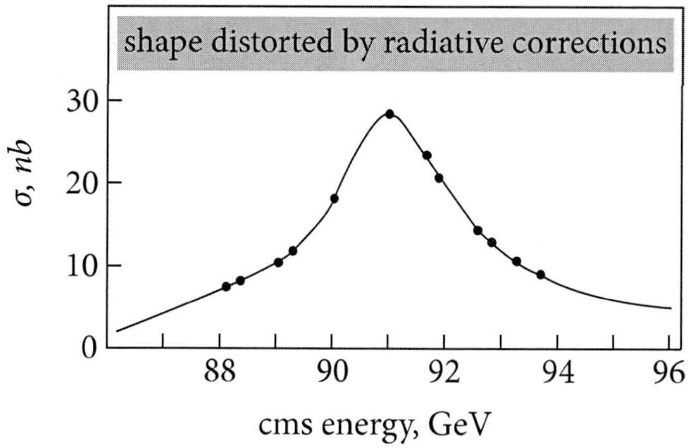

■ Weak decays of quarks, The GIM model and CKM matrix

※ GIM: Glashow-Iliopoulos-Maiani mechanism by which FCNC are suppressed

$n \to p + e^- + \bar{\nu}_e$	$\Sigma^- \to n + e^- + \bar{\nu}_e$
$(\Delta s = 0)$	$(\Delta s = 1)$
$u \;-\; u$	$d \;-\; d$
$d \;-\; d$	$d \;-\; d$

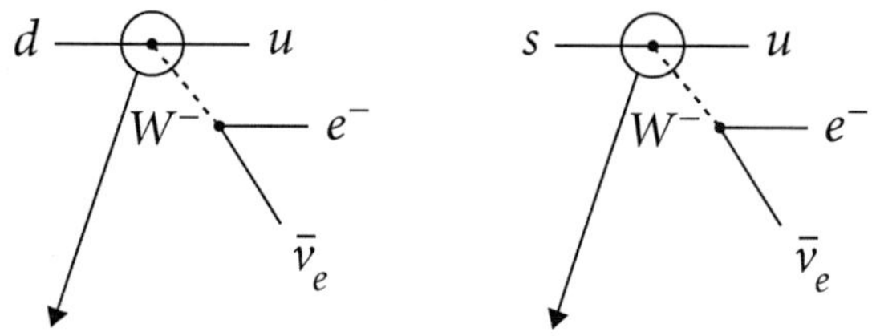

effective coupling in $g_W cos\theta_c$ $\qquad$ $g_W sin\theta_c$

① $\Delta s = 1$ weak decays are suppressed by a factor of about 20 compared with nuclear β-decays($\Delta s = 0$)

② Fermi constant(G_F) deduced from neutron β-decay is slightly smaller than G_F deduced from muon decay ($\mu^+ \rightarrow e^+ + \nu_e + \overline{\nu}_\mu$).

Can these facts be explained?

↳ Cabibbo theory in 1963

③ d and s quark states are not pure flavor eigenstates but rotated by a maxing angle, θ_c (Cabibbo angle).

→ quark mass eigenstate ≠ weak eigenstate

④ quark doublet

$\rightarrow \begin{pmatrix} u \\ d' \end{pmatrix} = \begin{pmatrix} u \\ d\cos\theta_c + s\sin\theta_c \end{pmatrix}$, where experiment yields $\theta_c \simeq 12°$ ($\sin\theta_c \approx 0.22$)

※ These three quarks (u, d, s) were the only quarks at the time Cabibbo wrote his theory.

⑤ with this quark doublet assumption,

lepton doublets $\begin{pmatrix} e \\ \nu_e \end{pmatrix}$, $\begin{pmatrix} \mu \\ \nu_\mu \end{pmatrix}$

- Flavor-changing neutral currents(FCNCs)

$K^+ \rightarrow \pi^+ + \nu + \overline{\nu}$ $\qquad$ $K^+ \rightarrow \pi^0 + \mu^+ + \nu_\mu$

$u - u$ $\qquad$ $d - d$

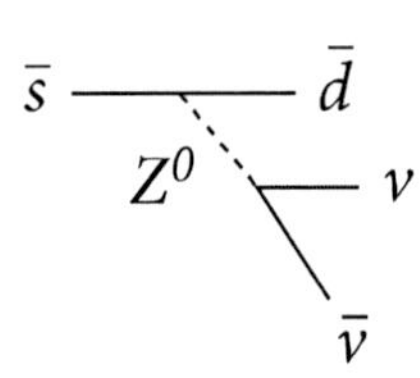

<neutral current>

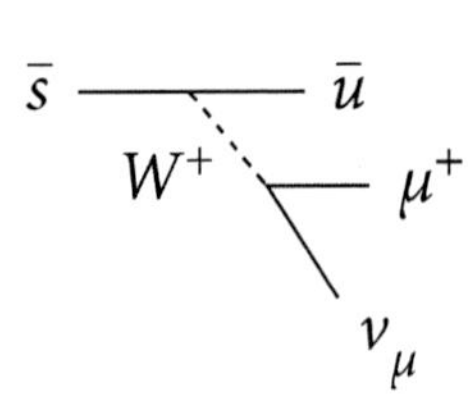

<charged current>

$$\frac{K^+ \rightarrow \pi^+ + \nu + \bar{\nu}}{K^+ \rightarrow \pi^0 + \mu^+ + \nu_\mu} = \frac{\text{NC}}{\text{CC}} \leq 10^{-5}$$

- Neutral current processes are non-flavor-changing process and are characterized by selection rule $\Delta S = 0$
- In Cabibbo model, it was hard to understand the absence of $\Delta S = 1$ in NC decay
- Try to understand this, Z^0 coupling to u quark and to 'Cabibbo-rotated' d and s quark combination:

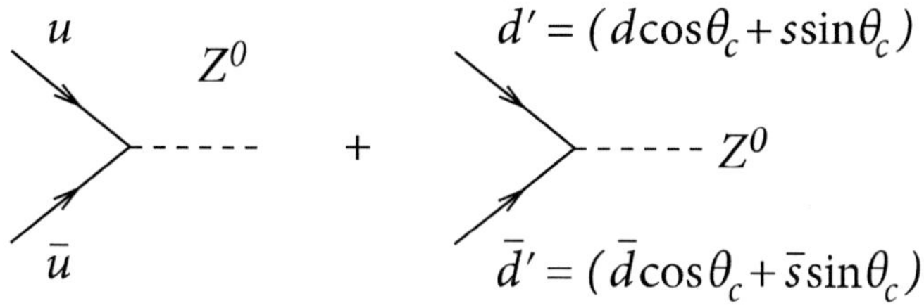

- NC interaction is proportional to product of wave functions of participating particles,

$$\underbrace{u\bar{u} + (d\bar{d}\cos^2\theta_c + s\bar{s}\sin^2\theta_c)}_{\Delta S = 0} + \underbrace{(s\bar{d} + d\bar{s})\sin^2\theta_c\cos\theta_c}_{\Delta S = 1}$$

- so that $\Delta S = 1$ NC are allowed (so, flavor-changing neutral interactions should be allowed)
- why were they so rare?

⇒ The problem was solved by introducing the charm quark in Glashow, Iliopoulos and Maiani (GIM) mechanism(1970).

→ i.e. introduce fourth quark("charm" quark)

→ $J/\Psi\ (c\bar{c})$ found in 1974

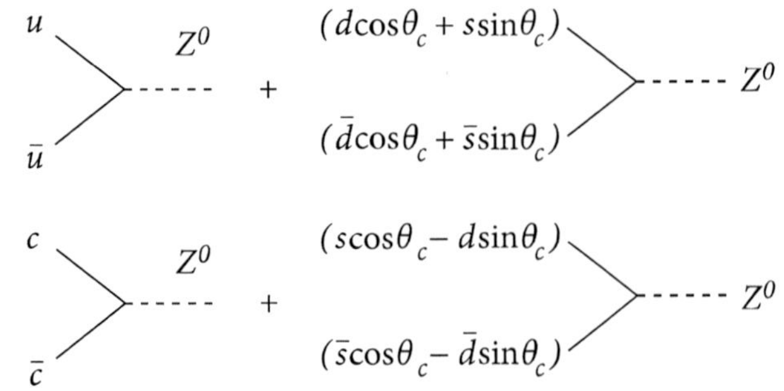

- GIM proposed that a new charge $+\frac{2}{3}$ quark with label c for "charm"

  "1st quark doublet" "second quark doublet"

  $\begin{pmatrix} u \\ d' \end{pmatrix} = \begin{pmatrix} u \\ d\cos\theta_c + s\sin\theta_c \end{pmatrix}$, $\begin{pmatrix} c \\ s' \end{pmatrix} = \begin{pmatrix} c \\ s\cos\theta_c - d\sin\theta_c \end{pmatrix}$

  └> orthogonal to 1st quark doublet

- Neutral current interaction becomes

$$\underbrace{u\bar{u} + c\bar{c} + (d\bar{d} + s\bar{s})\cos^2\theta_c + (s\bar{s} + d\bar{d})\sin^2\theta_c}_{\Delta S = 0} + \underbrace{(s\bar{d} + \bar{s}d - \bar{s}d - s\bar{d})\sin\theta_c\cos\theta_c}_{\Delta S = 1}$$

- Thus, $\Delta S = 1$ is cancelled by introducing a new quark and second quark doublet.

In Cabibbo-GIM formalism, Cabibbo angle represents the rotation of the mass eigenstate into the weak eigenstate.

- quark mixing

$$\begin{pmatrix} u \\ d' \end{pmatrix} = \begin{pmatrix} u \\ d\cos\theta_c + s\sin\theta_c \end{pmatrix}$$

$$\begin{pmatrix} c \\ s' \end{pmatrix} = \begin{pmatrix} c \\ s\cos\theta_c - d\sin\theta_c \end{pmatrix}$$

can be expressed in matrix form

$$\begin{pmatrix} d' \\ s' \end{pmatrix} = \begin{pmatrix} \cos\theta_c & \sin\theta_c \\ -\sin\theta_c & \cos\theta_c \end{pmatrix} \begin{pmatrix} d \\ s \end{pmatrix}$$

└> the weak interactions involving the four light quark are consistent with this model, being defined by unique value of θ_c

With six quark flavors

※ observing that CP-violation could not be explained in a four-quark model. KM(Kobayashi and Maskawa) generalized the Cabibbo matrix to keep track of weak decays of three generations of quark

$$\begin{pmatrix} d' \\ s' \\ b' \end{pmatrix} = V \begin{pmatrix} d \\ s \\ b \end{pmatrix}, \text{ where } V = \begin{vmatrix} V_{ud} & V_{us} & V_{ub} \\ V_{cd} & V_{cs} & V_{cb} \\ V_{td} & V_{ts} & V_{tb} \end{vmatrix}$$

$|V_{ij}|^2$ represent probability that the quark of j flavor decays into a quark of i flavors, V is unitary ($V^+ = V^{-1}$), $V_{ud}V_{ub}^* + V_{cd}V_{cb}^* + V_{td}V_{tb}^* = 0$

$V_{ud}(= \cos\theta_c)$ determined by comparing nuclear β-decay and μ-decay rates

V_{us} determined from semi-leptonic decays of strange particles

(i.e. $\Sigma^- \rightarrow n + e^- + \bar{\nu}_e$)

V_{ub} measured by selecting B semi-leptonic decays to non-strange particles at the end-point ($V_{ub}/V_{cb} \approx 0.06$)

V_{cb} from $B \rightarrow D\ell\nu$: $\Gamma(b \rightarrow c\ell\nu) = \dfrac{R(b \rightarrow c\ell\nu)}{\tau_B} = \dfrac{G_F^2 m_b^5}{192\pi^3}|V_{cb}|^2 f$

($V_{cb} = 0.040 \pm 0.005$)

V_{cd} from ν_μ di-μ events: $\nu_\mu + d \rightarrow \mu^- + c$

$\quad\quad\quad\quad\quad\quad\quad\quad\quad\quad$ ↳ $s + \mu^+ + \nu_\mu$,

so that μ^-, μ^+: two muons have opposite sign

($V_{cd} = 0.221 \pm 0.003$)

V_{cs} from $D^+ \rightarrow K^0 e^+ \nu_e$

or unitarity condition: $|V_{cd}|^2 + |V_{cs}|^2 + |V_{cb}|^2 = 1$

V_{td}, V_{ts} from B mixing measurements

$$\rightarrow \quad V_{CKM} = \begin{vmatrix} V_{ud} = 0.975 & V_{us} = 0.221 & V_{ub} = 0.005 \\ V_{cd} = 0.221 & V_{cs} = 0.974 & V_{cb} = 0.040 \\ V_{td} = 0.010 & V_{ts} = 0.041 & V_{tb} = 0.999 \end{vmatrix}$$

→ diagonal terms are close to unity and off-diagonal terms are small.

※ $N \times N$ matrix has $\frac{N(N+1)}{2}$ real parameters (Euler angles) and $\frac{(N-1)(N-2)}{2}$ non-trivial phase angles.

→ for 3×3 matrix, 3 Euler angles and 1 phase angle. This phase angle results in CP-violation(or T violation).

for V_{cb},

$$\Gamma(b \rightarrow c\ell\nu) = \frac{R(b \rightarrow c\ell\nu)}{\tau_B} = \frac{G_F^2 m_b^5}{192\pi^3} |V_{cb}|^2 f \quad \text{(analogy for } \mu \rightarrow e\nu_e\bar{\nu}_\mu)$$

← f: correction factor

① for b quark is not free but bound in a meson

② phase space is reduced compared with that for massless products

$\Rightarrow V_{cb} = 0.040 \pm 0.005$

for V_{cd},

$\nu_\mu + d \rightarrow \mu^- + c$

← μ^-, μ^+ : two muons of opposite sign

$c \rightarrow s + \mu^+ + \nu_\mu$

$\Rightarrow V_{cd} = 0.221 \pm 0.003$

$$V = \begin{vmatrix} V_{ud} & V_{us} & V_{ub} \\ V_{cd} & V_{cs} & V_{cb} \\ V_{td} & V_{ts} & V_{tb} \end{vmatrix}$$

① CKM matrix for 3 families may be represented by three rotations

3 Euler angles θ_{12}, θ_{23}, θ_{13}(Cabibbo angles) and one CP-violating phase δ_{13}

$$V_{CKM} = U_{23}U_{13}U_{12}$$

$$= \begin{pmatrix} 1 & 0 & 0 \\ 0 & c_{23} & s_{23} \\ 0 & -s_{23} & c_{23} \end{pmatrix} \begin{pmatrix} c_{13} & 0 & s_{13}e^{-i\delta_{13}} \\ 0 & 1 & 0 \\ -s_{13}e^{i\delta_{13}} & 0 & c_{13} \end{pmatrix} \begin{pmatrix} c_{12} & s_{12} & 0 \\ -s_{12} & c_{12} & 0 \\ 0 & 0 & 1 \end{pmatrix}$$

(※ "Standard" parameterization of CKM matrix (PDG), Chau and Keung, 1984 by three mixing angles and CP-violating KM phase)

$$V = \begin{pmatrix} c_{12}c_{13} & s_{12}c_{13} & s_{13}e^{-i\delta_{13}} \\ -s_{12}c_{23} - c_{12}s_{23}s_{13}e^{i\delta_{13}} & c_{12}c_{23} - s_{12}s_{23}s_{13}e^{i\delta_{13}} & s_{23}c_{13} \\ s_{12}s_{23} - c_{12}c_{23}s_{13}e^{i\delta_{13}} & -c_{12}s_{23} - s_{12}c_{23}s_{13}e^{i\delta_{13}} & c_{23}c_{13} \end{pmatrix},$$

where $c_{ij} = \cos\theta_{ij}$, $s_{ij} = \sin\theta_{ij}$, $s_{13} \ll s_{23} \ll s_{12}$ (experimentally known)

② Wolfenstein Parameterization (1983), expansion in $\lambda = |V_{us}|$

four Wolfenstein parameters: $\lambda \equiv s_{12}$, introduce A, ρ, η

※ $A \equiv \dfrac{s_{23}}{\lambda^2}$, $A\lambda^3(\rho - i\eta) \equiv s_{13}e^{-i\delta_{13}}$

$$V_{CKM} = \begin{vmatrix} 1-\dfrac{\lambda^2}{2} & \lambda & A\lambda^3(\rho - i\eta) \\ -\lambda & 1-\dfrac{\lambda^2}{2} & A\lambda^2 \\ A\lambda^3(1-\rho-i\eta) & -A\lambda^2 & 1 \end{vmatrix} + O(\lambda^4),$$

※ CP-violation can be determined by measuring $\rho - i\eta$

$$\bar{\rho} + i\eta = -\frac{V_{ud}V_{ub}^*}{V_{cd}V_{cb}^*}$$

$$c_{12}c_{13} = \sqrt{1-s_{12}^2}\,c_{13} = \sqrt{1-\lambda^2} = (1-\lambda^2)^{\frac{1}{2}} \approx 1 - \frac{1}{2}\lambda^2$$

As of 2022,

$$A = 0.826^{+0.018}_{-0.015},\ \lambda = 0.22500 \pm 0.00067,$$

$$\bar{\rho} = 0.159 \pm 0.010,\ \bar{\eta} = 0.348 \pm 0.010$$

■ Neutral K mesons

K mesons are in the form of two isospin doublet $\left(I_3 = \pm\frac{1}{2}\right)$

- $s = +1$ doublet consists of K^+ and K^0
- $s = -1$ doublet consists of K^- and $\overline{K}^0$

$\begin{cases} K^+,\ K^- \\ K^0,\ \overline{K}^0 \end{cases}$: related by charge conjugation

Meson	quark combinations	I	I_3	s	MeV
K^+	$u\overline{s}$	$\frac{1}{2}$	$\frac{1}{2}$	+1	494
K^0	$d\overline{s}$	$\frac{1}{2}$	$-\frac{1}{2}$	+1	498
K^-	$\overline{u}s$	$\frac{1}{2}$	$-\frac{1}{2}$	-1	494
$\overline{K}^0$	$\overline{d}s$	$\frac{1}{2}$	$\frac{1}{2}$	-1	498

- particle versus antiparticle
  - (i) reversal of I_3
  - (ii) change of strangeness $\Delta s = 2$
- These particle are clearly distinctive in strong interaction since strong interaction conserves isospin and strangeness.

All decay to non-strange particles and/or leptons through weak interaction is ($\Delta s = \pm 1$)

ex quark diagrams for $K \rightarrow 2\pi$ decay

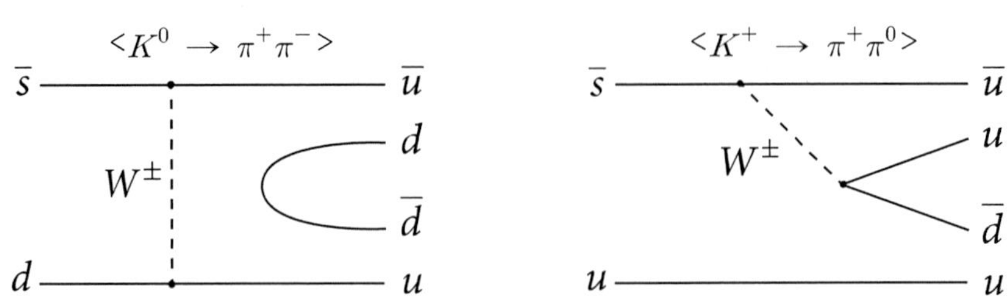

Pure K^0 or $\overline{K}^0$ states are produced in $p\bar{p}$ annihilation;

$$\begin{cases} p\bar{p} \to K^- \pi^+ K^0 \\ p\bar{p} \to K^+ \pi^- \overline{K}^0 \end{cases}$$

→ nature of neutral kaon produced is identified via charged K and π secondaries produced in association

K^0 and $\overline{K}^0$ can decay to pions with $|\Delta s| = 1$ by the weak interaction.

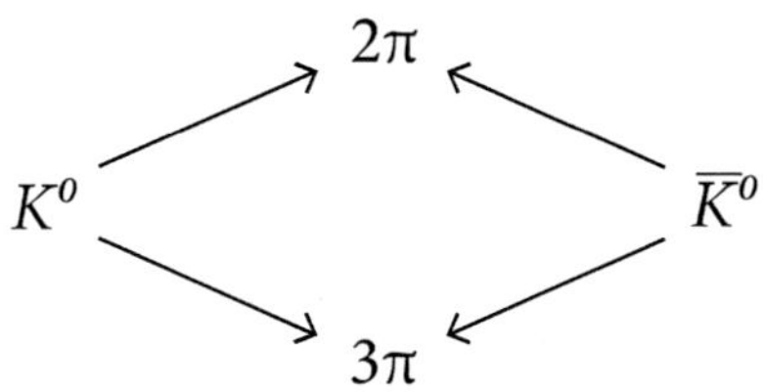

Thus, mixing can occur via (virtual) intermediate pion states.

If one has a pure K^0 state at $t = 0$, one can have a superposition (mixing state) of both K^0 and $\overline{K}^0$ at later time t:

$$|\, K(t) > \; = \alpha(t) \,|\, K^0 > + \beta(t) \,|\, \overline{K}^0 >$$

CP eigenstates can be defined for K^0 mesons:

$$CP \,|\, K^0 > \; = \eta \,|\, \overline{K}^0 >$$

$$CP \,|\, \overline{K}^0 > \; = \eta' \,|\, K^0 >,$$

where η and η' are the arbitrary phase factors

→ define $\eta = \eta' = 1$

and $|\, K^0 >$ and $\overline{K}^0 >$ are not CP eigenstates.

But we can form linear combinations:

$$|\, K_S > \equiv \frac{1}{\sqrt{2}}\left(|\, K^0 > + \,|\, \overline{K}^0 >\right),\; CP = +1 \xrightarrow[\text{in}]{\text{result}} CP \,|\, K_S > \to \,|\, K_S >$$

└ short lifetime for decay

$$|\, K_L > \equiv \frac{1}{\sqrt{2}}\left(|\, K^0 > - \,|\, \overline{K}^0 >\right),\; CP = -1 \xrightarrow[\text{in}]{\text{result}} CP \,|\, K_L > \to - \,|\, K_L >$$

※ Consider K^0 production and subsequent decay

Why don`t the $K^0$`s decay away with a single exponential?

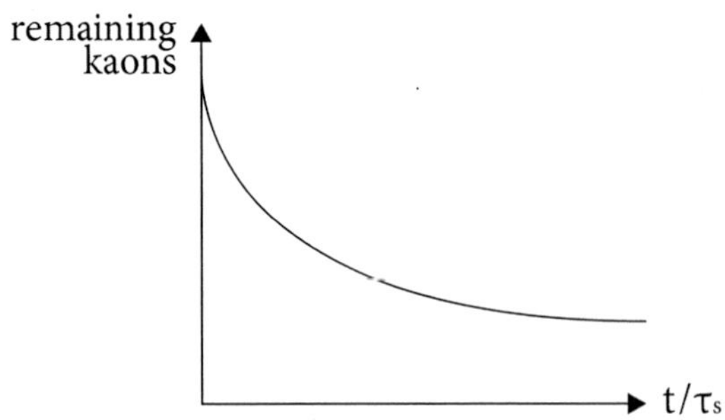

Because K^0`s are mixtures of K_S and K_L

K_S and K_L are distinguished by their decay modes

Consider 2π and 3π decay modes

(i) $\pi^0\pi^0$, $\pi^+\pi^-$

Total wave function is symmetric under two particle interchange by Bose symmetry.

$\rightarrow$ $CP = +1$ (no spin involved)

(ii) $\pi^+\pi^-\pi^0$ [three pions are in relative S-state due to small Q-value (70 MeV of decay)]

- 2π의 $CP = +1$
- π^0 has $C = +1$ (since it decays to 2γ)
- $P(\pi^0) = -1$

$\Rightarrow$ Combining π^0 with $\pi^+\pi^-$ system, $CP = -1$

(iii) $\pi^0\pi^0\pi^0$

By Bose symmetry, two pions must be even.

$\rightarrow$ ℓ-value of remaining pion about the di-pion is also even, since $J_K = 0$

$\rightarrow$ overall parity is product of intrinsic pion particles $\Rightarrow P = -1$ and neutral pion has $C = +1$

$\Rightarrow CP = -1$ for any orbital angular momentum

$\Rightarrow$ From (ii) and (iii), 3π state has $CP = -1$

So,

$$K_S = \frac{1}{\sqrt{2}}(K^0 + \overline{K}^0) \xrightarrow[\text{conserved}]{\text{if CP}} 2\pi \quad (CP = +1), \quad \tau_S = 0.893 \times 10^{-10} \text{ sec}$$

※ due to different Q-values, decay rates are different, so $CP = +1$ state is called K-short (K_S) and $CP = -1$ state is called K-long(K_L)

$$K_L = \frac{1}{\sqrt{2}}(K^0 - \overline{K}^0) \rightarrow 3\pi \quad (CP = -1), \ \tau_L = 0.517 \times 10^{-7} \text{ sec}$$

※ which is the reason why the $CP = +1$ state is called K-short (K_S) and $CP = -1$ state is called K-long(K_L)

• Strangeness oscillations

Since K_S and K_L have different lifetimes and decay modes, they can be expected to have a difference in mass, which will alter the relative phases of these two eigenstates.

※ wave functions describing a non-stationary decaying state of central angular frequency $\omega_R = \frac{E_R}{\hbar}$, where E_R is the resonance energy, and lifetime $\tau = \frac{\hbar}{\Gamma}$ is $\Psi(t) = \Psi(0)e^{-i\omega_R t}e^{-\frac{t}{2\tau}} = \Psi(0)e^{-\left(iE_R t + \frac{\Gamma}{2}t\right)}$ in $\hbar = c = 1$ units

Amplitude of $\begin{pmatrix} K_S \\ K_L \end{pmatrix}$ as a function of time, measured in the particle rest frame,

$$\begin{cases} K_S(t) = K_S(0)e^{-(\Gamma_S/2 + im_S)t} \\ K_L(t) = K_L(0)e^{-(\Gamma_L/2 + im_L)t}, \end{cases} \text{ where } \Gamma_S = \frac{\hbar}{\tau_S} \text{ and } \Gamma_L = \frac{\hbar}{\tau_L}$$

Suppose we start off with a pure K^0 beam of unit intensity,

$$\begin{cases} | K_S > = \dfrac{1}{\sqrt{2}}(| K^0 >+ | \overline{K}^0 >) \\ | K_L > = \dfrac{1}{\sqrt{2}}(| K^0 >- | \overline{K}^0 >) \end{cases} \quad \cdots\cdots (1)$$

$$\Rightarrow | K^0 > = \frac{1}{\sqrt{2}}(| K_S >+ | K_L >)$$

(※ K^0 intensity after time t is

$$I(K^0) = \frac{1}{2}[K_S(t)+K_L(t)][K_S^*(t)+K_L^*(t)] \quad \cdots\cdots (2)$$

$$\Rightarrow \begin{cases} | K_S(0) > = \dfrac{1}{\sqrt{2}}(| K^0(0) >+ | \overline{K}^0(0) >) = \dfrac{1}{\sqrt{2}} \\ | K_L(0) > = \dfrac{1}{\sqrt{2}}(| K^0(0) >- | \overline{K}^0(0) >) = \dfrac{1}{\sqrt{2}} \end{cases},$$

where $| K^0(0) > = 1, \quad | \overline{K}^0(0) > = 0$

Substitute Eq. (1) into (2),

$$I(K^0) = \frac{1}{2}[K_S^2(0)e^{-\Gamma_S t} + K_S(0)K_L^*(0)e^{i(m_L - m_S)t}e^{-\frac{\Gamma_S+\Gamma_L}{2}t} + K_S^*(0)K_L(0)e^{-i(m_L - m_S)t}e^{-\frac{\Gamma_S+\Gamma_L}{2}t} + K_L^2(0)e^{-\Gamma_L t}]$$

$$= \frac{1}{2}\left[\frac{1}{2}e^{-\Gamma_S t} + \frac{1}{2}(e^{i(m_L - m_S)t} + e^{-i(m_L - m_S)t})e^{-\frac{\Gamma_S+\Gamma_L}{2}t} + \frac{1}{2}e^{-\Gamma_L t}\right]$$

$$= \frac{1}{4}\left[e^{-\Gamma_S t} + e^{-\Gamma_L t} + 2e^{-\frac{\Gamma_S+\Gamma_L}{2}t}\cos\Delta mt\right] \cdots\cdots (3),$$

where $\Delta m = m_L - m_S$

Similarly, $\overline{K}^0$ intensity after time t is

$$I(\overline{K}^0) = \frac{1}{4}\left[e^{-\Gamma_S t} + e^{-\Gamma_L t} - 2e^{-\frac{\Gamma_S+\Gamma_L}{2}t}\cos\Delta mt\right] \quad \cdots\cdots (4)$$

From Eqs. (3) and (4), K^0 and $\overline{K}^0$ intensities oscillate with frequency Δm.

※ 1 sec = 1.52×10^{24} GeV^{-1}, $\tau_s = 0.9\times 10^{-10}$ sec = 1.37×10^{14} GeV^{-1}

Δm can be obtained from Δmt.

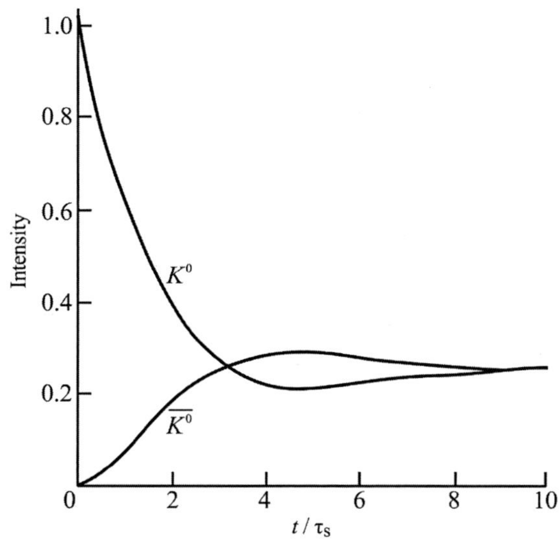

→ This mass difference corresponds to rate at which K^0 oscillates into $\overline{K}^0$.

→ gives $\Delta m = (3.191 \pm 0.009) \times 10^{-12}$ MeV

(I) Δm estimation

Since Fermi constant in the weak interaction is $\frac{g^2}{8M_W^2} = \frac{G_F}{\sqrt{2}}$, G_F^2 has dimension M^{-4}.

→ [mass difference] $= G_F^2 \times$ [kaon mass]^5

→ $\Delta m \sim G_F^2 m_K^5 \sim 10^{-9}$ MeV (c.f. for $M = m_\pi$, $\Delta m \sim 10^{-12}$ MeV)

(II) Actual Δm calculation comes from box diagram

Box diagram for $K^0 \to \overline{K}^0$ transition

※ this oscillation proceeds through the second-order weak interaction

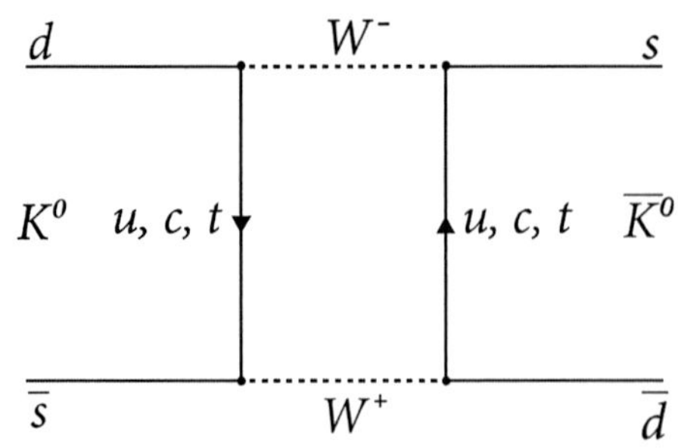

(Fig. 7.20, p. 231, Introduction to High Energy Physics 4th Edition by Donald H. Perkins)

→ initial c quark will dominate over a u quark (or t quark) and will introduce coupling $G_F^2 \cos^2\theta_c \sin^2\theta_c$ (this is due to CKM matrix)

→ actual expression for Δm is

※ Δm measurement allows that mass of charmed quark before it was discovered ($m_c \approx 1.5$ GeV).

$$\Delta m = \frac{G_F^2}{4\pi} m_K f_K^2 m_c^2 \cos^2\theta_c \sin^2\theta_c,$$

where $f_K \approx 1.2 m_\pi$ is the kaon decay constant, m_K is the kaon mass,

m_c is the charmed quark mass,

$m_K f_K^2$ is relates to $|\Psi(0)|^2$ of quark-antiquark($d\bar{s}$) wave function at origin or the point at which the quark interact $|\Psi(0)|^2 \simeq V^{-1}$, where $V \simeq m_h^{-3}$ and m_h is the some typical hadron mass.

- K^0 regeneration

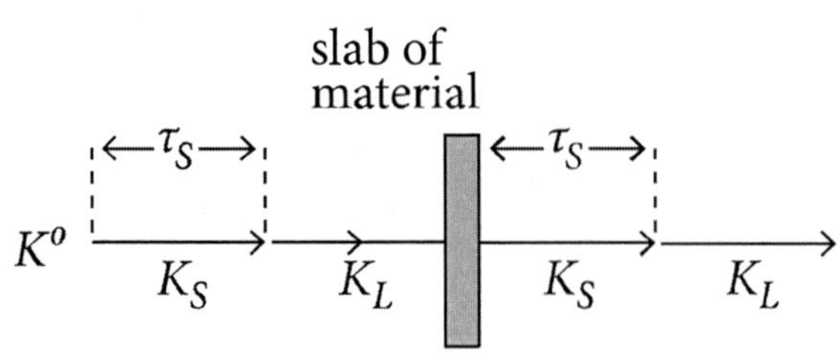

initially pure K^0 beam ($S = +1$)

$\Rightarrow$ all K_S component decayed and only K_L remains at a long way from source.

$\Rightarrow K_L = \frac{1}{\sqrt{2}}(K^0 - \overline{K}^0)$, generation of particles with $S = -1$

where $\overline{K}^0 \Rightarrow S = -1$

※ This was confirmed in early experiments by observation of hyperon production, $\overline{K}^0 + p \rightarrow \Lambda + \pi^+$

[decay $K^0 + p \rightarrow \Lambda + K^0 + K^+$ is unlikely because $(K^0 + K^+)$ leaves little free energy]

$\Rightarrow$ Both K^0 and $\overline{K}^0$ can undergo charge-exchange scattering but $\overline{K}^0$ can also transform to hyperons, $\overline{K}^0 + p \rightarrow \Lambda + \pi^+$.

$\Rightarrow$ After emerging from the slab, amplitude of $\overline{K}^0$ will be $\overline{f}$ and amplitude of K^0 will be f, where $\overline{f} < f < 1$

($\because$ $\overline{K}^0$ is transformed as hyperon)

$\Rightarrow$ K^0 and $\overline{K}^0$ are absorbed differently.

$$\Rightarrow \frac{1}{\sqrt{2}}(fK^0 - \overline{f}\overline{K}^0) = \frac{1}{2}(f+\overline{f})K_L + \frac{1}{2}(f-\overline{f})K_S$$

$$\because \begin{cases} K^0 = \frac{1}{\sqrt{2}}(K_S + K_L) \\ \overline{K}^0 = \frac{1}{\sqrt{2}}(K_S - K_L) \end{cases}$$

$\Rightarrow$ Since $f \neq \overline{f}$, some of K_S has been regenerated when K_L beam transverse a slab of material as shown in the above Fig.

■ CP violation in the neutral kaon system

① K_L is the $CP = -1$ eigenstate, and is not supposed to decay to two pions, just to three pions.

→ This violates CP since K_L is $CP = -1$ state and CP of two pions is $+1$.

② 1964-experiment finds that K_L does decays to two pions with the branching ratio(BR) of 0.003; $BR(K_L \rightarrow \pi^+\pi^-) = 0.002$, $BR(K_L \rightarrow \pi^0\pi^0) = 0.001$

CP violation in the neutral kaon system = K_L could also decay to $\pi^+\pi^-$ with a branching ratio of order 10^{-3}

Let K_1 and K_2 be amplitudes for $CP = +1$ and $CP = -1$, respectively.

⇒ K_L and K_S are their admixture:

$$\begin{cases} K_L = \dfrac{1}{\sqrt{1+|\epsilon|^2}}(K_2 + \epsilon K_1) \\ K_S = \dfrac{1}{\sqrt{1+|\epsilon|^2}}(K_1 - \epsilon K_2) \end{cases},$$

where small parameter ϵ is the CP violation parameter, which must be determined experimentally

⇒ degree of CP violation is usually quoted as the amplitude ratio,

$$|\eta_{+-}| = \frac{\text{amplitude of } K_L \rightarrow \pi^+\pi^-}{\text{amplitude of } K_S \rightarrow \pi^+\pi^-} = (2.29 \pm 0.02) \times 10^{-3} \quad \cdots\cdots (1),$$

where $+-$ in η_{+-} is for charges of pions,

$$\eta_{+-} = |\eta_{+-}| e^{i\phi_{+-}}$$

※ If CP violation is only due to the wave function impurity,

$$|\epsilon|^2 = |\eta_{+-}|^2$$

Similary,

$$|\eta_{00}| = \frac{\text{amplitude of } K_L \rightarrow \pi^0\pi^0}{\text{amplitude of } K_S \rightarrow \pi^0\pi^0} = (2.28 \pm 0.02) \times 10^{-3} \cdots\cdots (2)$$

- Interference between K_S and K_L amplitudes

Since both K_L and K_S can decay to two pions, interference effects in $\pi^+\pi^-$ signal are expected as a function of time development of an initially pure

$K^0(\overline{K}^0)$ beam.

$$\frac{I_{2\pi}(t)}{I_{2\pi}(0)} = e^{-\Gamma_S t} + |\eta_{+-}|^2 e^{-\Gamma_L t} + 2|\eta_{+-}|^2 e^{-\frac{\Gamma_L+\Gamma_S}{2}t}\cos(\Delta m t + \phi_{+-}),$$

where ϕ_{+-} is the phase angle between $K_S \to 2\pi$ and $K_L \to 2\pi$ amplitudes

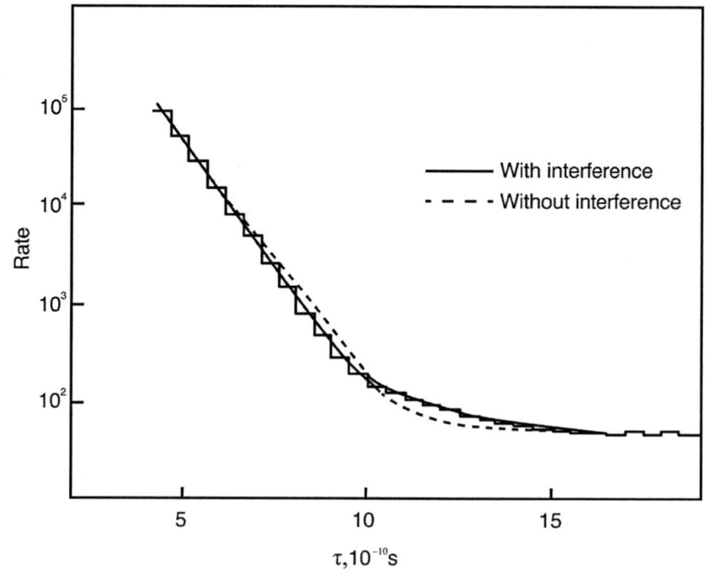

- Measured asymmetry between $K^0 \rightarrow \pi^+\pi^-$ and $\overline{K}^0 \rightarrow \pi^+\pi^-$;

※ CPLEAR experiment at CERN for measurement of CP, T and CPT in the neutral kaon system: $p\bar{p} \rightarrow K^-\pi^+K^0$, $p\bar{p} \rightarrow K^+\pi^-\overline{K}^0$

$$A_{+-}(t) = \frac{2|\eta_{+-}|e^{\frac{\Gamma_S-\Gamma_L}{2}t}\cos(\Delta m t + \phi_{+-})}{1+|\eta_{+-}|^2 e^{(\Gamma_S-\Gamma_L)t}},$$

※ denominator: $\left[e^{-\Gamma_S t} + |\eta_{+-}|^2 e^{-\Gamma_L t} + 2|\eta_{+-}|^2 e^{-\frac{\Gamma_L+\Gamma_S}{2}t}\cos(\Delta m t + \phi_{+-})\right]$

$$+\left[e^{-\Gamma_S t} + |\eta_{+-}|^2 e^{-\Gamma_L t} - 2|\eta_{+-}|^2 e^{-\frac{\Gamma_L+\Gamma_S}{2}t}\cos(\Delta m t + \phi_{+-})\right]$$

numerator: subtract the above two terms

where ϕ_{+-} and η_{+-} can be extracted taking Δm from other experiments

$\rightarrow\ \phi_{+-} = 43.7° \pm 0.6°,\quad \phi_{00} = 43.5° \pm 1.0°$ …… (3)

Then, why does the described CP violation occurs?

$\Rightarrow$ There two possible CP violation sources in the K^0 decay.

① indirect CP violation(admixture of ϵ): weakly decaying eigenstates of definite lifetime, K_S and K_L, are each an admixture of the wrong CP, to a degree ϵ, via the box diagram of mixing.

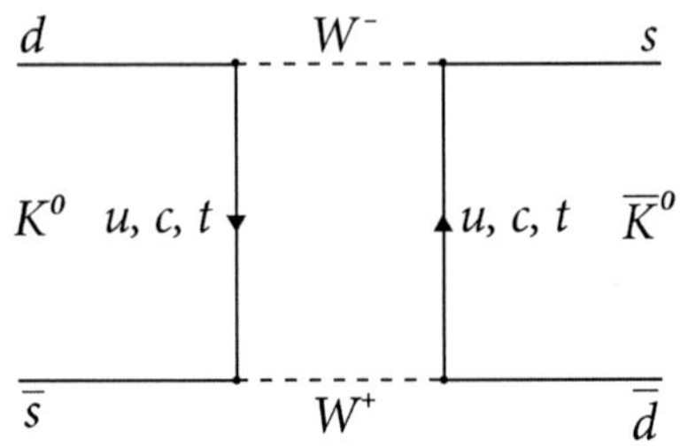

<mass eigenstate themselves are mixed state of even and odd *CP*>

② direct CP violation: violation ϵ' occurs in the decay process itself

※[FLAL: kTeV, CERN: NA48, LNFDaΦ_{ne}(ϕ Factory)]

Weak hadronic decays generally obey $\Delta I = \frac{1}{2}$ rule.

$$\frac{\Gamma_{(\Lambda \rightarrow n\pi^0)}}{\Gamma_{(\Lambda \rightarrow p\pi^-)}} = \frac{1}{2},$$ where $\Lambda\,(I = 0)$ hyperon

└ by Clebsch-Gordon coefficient

But $\Delta I = \frac{3}{2}$ transitions, although strongly suppressed, also occur.

$K \rightarrow \pi^+\pi^-$ or $\pi^0\pi^0$, pions can have $I = 0$ or less probably $I = 2$

(※ $I = 1$ is forbidden by Bose symmetry)

$\rightarrow$ corresponding decay amplitudes are A_0 or A_2

→ corresponding with phase factors $A_0 e^{i\delta_0}$ or $A_2 e^{i\delta_2}$ which arise from the final-state strong pion-pion interaction

→ $\delta_0 - \delta_2$ change sign under time reversal

→ therefore, equivalent to CP violation

Taking account of the Clebsch-Gordon coefficients in the coupling of $\pi^0\pi^0$ and $\pi^+\pi^-$ to $I=0$ or $I=2$ states, it is found that, taking A_0 as real,

$$\rightarrow \begin{cases} \eta_{+-} = |\eta_{+-}| e^{i\phi_{+-}} \simeq \epsilon + \epsilon' \\ \eta_{00} = |\eta_{00}| e^{i\phi_{+-}} \simeq \epsilon - 2\epsilon', \end{cases}$$

$$\text{where } \epsilon' = \frac{i}{\sqrt{2}}\left(\frac{Im A_2}{A_0}\right) e^{i(\delta_2 - \delta_0)}$$

If $\Delta I = \frac{1}{2}$ rule held strictly, we expect $A_2 = 0$

→ $|\eta_{+-}| = |\eta_{00}|$ as shown in Eqs. (1) and (2) in the previous pages,

$\phi_{+-} = \phi_{00}$ as shown in Eq. (3) in the previous page,

$\epsilon' = 0$

→ The experimental results are in good agreement with this prediction within two-standard-deviation level.

A total of four experiments to data (1999) find

$\frac{\epsilon'}{\epsilon} = (2.2 \pm 0.4) \times 10^{-3}$: direct CP violation appears to have been established.

→ This depends on the quark masses and for the top quark mass is extremely small.

→ This is a puzzle for the SM because the interference(ϵ') should be small with $m_t = 175 \geq V$.

In SM, direct CP violation is due to interference between the “tree” diagram and “penguin” diagram.

→ interference between these two diagrams give rise to non-trivial phase factor and direct CP violation

<$W^{\pm}$ exchange>

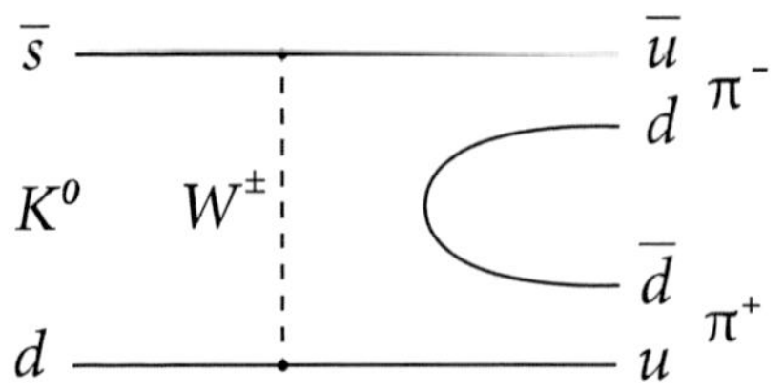

<penguin diagram>

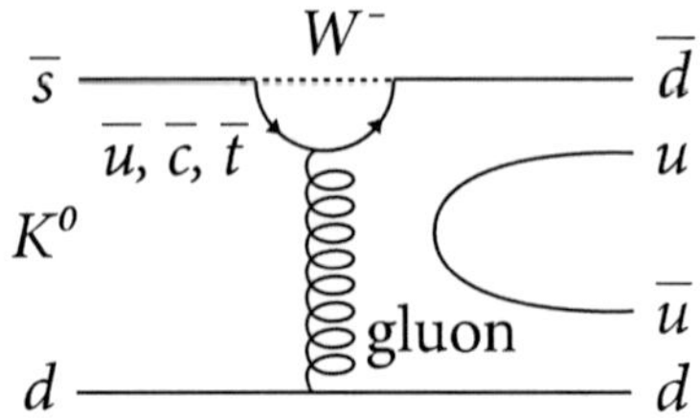

CP non-invariance is also demonstrated in the leptonic decay modes of K_L;

$$\begin{cases} K_L \rightarrow e^+ + \nu_e + \pi^- \\ K_L \rightarrow e^- + \bar{\nu}_e + \pi^+ \end{cases}$$ if CP invariance is violated, a small charge asymmetry expected.

- There decays transform into one another under CP operation, so CP invariance demands they be the same.

$$\rightarrow \Delta = \frac{\text{rate}(K_L \rightarrow e^+ + \nu_e + \pi^-) - \text{rate}(K_L \rightarrow e^- + \bar{\nu}_e + \pi^+)}{\text{rate}(K_L \rightarrow e^+ + \nu_e + \pi^-) + \text{rate}(K_L \rightarrow e^- + \bar{\nu}_e + \pi^+)}$$

$$= (0.327 \pm 0.012) \times 10^{-2} \quad \cdots\cdots (1)$$

- Since e^+ must come from K^0 and e^- from $\overline{K}^0$, we expect

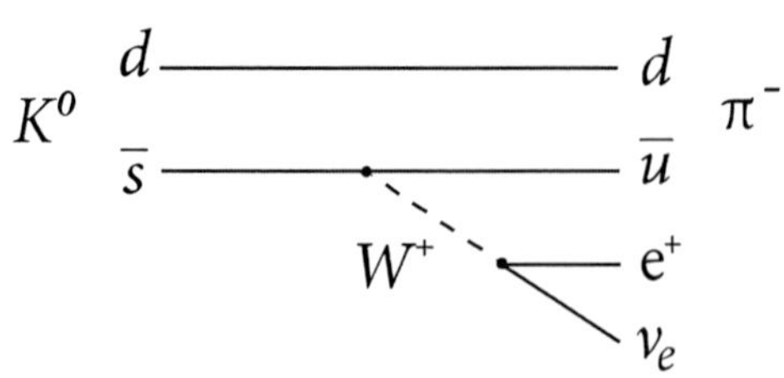

$$\Delta = \frac{|1+\epsilon|^2 - |1-\epsilon|^2}{|1+\epsilon|^2 + |1-\epsilon|^2} \simeq 2R_e\epsilon \quad \cdots\cdots (2)$$

if $\epsilon' = 0$, $\epsilon = \eta_{+-} = \eta_{00} \equiv \eta$ and $\phi_{+-} = \phi_{00} \equiv \phi$

$$\Delta(\epsilon' = 0) = 2\eta\cos\phi = (0.332 \pm 0.004) \times 10^{-2} \cdots\cdots (3)$$

→ From Eqs. (1) and (2), we obtain ϵ

→ Eq. (1) agrees with Eq. (3).

■ **$D^0 - \overline{D}^0$ and $B^0 - \overline{B}^0$ mixing:** mixing which is seen for kaons, can also occur in charm and bottom mesons

<(i) box diagrams for $D^0 - \overline{D}^0$ mixing>

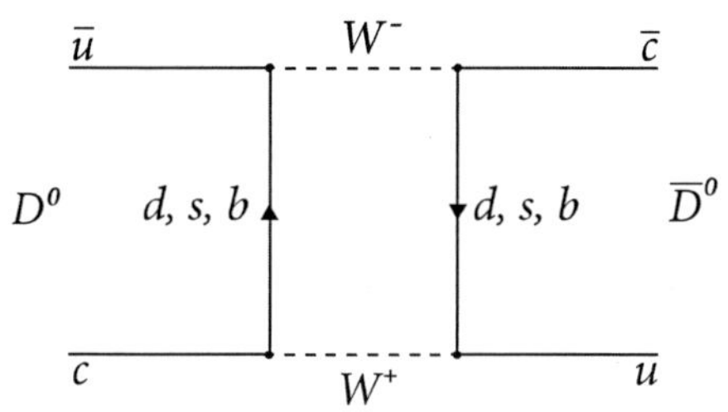

<(ii) box diagrams for $B^0 - \overline{B}^0$ mixing>

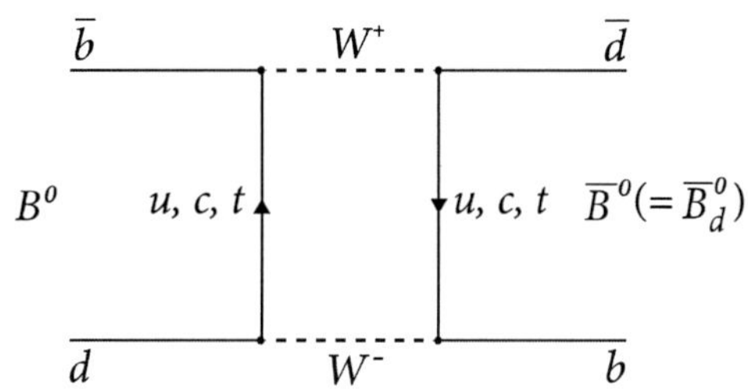

(i) box diagrams for $D^0 - \overline{D}^0$ mixing

→ contributions to rate vary as square of quark mass and as production of the vertex coupling

→ dominant contribution is b quark and s quark diagram is suppressed by small value of s quark mass

→ but product $V_{cb}V_{ub}$ is very small: $|V_{cb}V_{ub}|^2 \approx |0.04 \times 0.005|^2 \approx 4 \times 10^{-8}$

→ predicted $D^0 - \overline{D}^0$ mixing is minute

(ii) box diagrams for $B^0 - \overline{B}^0$ mixing

→ dominant contribution is t quark: $|V_{tb}V_{td}|^2 \approx |0.999 \times 0.01|^2 \approx 10^{-4}$

→ expect substantial mixing level

$$\begin{cases} I(K^0) = \frac{1}{4}\left[e^{-\Gamma_S t} + e^{-\Gamma_L t} + 2e^{-\frac{\Gamma_S+\Gamma_L}{2}}\cos\Delta m t\right] \\ I(\overline{K}^0) = \frac{1}{4}\left[e^{-\Gamma_S t} + e^{-\Gamma_L t} - 2e^{-\frac{\Gamma_S+\Gamma_L}{2}}\cos\Delta m t\right] \end{cases}$$

$\Rightarrow$ probability that an initially pure K^0 state decays as K^0 or $\overline{K}^0$

$$P(t)dt = \frac{\Gamma e^{-\Gamma t}}{2}\left(\frac{e^{-y\Gamma t}}{2} + \frac{e^{+y\Gamma t}}{2} \pm \cos x\Gamma t\right)dt \quad \cdots\cdots (1),$$

where ① $\Gamma = \frac{\Gamma_S + \Gamma_L}{2} \approx \frac{\Gamma_S}{2} \quad (\because \ \tau_s \ll \tau_L)$

② $x = \frac{\Delta m}{\Gamma}$ and $y = \frac{\Delta\Gamma}{2\Gamma} \quad (\Delta\Gamma = \Gamma_S - \Gamma_L \approx \Gamma_S)$

③ $\Delta m = m_L - m_S$

for K^0 system, $y \simeq 1$ and $x \simeq 0.95$

$$[\because \ x = \frac{\Delta m}{\Gamma} = \frac{2\Delta m}{\Gamma_S} = \frac{2\times 3.5\times 10^{-12}\ \text{MeV}}{\Gamma_S}$$

$$= \frac{2\times 3.5\times 10^{-12}}{\hbar}\times 0.9\times 10^{-10}\ (\text{MeV sec})$$

$$= \frac{2\times 3.5\times 10^{-12}\times 0.9\times 10^{-10}\ (\text{MeV sec})}{6.58\times 10^{-22}\ (\text{MeV sec})}\ ]$$

For B^0 system, $\Delta\Gamma \ll \Delta m$ ($\because$ $\Delta\Gamma$ is produced only decay channels common to both B^0 and $\overline{B}^0$, and owing to large B^0 meson mass)

ex $B^0 \rightarrow J/\Psi K_S^0 < 10^{-3}$ or $B^0 \rightarrow \pi^+\pi^- < 10^{-3}$

$\Rightarrow y \approx 0$

then, time-integrated probability that an initial B^0 ends up as $\overline{B}^0$ is

$$\chi = \frac{x_B^2}{2(1+x_B^2)}$$

[apply $\int_0^{\infty} dt$ to Eq. (1), for this, change $\cos$ term to exponential form]

$$\chi_{mea} = 0.17 \Rightarrow x_B \left(= \frac{\Delta m_B}{\Gamma_B} \right) \simeq 0.72$$

There is essentially no indirect CP violation ⇒ quantity ϵ is then simply $e^{i\alpha}$.
In the B^0 system, 'direct' CP violation will dominate and could be appreciable.
(the problem is that very low branching ratio for B^0 and $\overline{B}^0$ decay modes)

↳ this is the motivation for "B factories"

- **CKM is unitary matrix;** $V^+ = V^{-1} \Rightarrow V^+ V = 1$

  unitarity condition, $V_{ud}V_{ub}^* + V_{cd}V_{cb}^* + V_{td}V_{tb}^* = 0$ (see "Supplement-8")

  ※ off-diagonal terms will vanish

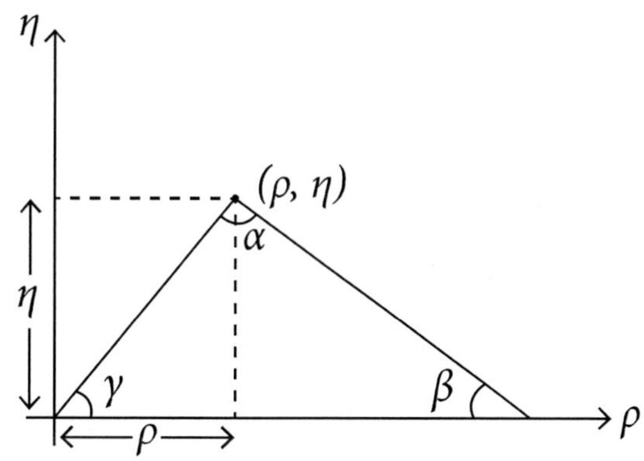

→ which corresponds to a (closed) triangle in $\rho\eta$-plane (unitarity triangle)

For $B_d^0(\overline{B}_d^0) \to J/\Psi K_S^0$,

decay rate are $R \propto e^{-\frac{t}{\tau}}(1 \pm \sin 2\beta \sin \Delta m t)$,

where $\sin 2\beta \sin \Delta m t$ measures CP violation asymmetry between B^0 and $\overline{B}^0$ decays

↳ in experiment, $\sin 2\beta = 0.741 \pm 0.067 \pm 0.034$ (with 88 million $B\overline{B}$ decay)

Sakharov(1966) showed that three conditions are required for baryon asymmetry of Universe to have developed.

① B-violating interactions

② a non-equilibrium situation

③ CP and C violation

CP violation observed in the K and B system is not sufficient to explain the baryon asymmetry

⇒ additional CP violation must occur

[Supplement-8]

Unitarity requires → any two different columns (formed by vectors of unit length) or any two different rows (ditto) are orthogonal:

$$|V_{ud}|^2+|V_{us}|^2+|V_{ub}|^2=1 \qquad V_{ud}^* V_{cd}+V_{us}^* V_{cs}+V_{ub}^* V_{cb}=0$$

$$|V_{cd}|^2+|V_{cs}|^2+|V_{cb}|^2=1 \qquad V_{ud}^* V_{td}+V_{us}^* V_{ts}+V_{ub}^* V_{tb}=0$$

$$|V_{td}|^2+|V_{ts}|^2+|V_{tb}|^2=1 \qquad V_{cd}^* V_{td}+V_{cs}^* V_{ts}+V_{cb}^* V_{tb}=0$$

$$|V_{ud}|^2+|V_{cd}|^2+|V_{td}|^2=1 \qquad V_{ud} V_{us}^*+V_{cd} V_{cs}^*+V_{td} V_{ts}^*=0$$

$$|V_{us}|^2+|V_{cs}|^2+|V_{ts}|^2=1 \qquad V_{ud} V_{ub}^*+V_{cd} V_{cb}^*+V_{td} V_{tb}^*=0$$

$$|V_{ub}|^2+|V_{cb}|^2+|V_{tb}|^2=1 \qquad V_{us} V_{ub}^*+V_{cs} V_{cb}^*+V_{ts} V_{tb}^*=0$$

vector in complex plane: $V_{cd} = |V_{cd}|e^{i\theta}$, $V_{cb}^* = |V_{cb}|e^{-i\phi}$

phase factor common to row or column can be eliminated.

Three vector define triangle:

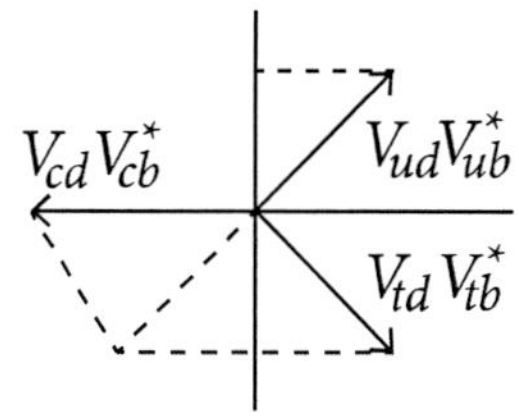

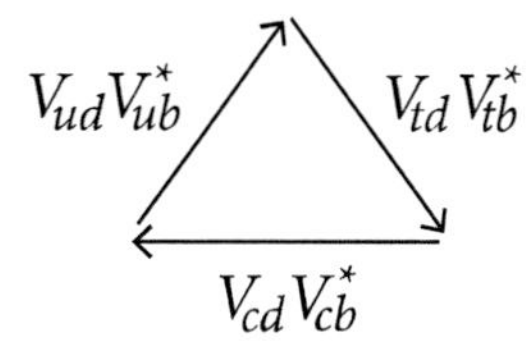

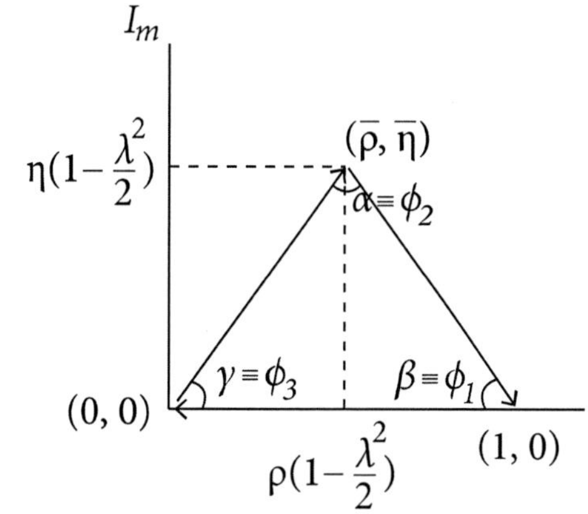

Use Wolfenstein third order:

$$V_{ud}V_{ub}^* = A\lambda^3\left(1-\frac{\lambda^2}{2}\right)(\rho+i\eta)$$

$$V_{cd}V_{cb}^* = -A\lambda^3$$

$$V_{td}V_{tb}^* = A\lambda^3(1-\rho-i\eta)$$

$$\alpha \equiv \arg\left[\frac{-V_{td}V_{tb}^*}{V_{ud}V_{ub}^*}\right], \quad \beta \equiv \arg\left[\frac{-V_{cd}V_{cb}^*}{V_{td}V_{tb}^*}\right], \quad \gamma \equiv \arg\left[\frac{-V_{ud}V_{ub}^*}{V_{cd}V_{cb}^*}\right]$$

Chapter 8

Electroweak Interactions and Standard Model

■ Introduction

To understand the interrelation between the different fields

- at some colossally high energy, four interactions are a single universal interaction
- at energies studied in particle physics, this symmetry is badly broken

ex unification of two different interactions

① electricity + magnetism ⇒ electromagnetic field (by Maxwell in 1865)

② electromagnetic interactions + weak interactions ⇒ electroweak interaction(by Glashow, Weinberg, and Salam in late 1960s)

※ These are different aspects of the same force, electroweak interaction. The symmetry between these two apparently different interactions would be clear at very large momentum transfers ($q \gg 10^4\ \mathrm{GeV}^2$).

※ electroweak interaction: four mediating vector bosons are

$$\begin{cases} \text{massless} \rightarrow \text{photon} \\ \text{massive} \rightarrow W^+,\ W^-,\ Z^0 \end{cases}$$

- massive bosons were observed in 1983 at CERN $p\bar{p}$ collider
- at low energy, the mass difference between photon and $W^{\pm}/Z^0$ break symmetry

- theory contains an arbitrary constant, $\sin^2\theta_w$ (weak mixing angle)

↳ determined by experiment, not given by the theory

• Divergences in the weak interactions

QED contains divergence terms associated with integrals over intermediate states.

→ These divergence can be absorbed into a redefinition of "bare" lepton charges and masses as physically measured values.

→ Theory is renormalizable.

Weak interaction`s theory also involve divergences in higher orders that could be cancelled only at the price of introducing an indefinitely large number of arbitrary constants, thus losing essentially any predictive power.

In Fermi theory of neutrino scattering, for $\nu_e + e^- \rightarrow e^- + \nu_e$

the cross-section has pointlike form

$\frac{d\sigma}{dq^2} = \frac{G_F^2}{\pi}$, where q^2 is the momentum transfer squared

⇒ total cross-section is then

$$\sigma_{tot}(\nu e) = \frac{G_F^2}{\pi} q_{\max}^2 = \frac{G_F^2}{\pi} 2mE = \frac{G_F^2}{\pi} S, \quad (\because \text{ for } E \gg m, \; S \simeq 2mE)$$

※ e^- is assumed at rest in the calculation

where m is the electron mass, E is the neutrino energy and S is the cms energy squared

→ the cross-section increases with energy

$$= \frac{G_F^2}{\pi} 4p^{*2} \; (\because \; p^{*2} = \frac{S}{4}) \; \cdots\cdots (1)$$

- (Another way to look at this)

  In wave theory, the cross-section for such an elastic scattering process is

$$\sigma_{max} = \pi\lambda\!\!\!^{-2}\frac{2\ell+1}{2s+1} = \frac{\pi\lambda\!\!\!^{-2}}{2} \cdots\cdots (2),$$ where $\lambda\!\!\!^{-} = \frac{\hbar}{p^*}$ is the cms wavelength

$\leftarrow \ell = 0$ for pointlike scattering and $s = 1/2$ for target electron

$$= \frac{\pi}{2}\frac{\hbar^2}{p^{*2}}$$

For $\frac{4G_F^2 p^{*2}}{\pi} > \frac{\pi}{2p^{*2}}$, $\sigma_{tot} > \sigma_{max} \Rightarrow p^* > \left(\frac{\pi}{\sqrt{8}\,G_F}\right)^{1/2} \simeq 300\ \mathrm{GeV}/c$

→ at sufficiently large energy, Fermi theory predicts the cross-section exceeding wave-theory limit

↳ unitarity limit at high energy

※ scattered intensity can`t exceed incident intensity in any partial wave

⇒ frequently this is called "unitarity limit"

→ This "crisis" is resolved since the propagator replaces G_F^2 with $\frac{G_F^2}{1+q^2/M_W^2}$

※ intermediate vector boson $W^{\pm}$ → introduce propagator $\frac{1}{q^2+M_W^2}$ into scattering amplitude

→ When $q^2 \gg M_W^2$, $\sigma_{tot}(\nu_e) \to \frac{G_F^2 M_W^2}{\pi}$: constant value at high energy

→ The divergences are fixed by the finite mass of the W for $\nu_e + e^- \to e^- + \nu_e$.

→ quadratic divergences still appear in other processes

→ need a mechanism to cancel the weak interaction divergences systematically and to all order in G_F

→ Introduction of Neutral Current

- divergences are fixed by finite mass of W for $\nu_e + e^- \to e^- + \nu_e$ but divergences persist for other processes.

  Consider $\nu\bar{\nu} \to W^+W^-$,

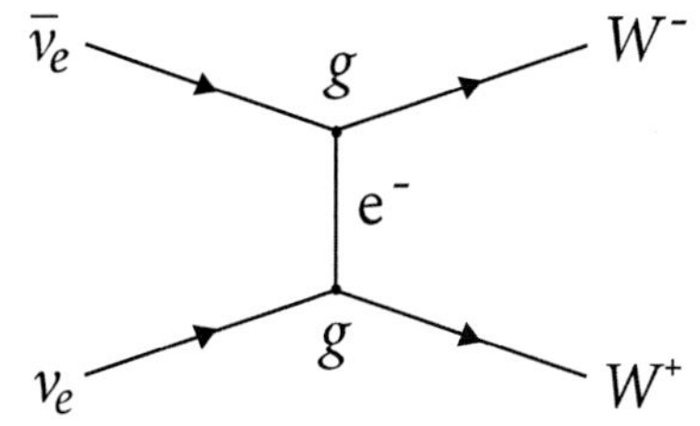

Since this diagram alone will diverge, we still need additional physics to control this cross section,

⇒

There are two possibilities:

① neutral boson

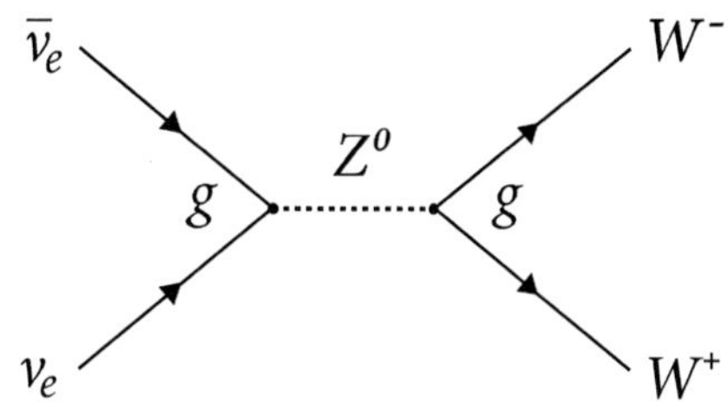

② new heavy lepton

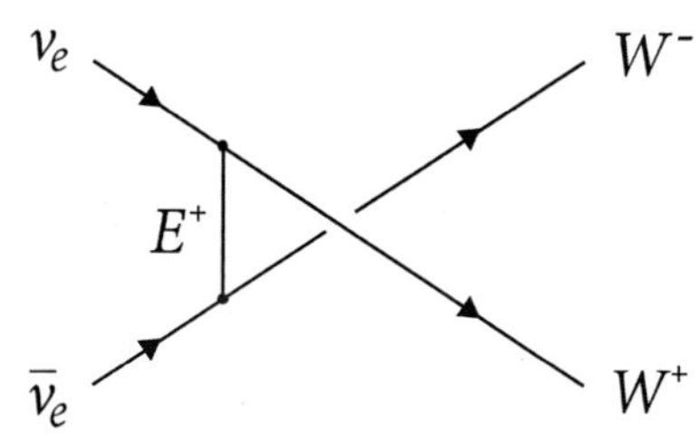

- there is no evidence for this solution.

→ we don`t consider this possibility further.

■ Introduction of neutral currents

For $e^+e^- \rightarrow W^+W^-$

① another divergent process is

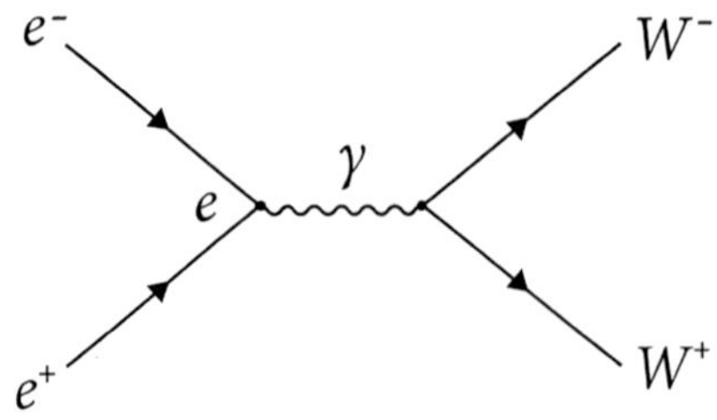

② Again, this divergent can be "saved" by introducing heavy neutral currents, which is exactly what Nature has chosen to do.

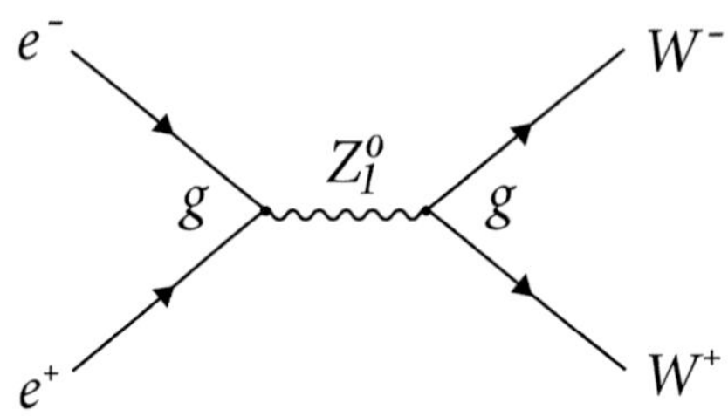

Therefore, EM process $e^+e^- \to W^+W^-$ is divergent and cancellation in this order is effected by introducing a neutral vector boson Z_1^0

※ The observation of neutral-current process in neutrino-electron and eutrino-nucleon scattering in 1973 gave the first experimental support for he electroweak theory and the first estimates for value of $\sin^2\theta_w$.

$$G_F = \frac{g^2}{M_W^2} \Rightarrow M_W = \frac{g}{\sqrt{G_F}}$$

$$\simeq \frac{e}{\sqrt{G_F}} \sim 90 \sim 100 \text{ GeV}$$

with $G_F \simeq 10^{-5}\ \text{GeV}^{-2}$, $e^2 = \dfrac{4\pi}{137}\left(\Leftarrow \alpha = \dfrac{e^2}{4\pi}\right)$

($\because$ Z^0 weak coupling g have to be similar in magnitude to the EM coupling e if two fields are unified.)

→ This is approximate mass if they are to have same coupling as in electromagnetism.

By introducing Z^0, the divergence in $e^+e^- \rightarrow W^+W^-$ disappear in which electron mass can be neglected.

For finite electron mass, a residual divergence exists and has to be cancelled by introduction of further scalar particles, Higgs scalars, with special couplings proportional to the lepton mass.

■ Weinberg-Salam Model

Weinberg-Salam and Glashow proposed the electroweak theory, $SU(2) \times U(1)$, in the late 1960s,

※ $SU(n)$: group of $n \times n$ unitary matrices with determinant 1

→ postulate four massless mediating bosons, arranged in a triplet and a singlet as members of multiples of 'weak isospin' I and 'weak hypercharge' Y.

→ (i) three of bosons, $\overrightarrow{W}_\mu = W_\mu^{(1)},\ W_\mu^{(2)},\ W_\mu^{(3)}$ are component of $I = 1$ triplet of group $SU(2)$. ※ In $W_\mu^{(i)}$, μ is the space-time components of 4-vector and (i) is the components in isospin space.

(ii) fourth of bosons, B_μ is the isoscalar($I = 0$) belonging to $U(1)$ group of weak hypercharge.

A process called "spontaneous symmetry breaking" endows the bosons with mass without spoiling the renormalizability of the theory.

⇒ This is achieved by $I = \frac{1}{2}$ doublet of complex fields called Higgs scalars (with four real components in total), which generate mass as a result of their self-interaction.

→ Higgs mechanism generates mass for three bosons leaving one (the photon) massless.

massive bosons are W_μ^+, W_μ^-, Z_μ^0
massless bosons is A_μ (photon),

where A_μ and Z_μ^0 are combinations of $W_\mu^{(3)}$ and B_μ and $W_\mu^{(3)}$ and B_μ are orthogonal

We can write the interaction energy of leptons and quarks with these mediating bosons in relativistically invariant form.

→ Lagrangian energy density;

$$\mathcal{L} = g\vec{J}_\mu \cdot \overrightarrow{W}_\mu + g' J_\mu^Y \cdot B_\mu,$$

where $\overrightarrow{W}_\mu$ and B_μ are 4-potentials associated with boson fields,

[※ W and B are boson fields, which transform into the physical boson fields, $W_\mu^\pm$, Z_μ and A_μ (photon)]

g and g' are the coupling constant,

$\vec{J}_\mu$ is the weak isospin currents of fermions,

J_μ^Y is the weak hypercharge current of fermions.

※ c.f. $e\vec{J}_\mu^{em} A_\mu$ is the interaction energy of electric current with EM field

Let`s define weak hypercharge, $Y = Q - I_3$

[※ c.f. for strong hypercharge: $Y = 2(Q - I_3)$]

then, $J_\mu^Y = J_\mu^{em} - J_\mu^{(3)}$,

where J_μ^{em} is the EM current coupling to Q,

$J_\mu^{(3)}$ is the third component of isospin current $\vec{J}_\mu$

definitions

$$W_\mu^\pm = \frac{1}{\sqrt{2}}\left[W_\mu^{(1)} \pm i\, W_\mu^{(2)}\right] \text{ and } W_\mu^{(3)} = \frac{gZ_\mu + g' A_\mu}{\sqrt{g^2 + g'^2}}$$

$$J_\mu^\pm = J_\mu^{(1)} \pm iJ_\mu^{(2)} \quad \text{and } B_\mu = \frac{-g' Z_\mu + gA_\mu}{\sqrt{g^2 + g'^2}}$$

produce the expected Lagrangian.

Hence,

$$\mathcal{L} = g[J_\mu^{(1)} W_\mu^{(1)} + J_\mu^{(2)} W_\mu^{(2)} + J_\mu^{(3)} W_\mu^{(3)}] + g'[J_\mu^{em} - J_\mu^{(3)}]B_\mu$$

$$= \frac{g}{\sqrt{2}}[J_\mu^- W_\mu^+ + J_\mu^+ W_\mu^-] + J_\mu^{(3)}[g W_\mu^{(3)} - g' B_\mu] + J_\mu^{em} g' B_\mu$$

$$※ \begin{cases} J_\mu^- W_\mu^+ = \frac{1}{\sqrt{2}}[J_\mu^{(1)} W_\mu^{(1)} + iJ_\mu^{(1)} W_\mu^{(2)} - iJ_\mu^{(2)} W_\mu^{(1)} + J_\mu^{(2)} W_\mu^{(2)}] \\ J_\mu^+ W_\mu^- = \frac{1}{\sqrt{2}}[J_\mu^{(1)} W_\mu^{(1)} - iJ_\mu^{(1)} W_\mu^{(2)} + iJ_\mu^{(2)} W_\mu^{(1)} + J_\mu^{(2)} W_\mu^{(2)}] \end{cases}$$

$$= \frac{g}{\sqrt{2}}[J_\mu^- W_\mu^+ + J_\mu^+ W_\mu^-] + \frac{g}{\cos\theta_w}[J_\mu^{(3)} - \sin^2\theta_w J_\mu^{em}]Z_\mu + g\sin\theta_w J_\mu^{em} A_\mu \quad \cdots\cdots (1),$$

↳ weak C.C ↳ weak N.C ↳ em N.C

where $e = g\sin\theta_w$, θ_w is the weak mixing (or Weinberg) angle

→ interaction energy contains the weak charge-changing currents, a weak neutral current, and the electromagnetic neutral current.

※ In Eq. (1), $J_\mu^{(3)}[g W_\mu^{(3)} - g' B_\mu] + J_\mu^{em} g' B_\mu$

(by inserting $W_\mu^{(3)} = \dfrac{gZ_\mu + g'A_\mu}{\sqrt{g^2 + g'^2}}$, $B_\mu = \dfrac{-g'Z_\mu + gA_\mu}{\sqrt{g^2 + g'^2}}$ and $g'/g \equiv \tan\theta_w$)

$$= J_\mu^{(3)}\left[g\left(\frac{gZ_\mu + g'A_\mu}{\sqrt{g^2 + g'^2}}\right) - g'\left(\frac{-g'Z_\mu + gA_\mu}{\sqrt{g^2 + g'^2}}\right)\right] + J_\mu^{em} g'\left(\frac{-g'Z_\mu + gA_\mu}{\sqrt{g^2 + g'^2}}\right)$$

$$= J_\mu^{(3)}\left[\frac{g(Z_\mu + \tan\theta_w A_\mu)}{\sqrt{1 + \tan^2\theta_w}} - \tan\theta_w \frac{g(A_\mu - \tan\theta_w Z_\mu)}{\sqrt{1 + \tan^2\theta_w}}\right]$$

$$+ J_\mu^{em} \tan\theta_w \frac{g(A_\mu - \tan\theta_w Z_\mu)}{\sqrt{1 + \tan^2\theta_w}}$$

$$= J_\mu^{(3)}\left[g\cos\theta_w Z_\mu + g\sin\theta_w A_\mu - g\sin\theta_w A_\mu + \frac{\sin^2\theta_w}{\cos\theta_w} g Z_\mu\right]$$

$$+ J_\mu^{em}\left[g\sin\theta_w A_\mu - \frac{\sin^2\theta_w}{\cos\theta_w} g Z_\mu\right]$$

$$= Z_\mu\left[J_\mu^{(3)} g\cos\theta_w + g\frac{\sin^2\theta_w}{\cos\theta_w}J_\mu^{(3)} - g\frac{\sin^2\theta_w}{\cos\theta_w}J_\mu^{em}\right] + g\sin\theta_w J_\mu^{em} A_\mu$$

$$= Z_\mu\left[\frac{g}{\cos\theta_w}J_\mu^{(3)} - g\frac{\sin^2\theta_w}{\cos\theta_w}J_\mu^{em}\right] + g\sin\theta_w J_\mu^{em} A_\mu$$

$$= \frac{g}{\cos\theta_w}\left[J_\mu^{(3)} - \sin^2\theta_w J_\mu^{em}\right] Z_\mu + g\sin\theta_w J_\mu^{em} A_\mu$$

■ Intermediate boson masses

Coupling g of a lepton to $W^\pm$ boson is related to Fermi constant G_F:

$$G_F = \frac{g^2}{M_W^2} \xrightarrow[(2)]{(1)} \frac{G_F}{\sqrt{2}} \equiv \frac{g^2}{8M_W^2};$$ Fermi constant is related to g and M_W

(1) G_F by $\frac{G_F}{\sqrt{2}}$: G_F was defined for Fermi (that is pure vector) transitions.

Now, we have in general V and A amplitudes.

(2) at time of $V-A$ theory, helicity operator was defined as $(1-\gamma_5)$ instead of $\frac{1}{2}(1-\gamma_5)$

→ applied at both vertices, this gives an overall factor of 4 by which g^2 should be divided.

$$\Rightarrow M_{W^\pm} = \left(\frac{g^2\sqrt{2}}{8G_F}\right)^{1/2} = \left(\frac{e^2\sqrt{2}}{8G_F\sin^2\theta_w}\right)^{1/2} \quad (\because\ e = g\sin\theta_w)$$

(by inserting $e^2 = \frac{4\pi}{137}$, $G_F \simeq 10^{-5}\ \mathrm{GeV}^{-2}$)

$$= \frac{37.4}{\sin\theta_w}\ [\mathrm{GeV}]$$ (※ Note: this is tree level expression)

⇒ Since we expected $M_W \sim 100\ \mathrm{GeV}$, this implies $\sin\theta_w \sim 0.4$ or $\sin^2\theta_w \sim 0.2$

※ $M_w \simeq \dfrac{g}{\sqrt{G}} \simeq \dfrac{e}{\sqrt{G}} \sim 100\ \mathrm{GeV}$ (rough guess)

• Let`s find Z^0 mass

※ empirical fact: ① photon is massless

② photon is orthogonal to Z particle

$$\begin{cases} W_\mu^{(3)} = \dfrac{Z_\mu + \tan\theta_w A_\mu}{\sqrt{1+\tan^2\theta_w}} = Z_\mu \cos\theta_w + A_\mu \sin\theta_w \\ B_\mu = \dfrac{\tan\theta_w Z_\mu + A_\mu}{\sqrt{1+\tan^2\theta_w}} = -Z_\mu \sin\theta_w + A_\mu \cos\theta_w \end{cases}$$

By inverting $W_\mu^{(3)}$ and B_μ Eqs.,

$$\Rightarrow \begin{cases} Z_\mu = W_\mu^{(3)} \cos\theta_w - B_\mu \sin\theta_w \\ A_\mu = W_\mu^{(3)} \sin\theta_w + B_\mu \cos\theta_w \end{cases}$$

$$\begin{pmatrix} Z_\mu^0 \\ A_\mu \end{pmatrix} = \frac{1}{\sqrt{g^2+g'^2}} \begin{pmatrix} g & -g' \\ g' & g \end{pmatrix} \begin{pmatrix} W_\mu^{(3)} \\ B_\mu \end{pmatrix} = \begin{pmatrix} \cos\theta_w & -\sin\theta_w \\ \sin\theta_w & \cos\theta_w \end{pmatrix} \begin{pmatrix} W_\mu^{(3)} \\ B_\mu \end{pmatrix}$$

$$\Rightarrow \begin{cases} M_Z^2 = M_W^2 \cos^2\theta_w + M_B^2 \sin^2\theta_w - 2M_{BW}^2 \cos\theta_w \sin\theta_w \quad \cdots\cdots (1) \\ M_\gamma^2 = 0 = M_W^2 \sin^2\theta_w + M_B^2 \cos^2\theta_w + 2M_{BW}^2 \cos\theta_w \sin\theta_w \quad \cdots\cdots (2) \\ M_{Z\gamma}^2 = 0 = (M_W^2 - M_B^2)\sin\theta_w \cos\theta_w + M_{BW}^2(\cos^2\theta_w - \sin^2\theta_w) \quad \cdots\cdots (3) \end{cases},$$

where M_{BW} and $M_{Z\gamma}$ are the off-diagonal mass terms

($\because$ photon is orthogonal to Z particle $\rightarrow$ $M_{Z\gamma}^2 = 0$)

From Eqs. (2) and (3), $M_B^2 = \dfrac{\sin^2\theta_w}{\cos^2\theta_w} M_W^2$

From Eqs. (1) and (2), $M_Z^2 = M_W^2 + M_B^2 = \dfrac{1}{\cos^2\theta_w} M_W^2$

$$\Rightarrow M_{Z^0} = \frac{M_{W^\pm}}{\cos\theta_w} = \frac{75}{\sin 2\theta_w}\ \mathrm{GeV} \ \left(\because\ M_{W^\pm} = \frac{37.4}{\sin\theta_w}\ \mathrm{GeV}\right)$$

※ Note: this is tree level expression

→ For $\sin^2\theta_w \sim 0.2$, Z mass is about 10% larger than W mass.

① These predictions of boson masses are for the simplest model of spontaneous symmetry breaking, which involves one weak isospin doublet of scalar (Higgs) particles.

→ Many other models are possible

↓

② We can limit the possible alternative models since a factor ρ can be different from 1 in other models,

$M_{Z^0}^2 = \dfrac{M_W^2}{\rho \cos^2\theta_w}$; Data to date indicates $\rho = 1$, consistent with the simplest model, as well as some others (i.e. SUSY with two doublets)

⇒ Many other models are possible

■ Electroweak Couplings of leptons and quarks

Let`s assign the lepton states

$$\Psi_L = \begin{pmatrix} \nu_e \\ e^- \end{pmatrix}_L, \quad \text{LH doublet} \begin{cases} I = \frac{1}{2},\ I_3 = +\frac{1}{2},\ Q = 0 \\ I = \frac{1}{2},\ I_3 = -\frac{1}{2},\ Q = -1 \end{cases} \Rightarrow Y = -\frac{1}{2}$$

$$\Psi_R = (e^-)_R \quad \text{RH singlet} \quad I = 0,\ Q = -1,\ Y = -1$$

(∵ SM assigns leptons to LH doublet and RH singlet in weak isospin.)

※ Quarks are also assigned to left-handed doublets and right-handed singlet in weak isospin:

$$\Psi_L = \begin{pmatrix} u \\ d \end{pmatrix}_L \quad \begin{cases} I = \frac{1}{2},\ I_3 = +\frac{1}{2},\ Q = \frac{2}{3} \\ I = \frac{1}{2},\ I_3 = -\frac{1}{2},\ Q = -\frac{1}{3} \end{cases} \Rightarrow Y = \frac{1}{6}$$

$$\Psi_R = (u)_R \qquad I = 0,\ Q = \frac{2}{3},\ Y = \frac{2}{3}$$

$$\Psi_R = (d)_R \qquad I = 0,\ Q = -\frac{1}{3},\ Y = -\frac{1}{3}$$

The coupling to the Z is $\left(Z_\mu^{(3)} - \sin^2\theta_w J_\mu^{em}\right)$:

- left-handed coupling of fermions (antifermions) to Z^0 is

$$g_L = I_3 - Q\sin^2\theta_w \quad (g_R)$$

- right-handed coupling of fermions (antifermions) to Z^0 is

$$g_R = -Q\sin^2\theta_w \quad (g_L)$$

Since vector interactions coupling LH and RH states with the same sign and axial vector interactions coupling LH and RH states with the opposite sign, this means that the vector and axial vector coupling to the Z are

V coefficient: $C_V = g_L + g_R = I_3 - 2Q\sin^2\theta_w$

A coefficient: $C_A = g_L - g_R = I_3$

(※ for antifermion, $C_V \to C_V$, $C_A \to -C_A$)

then,

(i) ν_ℓ: $2C_V = 2I_3 - 4Q\sin^2\theta_w = 2 \cdot \frac{1}{2} = 1$, where ν_ℓ, $(\ell = e, \mu, \tau)$

$$2C_A = 2 \cdot \frac{1}{2} = 1$$

(ii) ℓ: $2C_V = 2\left(-\frac{1}{2}\right) + 4\sin^2\theta_w = -1 + 4\sin^2\theta_w$

$$2C_A = 2 \cdot \left(-\frac{1}{2}\right) = -1$$

(iii) u-type quarks: $2C_V = 2\left(\frac{1}{2}\right) - 4\left(\frac{2}{3}\right)\sin^2\theta_w = 1 - \left(\frac{8}{3}\right)\sin^2\theta_w$

$$2C_A = 2 \cdot \left(\frac{1}{2}\right) = 1$$

(iv) d-type quarks: $2C_V = 2\left(-\frac{1}{2}\right) - 4\left(-\frac{1}{3}\right)\sin^2\theta_w = -1 + \left(\frac{4}{3}\right)\sin^2\theta_w$

$$2C_A = 2 \cdot \left(-\frac{1}{2}\right) = -1$$

Summarizing,

(I)

fermion	$2C_V$	for $\sin^2\theta_w \sim 0.23$	$2C_A$
ν_ℓ	1	1	1
ℓ	$-1+4\sin^2\theta_w$	-0.08	-1
u-type	$1-\frac{8}{3}\sin^2\theta_w$	0.39	1
d-type	$-1+\frac{4}{3}\sin^2\theta_w$	-0.69	-1

(II) $L = \frac{g}{\sqrt{2}}[J_\mu^- W_\mu^+ + J_\mu^+ W_\mu^-] + \frac{g}{\cos\theta_w}[J_\mu^{(3)} - \sin^2\theta_w J_\mu^{em}]Z_\mu + g\sin\theta_w J_\mu^{em} A_\mu$

↳ weak C.C ↳ weak N.C ↳ em N.C

	photon		$W^+,\ W^-$		Z^0	
fermion	g_L^{em}	g_R^{em}	L	R	g_L	g_R
$\nu_e,\ \nu_\mu,\ \nu_\tau$	0	0	1	0	0.5	0
$e,\ \mu,\ \tau$	-1	-1	1	0	-0.27	0.23
$u,\ c,\ t$	0.67	0.67	1	0	0.35	-0.15
$d,\ s,\ b$	-0.33	-0.33	1	0	-0.42	0.08
	$\times e = g\sin\theta_w$		$\times g/\sqrt{2} = e/\sqrt{2}\sin\theta_w$		$\times g/\cos\theta_w = e/\cos\theta_w\sin\theta_w$	
$\sin^2\theta_w \sim 0.23,\quad \sin\theta_w \sim 0.48,\quad \cos\theta_w \sim 0.88$						

where $g_L = I_3 - Q\sin^2\theta_w,\ \ g_R = -Q\sin^2\theta_w,\quad \sin^2\theta_w \approx 0.23$

■ Neutrino scattering via Z exchange

- value of $\sin^2\theta_w$ can be measured in neutrino scattering.

CC scattering (via $W^{\pm}$ exchange)	NC cross-section
$(LL \to LL)$ $\frac{d\sigma}{dy}(\nu_e e \to \nu_e e)_{C.C.} = \frac{G_F^2 S}{\pi}$	$\frac{d\sigma}{dy}(\nu_e e \to \nu_e e)_{N.C.} = \frac{G_F^2 S}{\pi}[g_L^2 + g_R^2(1-y)^2]$ …… (1)
$(RL \to RL)$ $\frac{d\sigma}{dy}(\bar{\nu}_e e \to \bar{\nu}_e e)_{C.C.} = \frac{G_F^2 S}{\pi}(1-y)^2$	$\frac{d\sigma}{dy}(\bar{\nu}_e e \to \bar{\nu}_e e)_{N.C.} = \frac{G_F^2 S}{\pi}[g_R^2 + g_L^2(1-y)^2]$ …… (2)
where $y = \frac{E_e}{E_\nu}$ is the fractional energy acquired by electron	

(i) For ν_μ and $\bar{\nu}_\mu$ scattering on electron, they can only proceed through Z^0 exchange:

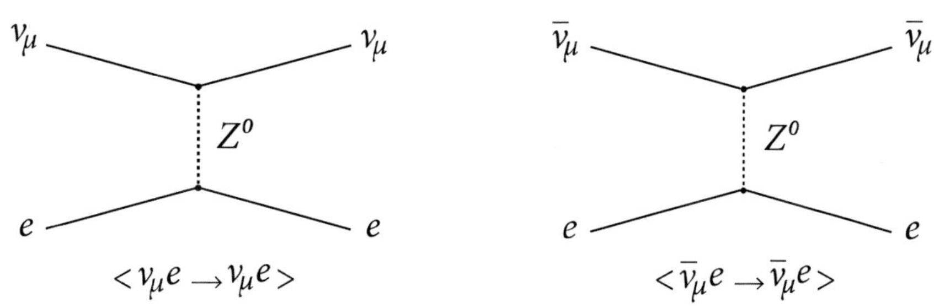

$$g_L = I_3 - Q\sin^2\theta_w = -\frac{1}{2} + \sin^2\theta_w$$

$$g_R = -Q\sin^2\theta_w = \sin^2\theta_w$$

※ scattering from quarks,

	u	d
$g_L = I_3 - Q\sin^2\theta_w$	$\frac{1}{2} - \frac{2}{3}\sin^2\theta_w$	$-\frac{1}{2} + \frac{1}{3}\sin^2\theta_w$
$g_R = -Q\sin^2\theta_w$	$-\frac{2}{3}\sin^2\theta_w$	$\frac{1}{3}\sin^2\theta_w$

(ii) For ν_e scattering on electron, we have to add Z^0 and $W^{\pm}$ exchange amplitudes:

$<\nu_\mu e \rightarrow \nu_\mu e>$

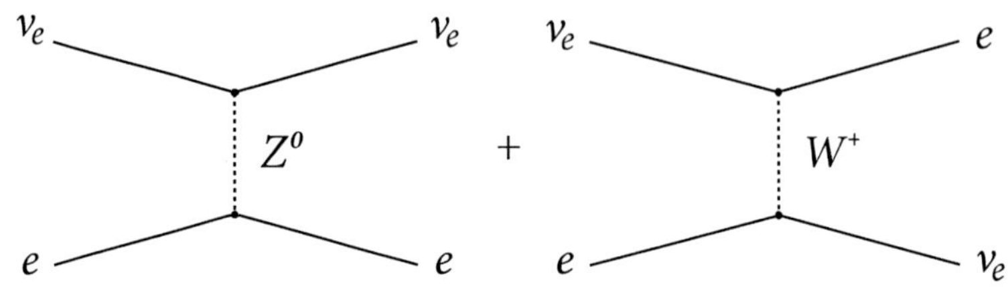

$$g_L = 1 + \left[-\frac{1}{2} + \sin^2\theta_w\right] = 1 - \frac{1}{2} + \sin^2\theta_w, \quad \text{where } g_L = 1 \text{ for } CC$$

$$g_R = 0 + \sin^2\theta_w, \text{ where } g_R = 0 \text{ for } CC$$

(iii) For $\bar{\nu}_e$ scattering on electron, it is obtained by interchanging g_L and g_R in Eq.(1).

$<\bar{\nu}_\mu e \rightarrow \bar{\nu}_\mu e>$

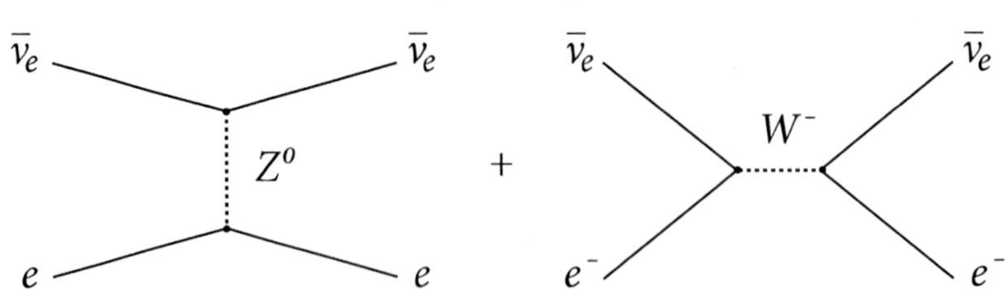

- Separate neutral current(NC) and charged current(CC) events.
  - NuTeV experiment at FNAL

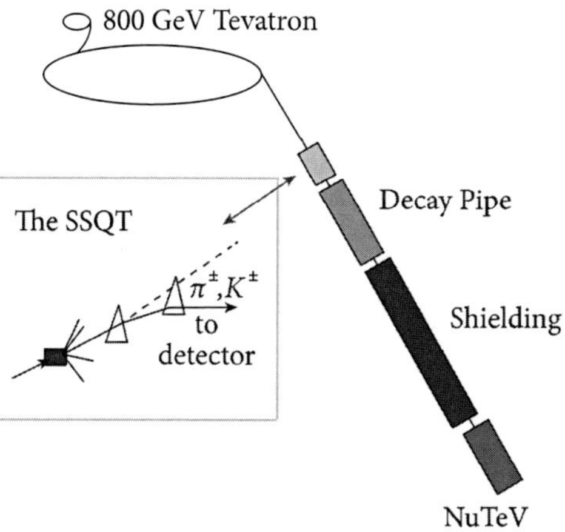

(i) 800 GeV proton produce π, K from interactions in BeO target and decay of π, K produce ν and $\bar{\nu}$

(ii) only νs ($\bar{\nu}$s) penetrate 18 m long shielding

- determine $\nu/\bar{\nu}$ type from sign of charged meson by dipole

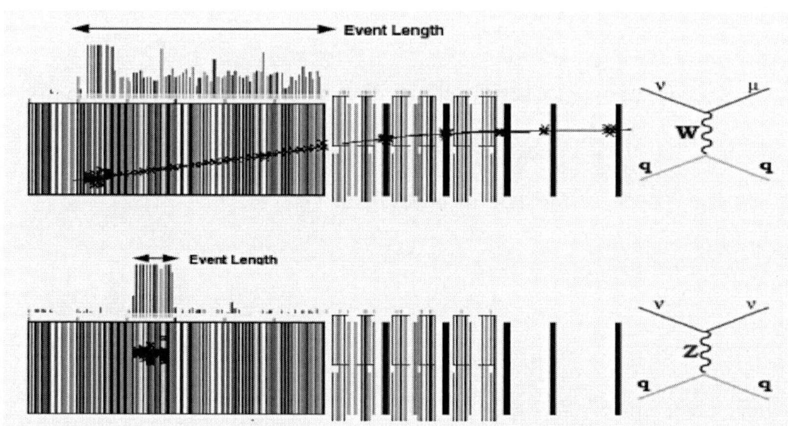

NuTeV(neutrino collaboration), G. Zeller etal, PRL 88, 091802 (02); PR D65, 111103 (02).

18 m long steel plates interspersed with liquid scintillator and drift chambers.
- CC events: track through several plates and large visible energy deposit
- NC events: short visible track and large missing energy

$\Rightarrow$ 1.62×10^6 ν and 351×10^3 $\bar{\nu}$

$$\Rightarrow \begin{cases} R = \dfrac{\sigma^{\nu N}(NC)}{\sigma^{\nu N}(CC)} = \dfrac{1}{2} - \sin^2\theta_w + \dfrac{20}{27}\sin^4\theta_w \\ \overline{R} = \dfrac{\sigma^{\bar{\nu} N}(NC)}{\sigma^{\bar{\nu} N}(CC)} = \dfrac{1}{2} - \sin^2\theta_w + \dfrac{20}{9}\sin^4\theta_w \end{cases}$$

Measurements of the cross-sections for $\nu_\mu e$ and $\overline{\nu}_\mu e$ give values of g_L^2 and g_R^2 in Eqs. (1) and (2).

$\Rightarrow$ g_L and g_R are the functions of $\sin^2\theta_w$

$\Rightarrow$ There is four-fold ambiguity on magnitudes of g_L and g_R

$\Rightarrow$ independent data, $e^+e^- \rightarrow \mu^+\mu^-$ forward-backward asymmetry at LEP, resolve this ambiguity and give a value for $\sin^2\theta_w \rightarrow \sin^2\theta_w = 0.222 \pm 0.006$

$\Rightarrow$ estimate of expected $W^\pm$ and Z^0 masses (several years before their discovery)

Since M_W and M_Z are related through the value of $\sin^2\theta_w$ and M_Z is very precisely known from LEP, one can compare $\sin^2\theta_w$ measurement of neutrino scattering to the direct M_W measurements:

$$M_{Z^0} = \frac{M_{w^\pm}}{\cos\theta_W} \Rightarrow \sin^2\theta_w = 1 - \frac{M_w^2}{M_Z^2}$$

(※ The calculations that led to the above results.)

$$\left.\frac{d\sigma}{dy}(\nu_\mu q \rightarrow \nu_\mu q)\right|_{NC} = \frac{G_F^2 S}{\pi}[g_L^2 + g_R^2(1-y)^2]$$

$$\Rightarrow \left(\frac{1}{2} - \frac{2}{3}\sin^2\theta_w\right)^2 + \left(-\frac{1}{2} + \frac{1}{3}\sin^2\theta_w\right)^2 + \frac{1}{3}\left(-\frac{2}{3}\sin^2\theta_w\right)^2 + \frac{1}{3}\left(\frac{1}{3}\sin^2\theta_w\right)^2$$

$$= \frac{1}{2} - \sin^2\theta_w + \frac{20}{27}\sin^4\theta_w \qquad \left(\leftarrow \int_0^1 (1-y)^2 dy = \frac{1}{3}\right)$$

$$\left.\frac{d\sigma}{dy}(\nu_\mu q \rightarrow \nu_\mu q)\right|_{CC} = \frac{G_F^2 S}{\pi}$$

$$\frac{d\sigma}{dy}(\bar{\nu}_\mu q \to \bar{\nu}_\mu q)\Big|_{NC} = \frac{G_F^2 S}{\pi}[g_R^2 + g_L^2(1-y)^2]$$

$$\frac{d\sigma}{dy}(\bar{\nu}_\mu q \to \bar{\nu}_\mu q)\Big|_{CC} = \frac{G_F^2 S}{\pi}(1-y)^2 \quad \left[\leftarrow (1-y)^2 = \frac{1}{3}\right]$$

■ Asymmetries in the scattering of polarized electrons by deuterons

※ asymmetries: parity violation effects due to interference between Z^0 and γ xchange

(i) existence of neutral weak current ← neutrino experiment in 1973

(ii) asymmetries provided in 1978 at SLAC

※ experiment confirmed the neutral current measurements in neutrino scattering and provided a measurement of weak mixing angle

$$e^-_{L,R} + d_{\text{unpolarized}} \to e^- + X,$$

where γ exchange dominates but small contribution from Z^0 exchange,

X is for any final hadron states.

Difference in the cross-section for LH and RH electrons (i.e. parity-conserving asymmetry) is

$$A = \frac{\sigma_R - \sigma_L}{\sigma_R + \sigma_L} \simeq \frac{G_F q^2}{e^2} = \frac{137\times 10^{-5}}{4\pi} q^2 \simeq 10^{-4} q^2 \ (q^2 \text{ in GeV}^2)$$

ⓐ using unpolarized source (electron gun)

$$A = (-2.5 \pm 2.2)\times 10^{-5},$$

which is consistent with zero and proves the sensitivity of the method and reliability of beam monitoring

- electron beam dependence is consistent with $g-2$ precession

$$\theta_{\text{precession}} = \frac{E_0}{mc^2}\left(\frac{g-2}{2}\right)\theta_{\text{bend}}$$

ⓑ $\frac{A}{q^2} = -(9.5 \pm 1.6) \times 10^{-5} (\mathrm{GeV}/c)^{-2}$: final result showed clear asymmetry

→ fractional energy loss of electron in collision, $y = \frac{E_0 - E}{E_0}$

$\frac{A}{q^2} = -\frac{9G_F}{20\sqrt{2}\pi\alpha}\left[a_1 + a_2\frac{1-(1-y)^2}{1+(1-y)^2}\right]$; A was measured as a function of y

where $a_1 = 1 - \frac{20}{9}\sin^2\theta_w$, $a_2 = 1 - 4\sin^2\theta_w$

→ observed asymmetries gave ⇒ $\sin^2\theta_w = 0.22 \pm 0.02$

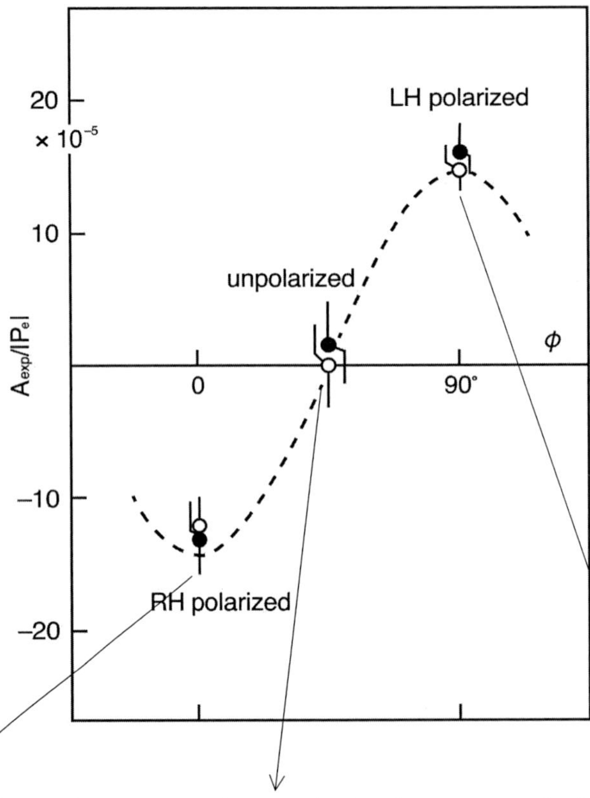

$\phi = 0^0$, RH polarized $\phi = 45^0$, unpolarized $\phi = 0^0$, RH polarized

■ **Observations on the Z resonance** at SLC(SLAC) and LEP(CERN) e^+e^- colliders LEP and SLC had produced Z^0 in the electron-position collisions and measured the properties:

※ LEP experiments (ALEPH, DELPHI, L3, OPAL)

SLC experiments [MarkII(initial), SLD(replaced MarkII)]

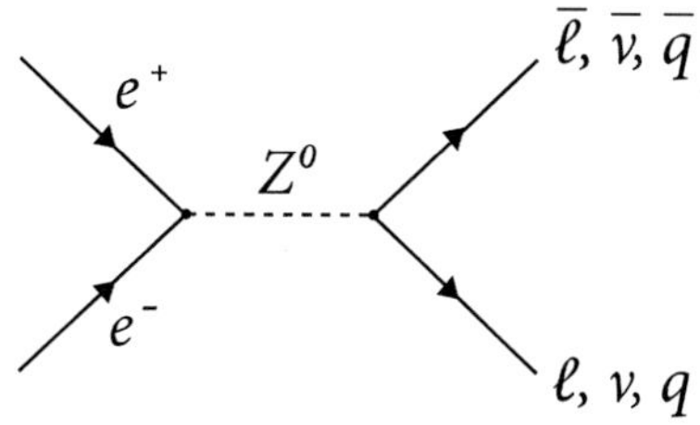

Measurements: - Z^0 mass, line shape

- branching ratios to leptons and hadrons
- angular asymmetries of Z^0 decay products
- cross section asymmetries with longitudinally polarized electron beams

⋮

- Total and partial widths of Z^0
  - total width Γ is made up of contributions from decay to charged lepton pairs, neutral lepton pairs, and quark-antiquark pairs.
  - partial width for any of these mode is

$$\Gamma_{\text{partial}} = \frac{G_F M_Z^3 \rho}{6\pi\sqrt{2}}(C_A^2 + C_V^2)F \text{ , where } \rho \text{ comes from } M_Z^2 = \frac{M_W^2}{\rho\cos^2\theta_w}$$

for $Z^0 \rightarrow \nu\bar{\nu},\ F = 1$

$\rightarrow \ell\bar{\ell},\ F = 1 + \dfrac{3\alpha}{4\pi}$

(※ F: QED corrections for photon radiation in the final state)

$\rightarrow q\bar{q},\ F = 3\left(1 + \dfrac{\alpha_s}{\pi}\right)$, where 3 for color factor

[※ α, α_s: running coupling constants at Z^0 mass

$\left(1+\dfrac{\alpha_s}{\pi}\right)$: QCD factor for gluon radiation by quarks]

- prediction (by inserting obtained values):

$$\Gamma_{\nu\bar{\nu}} = 0.166\ \text{GeV}, \qquad \Gamma_{u\bar{u}} = \Gamma_{c\bar{c}} = 0.29\ \text{GeV}$$

$$\Gamma_{\ell\bar{\ell}} = 0.084\ \text{GeV}, \qquad \Gamma_{d\bar{d}} = \Gamma_{s\bar{s}} = \Gamma_{b\bar{b}} = 0.38\ \text{GeV}$$

→ quark and lepton masses and radiative corrections have only small effects on width

- assuming $N_\nu = 3$ neutrino flavors,

$$\Gamma_{tot} = 3\times\Gamma_{\nu\bar{\nu}} + 2\times\Gamma_{u\bar{u}} + 3\times\Gamma_{\ell\bar{\ell}} + 3\times\Gamma_{d\bar{d}}$$

$$\rightarrow \begin{cases}\Gamma_{total}(\text{calculated}) = 2.49\ \text{GeV} \\ \Gamma_{\text{total}}(\text{observed}) = 2.50\ \text{GeV}\end{cases} \Rightarrow \text{good agreement}$$

<SLD at SLC>

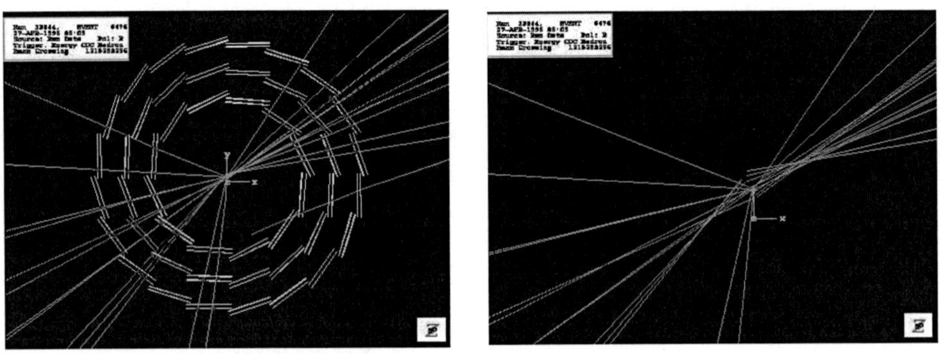

$Z^0 \rightarrow b\bar{b} \rightarrow$ two-jet events

- precise tracking makes possible to reconstruct the secondary vertices from B and D hadron decays and tagging of b- and c-quark flavored jets

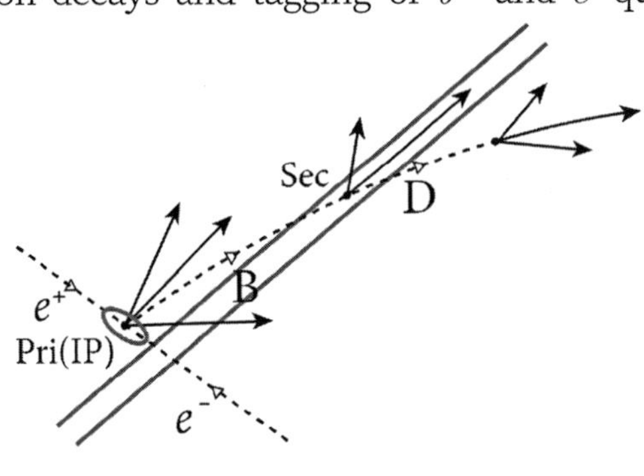

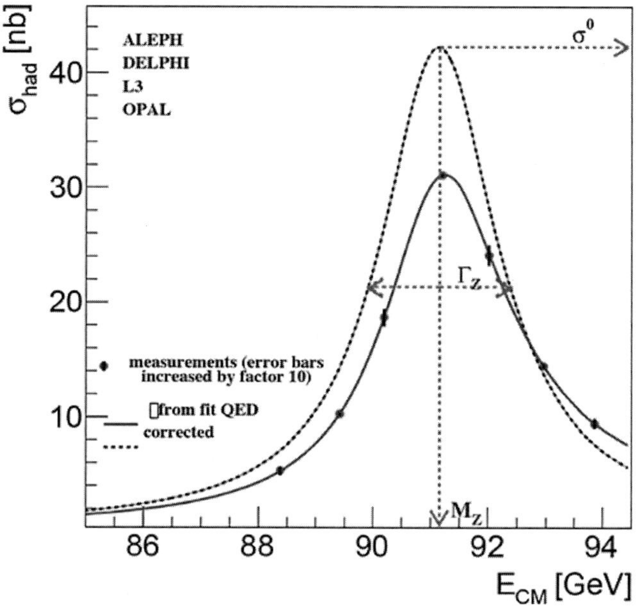

- gree line is the theoretical prediction for preferred values of M_Z and Γ_Z
- pink line represents the corresponding Breit-Wigner distribution,- unfolded for QED radiative corrections

$$\sigma = \frac{4\pi\lambda\!\!\!^{-2}(2J+1)}{(2S+1)^2}\frac{\Gamma_e\Gamma/4}{(E-E_0)^2+\Gamma^2/4}, \quad \text{where } E_0 = m_{Z^0}$$

- Forward-backward asymmetries

  Even well below Z^0 resonances, the interference between photon exchange and Z^0 exchange results in a forward-backward asymmetry.

  - Since Z^0 exchange involves both V and A couplings, parity-violating forward-backward asymmetries are expected.

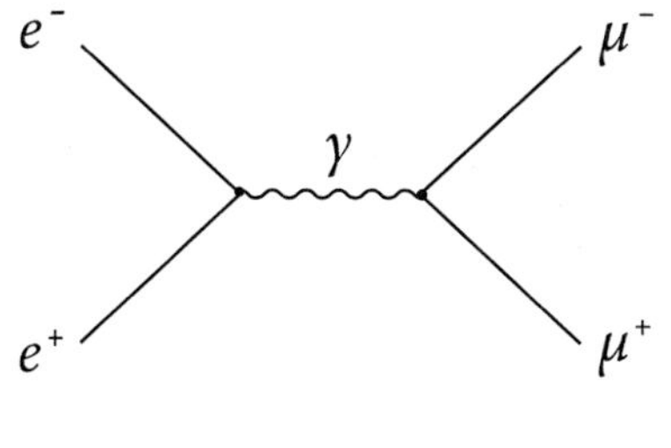

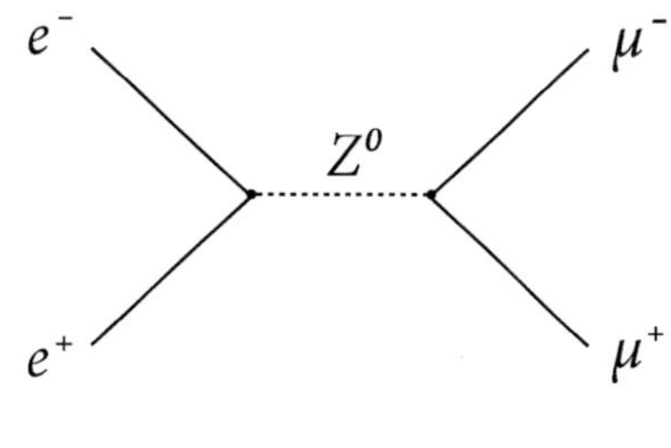

$$f \sim \frac{a_{weak}a_{em}}{a_{em}^2} \sim \frac{G_F S}{4\pi\alpha} \sim 10^{-4}S \text{ (interference)}$$

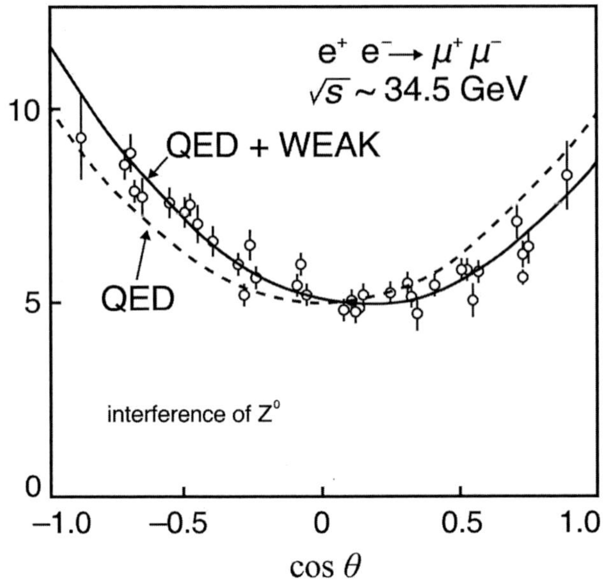

Total cross-section below the Z,

$$\frac{d\sigma}{d\Omega} = \left(\frac{d\sigma}{d\Omega}\right)_{QED} + \left(\frac{d\sigma}{d\Omega}\right)_{interf} + \left(\frac{d\sigma}{d\Omega}\right)_{weak}$$

$$\Rightarrow \frac{\alpha^2}{S} \qquad \Rightarrow \frac{G_F S}{\alpha} \qquad \Rightarrow G_F^2 S$$

Forward-backward asymmetry,

$$A_{FB} = \frac{F-B}{F+B} \simeq \frac{G_F S}{\alpha},$$

where F and B are the forward and backward intensities

- full formula for asymmetry is

$$A_{FB} = -\frac{3C_A^2}{4\sqrt{2}\pi}\frac{G_F S}{\alpha}$$: depend on C_A at small S and therefore is not sensitive to $\sin^2\theta_w$

On the Z^0 resonance ($e^+e^- \rightarrow Z^0 \rightarrow f\bar{f}$),

$$A^0_{FB} = 3A^0_e A^0_f, \quad \text{where } A^0 = \frac{C_V C_A}{C_V^2 + C_A^2} = \frac{(I_3 - 2Q\sin^2\theta_w)I_3}{(I_3 - 2Q\sin^2\theta_w)^2 + I_3^2}$$

Fermion	$2C_V$	$2C_A$
ν_e, ν_μ, ν_τ	1	1
e, μ, τ	$-1+4\sin^2\theta_w$	-1
u, c, t	$1-\frac{8}{3}\sin^2\theta_w$	1
d, s, b	$-1+\frac{4}{3}\sin^2\theta_w$	-1

By inserting the values in the above table,

$$\Rightarrow A_e = \frac{\frac{1}{4}(-1+4\sin^2\theta_w)(-1)}{\frac{1}{4}(-1+4\sin^2\theta_w)^2+\frac{1}{4}} = \frac{1-4\sin^2\theta_w}{1+(1-4\sin^2\theta_w)^2}$$

Fermion	A^0_{FB}	A^0_f
ν_ℓ	0.12	1
ℓ	0.02	0.16
u, c, t	0.08	0.67
d, s, b	0.11	0.93

- **Left-right polarization and asymmetries at SLAC**

  Using left or right circularly polarized electron beam

  $A^0_{LR} = \frac{\sigma_L - \sigma_R}{\sigma_L + \sigma_R} = A^0_e$: A_{LR} is a direct measurement of $A^0_e \Rightarrow$ gives $\sin^2\theta_w$

  In SLD experiment at SLAC,

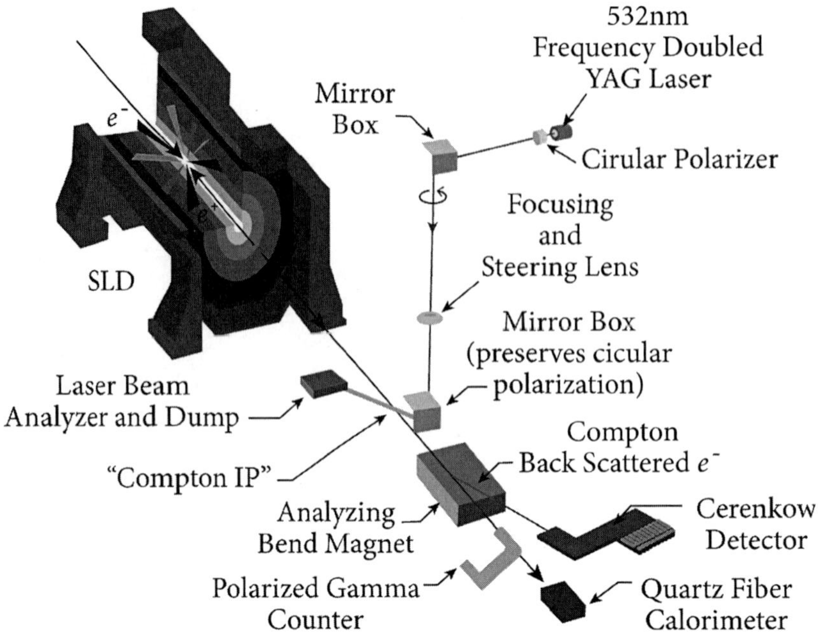

don`t measure A_{LR} directly but

$$A_{LR}^{mea} = p_e A_{LR} = p_e A_e \Rightarrow \text{so, we must know } p_e$$

$$A_e = \frac{2(1-4\sin^2\theta_w)}{1+(1-4\sin^2\theta_w)^2}$$

$$A_{LR}^0 = \frac{2va_e}{v_e^2+a_e^2} = \frac{2(1-4\sin^2\theta_w^{eff})}{1+(1-4\sin^2\theta_w^{eff})^2} = 0.1513 \pm 0.0021,$$

where $\sin^2\theta_w^{eff}$ is defined as the ratio of the EM to the weak isospin coupling constant

$$\rightarrow \sin^2\theta_w^{eff} = \frac{1}{4}\left(1-\frac{v_e}{a_e}\right),$$

$$\rightarrow \sin^2\theta_w^{eff} = 0.23098 \pm 0.00026 \equiv \frac{1-C_V/C_A}{4}$$

■ Fits to the SM and radiative corrections

The SM can only be tested precisely after effects of radiative corrections havebeen accounted for.

Initial state radiation of real photons prior to collision distorts the Breit-Wigner resonance shape.

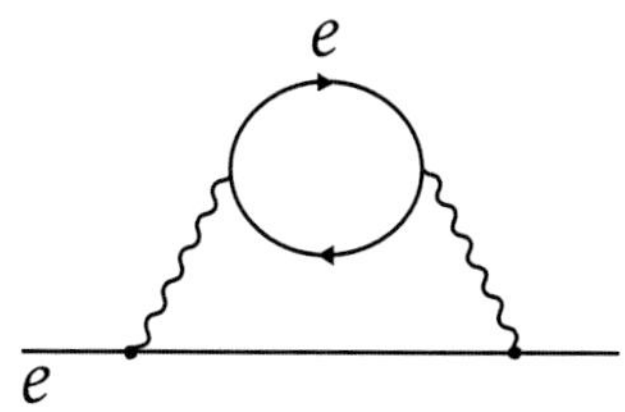

Other corrections:

- virtual photon emission $\Rightarrow$ running of α_{em}
- also, virtual gluon emission
- loop diagrams of virtual top quark and Higgs

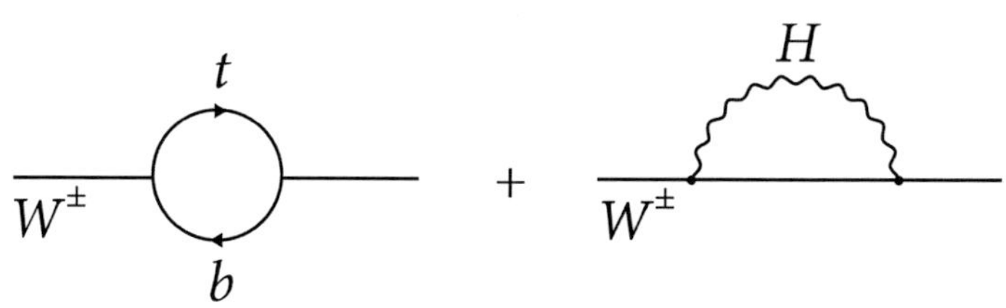

The loop diagrams affect W and Z differently, leading to an effect on the ratio of the neutral- to charged-current couplings:

$$M_{Z^0}^2 = \frac{M_W^2}{\rho \cos^2\theta_w}, \quad \sin^2\theta_w \equiv 1 - \frac{M_W^2}{M_Z^2}$$

$$※ \ \sin^2\theta_w(eff) \simeq \sin^2\theta_w(1 + \cot^2\theta_w \Delta\rho) \leftarrow \cot^2\theta_w \Delta\rho \approx 0.03$$

$$\text{where } \Delta\rho \simeq + \frac{3 G_F m_t^2}{8\pi^2\sqrt{2}} = 0.01 \quad \text{for } m_t = 176 \text{ GeV}$$

Now the measured value of $\sin^2\theta_w$ will depend on the input parameters of the model.

Normally we can take m_Z, m_W, α, α_s as known and allow variation in m_t and m_H.

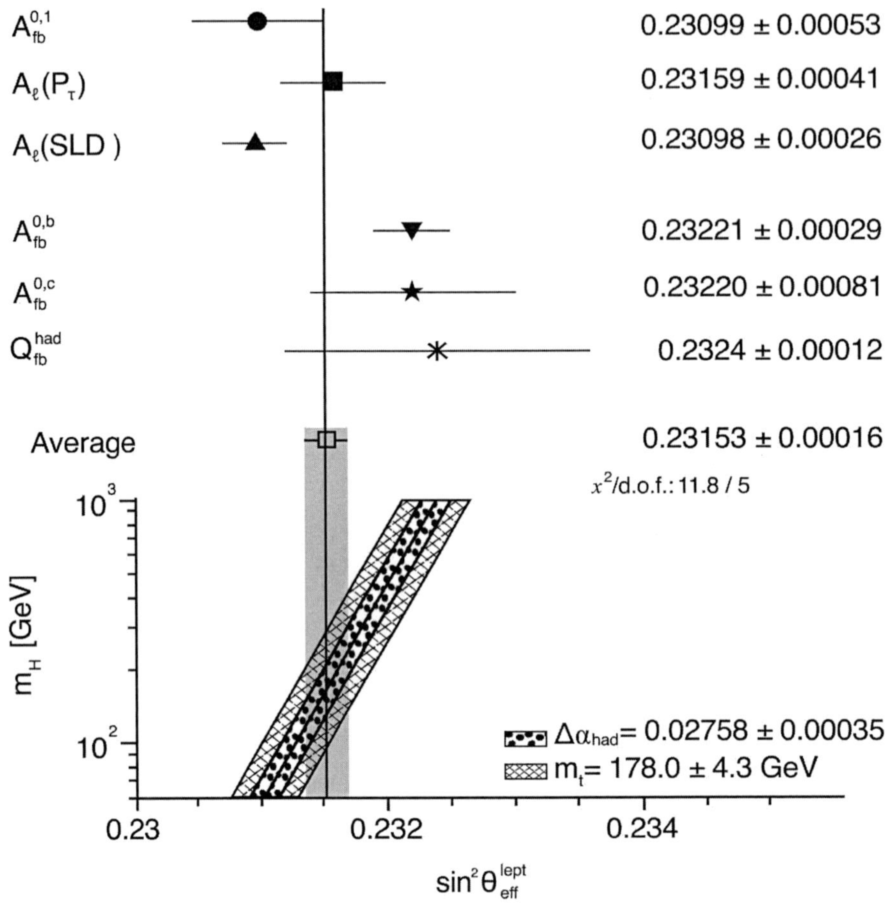

Fitted values of $\sin^2\theta_w(eff)$ as a function of top quark mass for $m_H = 300$ GeV (1996)

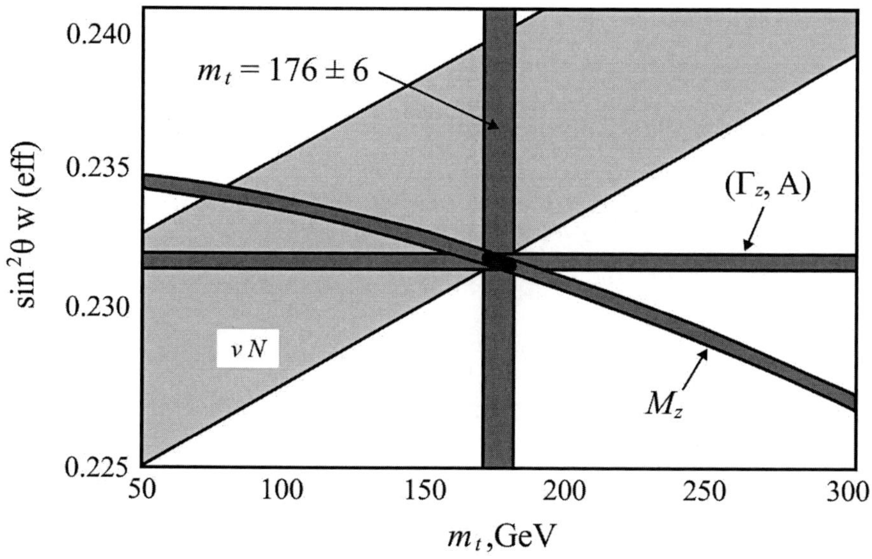

$m_t = 176 \pm 6$ is the direct measurement from FNAL

→ SM gives excellent account of all data, and agreement between m_t deduced indirectly from radiative corrections and that found from direct measurementis very impressive

→ While loop corrections from the top quark vary as m_t^2, they vary only logarithmically with m_H

→ Therefore, dependence on the Higgs mass is much weaker

■ W pair production in e^+e^- collisions

In 1996, the energy of LEP at CERN was increased above the W pair production threshold to begin the LEP2 program.

<triple boson couplings>

νWW coupling γWW coupling ZWW coupling

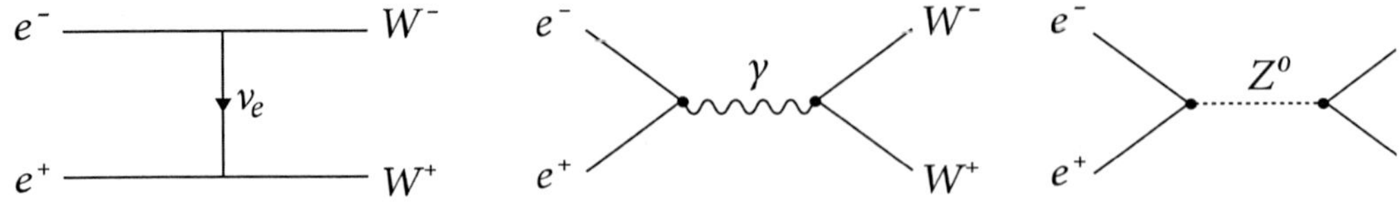

- Each of these diagrams is individually divergent (with $\sigma \propto S$)

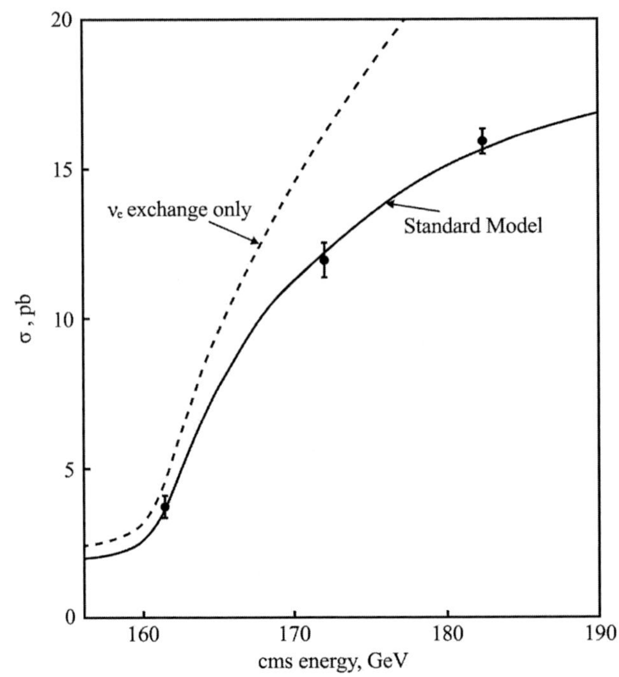

→ $\sigma_{max} \approx 17$ pb

- When these are combined, with the couplings given by the SM, the total cross-section remains finite, and falls as $\frac{\ln(S)}{S}$

■ Spontaneous symmetry breaking and the Higgs mechanism

Let`s outline the basic ideas of the Higgs mechanism.

The open issue of the electroweak symmetry breaking is "How is the symmetry broken?"

In other words, how do we move from the underlying Lagrangian energy density,

$$\mathcal{L} = g\vec{J}_\mu \cdot \overrightarrow{W}_\mu + g' J_\mu^Y \cdot B_\mu \text{ (electroweak field Lagrangian)}$$

to one in which the EM field remains massless and the weak neutral current acquires mass.

→ One problem was that the Weinberg-Salam model is dealing with massless bosons and arbitrarily adding mass for W and Z spoils gauge invariance and leads to divergences.

→ Solution to this problem is the local gauge symmetry and the Higgs mechanism which was invented to explain how the symmetry could be broken and endow the weak bosons with mass, and preserve the massless photon.

Lagrange equation with generalized coordinate q_i and $\dot{q}_i$ is

$$\frac{d}{dt}\left(\frac{\partial \mathcal{L}}{\partial \dot{q}_i}\right) - \frac{\partial \mathcal{L}}{\partial \dot{q}_i} = 0$$

$$\xrightarrow[\text{Mechanics}]{\text{Quantum}} \quad \partial_\mu \left[\frac{\partial \mathcal{L}}{\partial(\partial_\mu \phi)}\right] - \frac{\partial \mathcal{L}}{\partial \phi} = 0 \;\cdots\cdots\; (1),$$

where Q.M. means that (i) $\frac{\partial}{\partial x_\mu} \equiv \partial_\mu$: 4-vector space-time derivative

(ii) $q_i \rightarrow \phi$: field amplitude or continuously variable wave amplitude

let`s guess the Lagrangian for free scalar particles of mass μ

$$\mathcal{L} = K - U \equiv \frac{1}{2}(\partial_\mu \phi)^2 - \frac{1}{2}\mu^2\phi^2 \cdots ②$$

then, we obtain by substituting Eq. (2) into (1)

$$\partial_\mu^2 \phi + \mu^2 \phi = 0: \text{ Klein-Gordon equation}$$

[※ Substitute $\vec{p} = \frac{\hbar}{i}\vec{\nabla}$ and $E = i\hbar\frac{\partial}{\partial t}$ into Einstein equation, $E^2 = p^2c^2 + m^2c^4$,

we get $\frac{1}{c^2}\frac{\partial^2 \Psi}{\partial t^2} = \nabla^2 \Psi - \frac{m^2c^2}{\hbar^2}\Psi$. Since $\partial_\mu^2 = \frac{\partial^2}{\partial t^2} - \nabla^2$ and we put $\Psi \rightarrow \phi$

and $\hbar = c = 1$, we obtain Klein-Gordon equation.]

→ Therefore, Eq. (2) was obviously chosen to give this result.

• Higgs Lagrangian

Suppose that we are dealing with scalar particles that interact with each other. The most general Lagrangian for this field is

$$\mathcal{L} = \frac{1}{2}(\partial_\mu \phi)^2 - \frac{1}{2}\mu^2\phi^2 - \frac{1}{4}\lambda\phi^4 \text{ (Higgs field Lagrangian)} \cdots\cdots (1),$$

where ϕ is the boson field, μ is the particle mass, λ is the dimensionless constant representing coupling of 4-boson vector, and $-\frac{1}{4}\lambda\phi^4$ is for the self-interaction.

※ odd powers for the self-interaction are excluded by symmetry under transformation $\phi \rightarrow -\phi$, and higher powers are excluded by requirement of renormalizability.

→ Therefore, the potential energy of the Lagrangian is

$$U = \frac{1}{2}\mu^2\phi^2 + \frac{1}{4}\lambda\phi^4 \cdots\cdots (2)$$

For minimum value of the boson field,

$$\left.\frac{\partial U}{\partial \phi}\right|_{\phi \rightarrow \phi_{\min}} = 0 \Rightarrow \phi_{\min}\left(\mu^2 + \lambda\phi_{\min}^2\right) = 0$$

it depends on sign of μ^2

(i) $\mu^2 > 0$ (situation for a massive particle)

$\phi_{min} = 0$

→ This is normal situation for the lowest energy vacuum state with $U = 0$ [by substituting into Eq. (2)].

(ii) $\mu^2 < 0$

$$\phi_{min} = \pm\sqrt{\frac{-\mu^2}{\lambda}} \equiv \pm v;$$ non-zero vacuum expectation value

→ v is called the vacuum expectation value of ϕ and ϕ is called Higgs field.

By substituting into Eq. (2),

$U_{min} = -\frac{\mu^4}{4\lambda}$ is the energy where non-zero constant.

→ We have "spontaneous symmetry breaking" by randomly choosing $+v$ or $-v$

From (i) and (2),

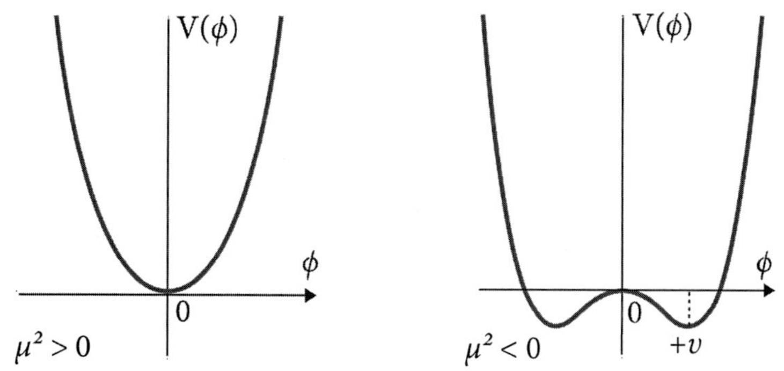

⇒ In either case, a symmetry curve results.

⇒ There are two minima, $+v$ and $-v$, for $\mu^2 < 0$.

⇒ We are concerned with evaluating small perturbations about the energy minimum in weak interactions.

So, let`s expand the field about the minium $+v$ (or $-v$)

$\phi = v + \sigma(x)$ ······ (3),

where $\sigma(x)$ is the variable value of the field over and above the constant and uniform value

Substituting Eq. (3) into Eq. (1),

$$\mathcal{L} = \frac{1}{2}(\partial_\mu \upsilon)^2 - \lambda v^2 \sigma^2 - \left(\lambda v \sigma^3 + \frac{1}{4}\lambda\sigma^4\right) + \text{constant}, \ (\because \ \mu^2 = -\lambda v^2)$$

(※ For detailed calculations, see [Supplement-9])

where $\frac{1}{2}(\partial_\mu \sigma)^2$ is the kinetic energy term, $\lambda v^2 \sigma^2$ is the field mass term, $\lambda v \sigma^3$ is the new term, $\frac{1}{4}\lambda\sigma^4$ is the self-interaction term, and constant is for the terms in v^2 and v^4

Let`s look at the mass term in Eq. (1). → form is $\frac{1}{2}(\text{mass})^2(\text{field})^2$

$$\rightarrow \frac{1}{2}(\text{mass})^2(\text{field})^2 = \lambda v^2 \sigma^2$$

so, $(\text{mass})^2$ is $2\lambda v^2 \Rightarrow m = \sqrt{2\lambda v^2} = \sqrt{-2\mu^2} \ \left(\because \ v = \pm\sqrt{\frac{-\mu^2}{\lambda}}\right)$

⇒ A real positive mass (as against an imaginary one $\mu^2 < 0$) is appeared by making a perturbation expansion of Eq. (3) about either of two minima $\pm v$.

⇒ When this is done, the symmetry in the above plot will be broken.

⇒ This behavior is called "spontaneous symmetry breaking."

[Supplement-9]

Substituting Eq. (3) into Eq. (1) $\mathcal{L} = \frac{1}{2}(\partial_\mu \phi)^2 - \frac{1}{2}\mu^2\phi^2 - \frac{1}{4}\lambda\phi^4$,

1st term: $\frac{1}{2}(\partial_\mu \phi)^2 = \frac{1}{2}[\partial_\mu(v+\sigma)]^2 = \frac{1}{2}(\partial_\mu \sigma)^2$ ($\because$ v is the minium → $\partial_\mu v = 0$)

2nd term: $-\frac{1}{2}\mu^2(v+\sigma)^2 = -\frac{1}{2}\mu^2 v^2 - \mu^2 v\sigma - \frac{1}{2}\mu^2\sigma^2 = \frac{1}{2}\lambda v^4 + \lambda v^3 \sigma + \frac{1}{2}\lambda v^2 \sigma^2$

$$= \text{constant} + \lambda v^3\sigma + \frac{1}{2}\lambda v^2\sigma^2$$

$$\left(\because\ v = \pm\sqrt{\frac{-\mu^2}{\lambda}} \rightarrow \mu^2 = -\lambda v^2\right)$$

3rd term: $-\frac{1}{4}\lambda(v+\sigma)^4 = -\frac{1}{4}\lambda(v^2+2v\sigma+\sigma^2)^2$

$$= -\frac{1}{4}\lambda(v^4+4v^2\sigma^2+\sigma^4+4v^3\sigma+4v\sigma^3+2v^2\sigma^2)$$

$$= \text{constant} - \lambda v^2\sigma^2 - \frac{1}{4}\lambda\sigma^4 - \lambda v^3\sigma - \lambda v\sigma^3 - \frac{1}{2}\lambda v^2\sigma^2$$

■ Gauge invariance in the electroweak model

① Lagrangian energy density in QED is invariant under arbitrary local phase transformation, $U(1): \phi(x) \rightarrow e^{ie\theta(x)}\phi(x)$, infinite set of phase transformations form the unitary group.

where $U(1)$ group is Abelian (commutative group), e is the particle charge, and $\theta(x)$ is the scalar quantity.

If these were accompanied by gauge transformations on electromagnetic potential A_μ, $[A_\mu \rightarrow A_\mu + \partial_\mu\theta(x)]$, to which particles were coupled.

→ This occurs automatically, if the derivative D_μ in the Lagrangian is replaced by a covariant derivative.

→ Therefore, $D_\mu \rightarrow \partial_\mu - ieA_\mu$

※ This is well-known local gauge symmetry in QED, $U(1)$.

For detailed calculations, see [Supplement-10]

② What about $SU(2) \times U(1)$?

Consider the gauge symmetries of the weak hypercharge $U(1)$ and the weak isospin $SU(2)$ interactions.

- The weak-hypercharge behaves under gauge transformations as electric charge since both are $U(1)$.

- The weak isospin will be invariant under a rotation in weak isospin space:

(※ For detailed calculations, see [Supplement-10])

$\Psi \rightarrow e^{ig\vec{\tau}\cdot\vec{\Lambda}}\Psi$, where Λ is an arbitrary vector in isospin space and g is the coupling constant

③ Yang and Mills considered gauge transformations by non-Abelian operators, specially in the group $SU(2)$ of isospin, involving non-commuting operators, $\vec{\tau} = \tau_1, \tau_2, \tau_3$, which are identical to Pauli spin operators

i.e, $\tau_1 = \begin{pmatrix} 0 & 1 \\ 1 & 0 \end{pmatrix}$, $\tau_2 = \begin{pmatrix} 0 & -i \\ i & 0 \end{pmatrix}$, $\tau_3 = \begin{pmatrix} 1 & 0 \\ 0 & -1 \end{pmatrix}$

→ Let`s apply these ideas to the groups of the weak hypercharge $U(1)$ and the weak isospin $SU(2)$.

- The weak hypercharge clearly behaves in the same way under gauge transformations as electric charge.
- Conservation of the weak isospin implies invariance under rotation in weak isospin space.

[Supplement-10]

• local gauge symmetry

$\mathcal{L} = \frac{1}{2}(\partial_\mu\phi)(\partial_\mu\phi)^* - \frac{1}{2}\mu^2\phi\phi^* - \frac{1}{4}\lambda(\phi\phi^*)^2$ is invariant under local gauge transformation, accompanied by gauge transformations on electromagnetic potential $A_\mu \rightarrow A_\mu + \partial_\mu\theta(x)$.

(pf)

(i) $(D_\mu^{'}\phi^{'})(D_\mu^{'}\phi^{'})^* = (\partial_\mu - ieA_\mu^{'})e^{ie\theta(x)}\phi(x)\,(\partial_\mu + ieA_\mu^{'})e^{-ie\theta(x)}\phi^*(x)$

$$= \left[(ie\partial_\mu\theta)e^{ie\theta(x)}\phi(x) + e^{ie\theta(x)}\partial_\mu\phi(x) - ie\,A_\mu^{'}e^{ie\theta(x)}\phi(x)\right](\partial_\mu + ieA_\mu^{'})e^{-ie\theta(x)}\phi^*(x)$$

$$= \left[ie\partial_\mu\theta + \partial_\mu - ieA_\mu{'}\right]e^{ie\theta(x)}\phi e^{-ie\theta(x)}\left[-ie\partial_\mu\theta + \partial_\mu + ieA_\mu{'}\right]\phi^*,$$

[where $ie\partial_\mu\theta + \partial_\mu - ieA_\mu{'} = ie\partial_\mu\theta + \partial_\mu - ie(A_\mu + \partial_\mu\theta)$

$= ie\partial_\mu\theta + \partial_\mu - ieA_\mu - ie\partial_\mu\theta = \partial_\mu - ieA_\mu$

similarly, $-ie\partial_\mu\theta+\partial_\mu+ieA_\mu{}'=\partial_\mu+ieA_\mu$]

$$=(\partial_\mu-ieA_\mu)\phi e^{ie\theta(x)}e^{-ie\theta(x)}(\partial_\mu+ieA_\mu)\phi^*$$

$$=(\partial_\mu-ieA_\mu)\phi(\partial_\mu+ieA_\mu)\phi^*=(D_\mu\phi)(D_\mu^*\phi^*)$$

(ii) and potential energy of the Lagrangian

$U=\frac{1}{2}\mu^2\phi\phi^*+\frac{1}{4}\lambda(\phi\phi^*)^2$ is invariant under $\phi\rightarrow e^{ie\theta}\phi$ transformation.

∴ From (i) and (ii), $\mathcal{L}$ is invariant.

We must introduce a massless isovector (ie. $I=1$) field, $\overrightarrow{W}_\mu$, which contains charged and neutral components to preserve the interaction.

Invariant under the transformation, $\Psi\rightarrow e^{ig\vec{\tau}\cdot\vec{\Lambda}}$, is obtained by introducing a covariant derivative of the form

$SU(2)$; $D_\mu=\partial_\mu-ig\vec{\tau}\cdot\overrightarrow{W}_\mu$,

[※ c.f. QED: $D_\mu=\partial_\mu-ieA_\mu$, $A_\mu\rightarrow A_\mu+\partial_\mu\theta(x)$]

where infinitesimal gauge transformation of field $\overrightarrow{W}_\mu$ is

$$\overrightarrow{W}_\mu\rightarrow\overrightarrow{W}_\mu+\partial_\mu\vec{\Lambda}-g\vec{\Lambda}\times\overrightarrow{W}_\mu$$

[※ $-g\vec{\Lambda}\times\overrightarrow{W}_\mu$ leads to local gauge invariance and implies an interaction of $\overrightarrow{W}_\mu$ with all particles carrying isospin and hence with itself]

$SU(2)\otimes U(1)$;

The effect of Higgs mechanism on the massless fields $\overrightarrow{W}_\mu$ and B_μ is found by substituting the covariant derivative for $\overrightarrow{W}$ and B fields, which involves appropriate weak charges g and g'.

$$D_\mu=\partial_\mu+\frac{ig}{2}\vec{\tau}\cdot\overrightarrow{W}_\mu+\frac{ig'}{2}YB_\mu\quad\cdots\cdots(1)$$

($\because$ The second term on the RHS of the above is originally igW_μ, but similarly to $\vec{S} = \frac{1}{2}\vec{\sigma}$, W_μ is replaced by $\left(\frac{\sigma}{2}\right)^\ell W_\mu^\ell$. The third term is originally $i\frac{g'}{2}A_\mu$, but it can be expressed like this because Y is strong hypercharge.)

Let`s calculate by substituting D_μ of Eq. (1) into ∂_μ in the Lagrangian;

$$\mathcal{L} = \frac{1}{2}(\partial_\mu\phi)^2 - \frac{1}{2}\mu^2\phi^2 - \frac{1}{4}\lambda\phi^4$$

$$\rightarrow\ \mathcal{L} = \frac{1}{2}(D_\mu\phi)^2 - \frac{1}{2}\mu^2\phi^2 - \frac{1}{4}\lambda\phi^4$$

The Higgs Lagrangian becomes

$$\mathcal{L}_\Phi = (D_\mu\Phi)^+(D_\mu\Phi) - \mu^2\Phi^+\Phi - \lambda(\Phi^+\Phi)^2 \quad \cdots\cdots (2)$$

The Higgs field is now a isospin doublet since D_μ is

[※ In the SM, the Higgs is a doublet in $SU(2)$]

$$\Phi = \begin{pmatrix}\phi^+\\ \phi^0\end{pmatrix} = \frac{1}{\sqrt{2}}\begin{pmatrix}\phi_1 + i\phi_2\\ \phi_3 + i\phi_4\end{pmatrix} \text{ with hypercharge}$$

$$Y = +1 = \begin{cases} I_3 & Y = 2(Q - I_3) \\ +\frac{1}{2} & +\frac{1}{2} \\ -\frac{1}{2} & -\frac{1}{2} \end{cases}$$

($\because$ electric charge of the upper and lower component of the doublet are chosen to ensure that the hypercharge $Y = +1$)

The minimum in the potential occurs at $\frac{dU}{d(\Phi^+\Phi)} = 0$.

$$\Rightarrow\ \mu^2 + 2\lambda(\Phi^+\Phi)_{min} = 0 \quad (\because\ U(\Phi^+\Phi) = \mu^2(\Phi^+\Phi) + \lambda(\Phi^+\Phi)^2\ )$$

$$\Rightarrow\ (\Phi^+\Phi)_{min} = -\frac{\mu^2}{2\lambda} = \frac{v^2}{2} \quad \cdots\cdots (3)\ (\because\ \mu^2 = -\lambda v^2)$$

$$\hookrightarrow (\Phi^+\Phi)_{min} = \frac{1}{2}(\phi_1^2 + \phi_2^2 + \phi_3^2 + \phi_4^2) \quad \cdots\cdots (4)$$

$$[\because\ \Phi^+ = \frac{1}{\sqrt{2}}(\phi_1 - i\phi_2,\ \phi_3 - i\phi_4),\ \Phi = \frac{1}{\sqrt{2}}\begin{pmatrix}\phi_1 + i\phi_2\\ \phi_3 + i\phi_4\end{pmatrix}\ ]$$

Choose $\phi_1 = \phi_2 = \phi_4 = 0$ in Eq. (4), then

$$\Phi_{\min} = \frac{1}{\sqrt{2}}\begin{pmatrix}0\\v\end{pmatrix} \quad (\because \text{ Eq. (3)})$$

$\rightarrow$ This vacuum is neutral because $I = \frac{1}{2}$ and $I_3 = -\frac{1}{2}$ with our choice of $Y = +1$ and then we have $Q = I_3 + \frac{1}{2}Y = 0$.

$$\tau_1 = \begin{pmatrix}0 & 1\\1 & 0\end{pmatrix}, \quad \tau_2 = \begin{pmatrix}0 & -i\\i & 0\end{pmatrix}, \quad \tau_3 = \begin{pmatrix}1 & 0\\0 & -1\end{pmatrix}$$

$$\Rightarrow \vec{\tau} \cdot \overrightarrow{W}_\mu = \tau_1 W_\mu^{(1)} + \tau_2 W_\mu^{(2)} + \tau_3 W_\mu^{(3)} = \begin{pmatrix} W_\mu^{(3)} & W_\mu^{(1)} - i W_\mu^{(2)} \\ W_\mu^{(1)} + i W_\mu^{(2)} & -W_\mu^{(3)} \end{pmatrix}$$

and

$$B_\mu = \begin{pmatrix}B_\mu & 0\\0 & B_\mu\end{pmatrix}$$

Since we are interested in the masses of the vector bosons, from Eq. (1),

$$D_\mu \Phi = \left[\frac{ig}{2}\begin{pmatrix} W_\mu^{(3)} & W_\mu^{(1)} - i W_\mu^{(2)} \\ W_\mu^{(1)} + i W_\mu^{(2)} & -W_\mu^{(3)} \end{pmatrix} + \frac{ig'}{2}Y\begin{pmatrix}B_\mu & 0\\0 & B_\mu\end{pmatrix}\right]\frac{1}{\sqrt{2}}\begin{pmatrix}0\\v\end{pmatrix}$$

$$= \frac{i}{\sqrt{8}}\begin{pmatrix} g W_\mu^{(3)} + g' Y B_\mu & g(W_\mu^{(1)} - i W_\mu^{(2)}) \\ g(W_\mu^{(1)} + i W_\mu^{(2)}) & -g W_\mu^{(3)} + g' Y B_\mu \end{pmatrix}\begin{pmatrix}0\\v\end{pmatrix}$$

$$= \frac{iv}{\sqrt{8}}\begin{pmatrix} g(W_\mu^{(1)} - i W_\mu^{(2)}) \\ -g W_\mu^{(3)} + g' Y B_\mu \end{pmatrix}$$

Similarly,

$$(D^\mu \Phi)^+ = -\frac{iv}{\sqrt{8}}\left(g(W_\mu^{(1)} + i W_\mu^{(2)}), \quad -g W_\mu^{(3)} + g' Y B_\mu\right)$$

So, we get the following expression for the kinetic part of the Lagrangian;

$$(D^\mu \Phi)^+ (D_\mu \Phi) = \frac{1}{8}v^2\left[g^2(W_1^2 + W_2^2) + (-g W_3 + g' Y B_\mu)^2\right] \quad \cdots\cdots (5),$$

where $W_\mu^{(i)} \equiv W_i$ for simplicity,

$$W_1 = \frac{1}{\sqrt{2}}(W^+ + W^-), \quad W_2 = \frac{1}{\sqrt{2}}(W^+ - W^-)$$

$$\rightarrow \quad W_1^2 + W_2^2 = W^{+2} + W^{-2}$$

In Eq. (5),

$$(-gW_3+g'YB_\mu)^2 = (W_3,\ B_\mu)\begin{pmatrix} g^2 & -gg'Y \\ -gg'Y & g'^2 \end{pmatrix}\begin{pmatrix} W_3 \\ B_\mu \end{pmatrix} \quad \cdots\cdots (6)$$

└> mixing matrix

(i) only if $Y \neq 0$, W_3 and B_μ fields mix

(ii) if $Y = \pm 1$, determinant of the mixing matrix vanishes and one of the combinations will be massless (the coefficient for that gauge field squared is 0)

In our choice of vacuum, we have $Y = +1$. If we choose $Y = -1$, the photon does not couple to the electric charge.

mixing matrix in Eq. (6)

eigenvalue	eigenvector	
$\lambda = 0$	$\frac{1}{\sqrt{g^2+g'^2}}\begin{pmatrix} g \\ g' \end{pmatrix} = \frac{1}{\sqrt{g^2+g'^2}}(g'W_B+gB_\mu) \equiv A_\mu$	photon(γ)
$\lambda = (g^2+g'^2)$	$\frac{1}{\sqrt{g^2+g'^2}}\begin{pmatrix} g \\ -g' \end{pmatrix} = \frac{1}{\sqrt{g^2+g'^2}}(gW_3-g'B_\mu) \equiv Z_\mu$	Z-boson(Z)

$$\Rightarrow \begin{cases} Z_\mu = \dfrac{gW_3-g'B_\mu}{\sqrt{g^2+g'^2}} \\ A_\mu = \dfrac{g'W_3+gB_\mu}{\sqrt{g^2+g'^2}} \end{cases} \Rightarrow \begin{cases} W_3 = \dfrac{gZ_\mu+g'A_\mu}{\sqrt{g^2+g'^2}} \\ B_\mu = \dfrac{gA_\mu-g'Z_\mu}{\sqrt{g^2+g'^2}} \end{cases} \quad \cdots\cdots (7)$$

Substituting Eq. (7) into (6),

$$(-gW_3+g'YB_\mu)^2 = (-gW_3+g'B_\mu)^2 = \left[\frac{(-g^2-g'^2)Z_\mu+(g'g-gg')A_\mu}{\sqrt{g^2+g'^2}}\right]^2$$

$$= (g^2+g'^2)Z_\mu^2+0\cdot A_\mu^2 \quad \cdots\cdots (8)$$

Substituting Eq. (8) into (5),

$$(D^\mu\Phi)^+(D_\mu\Phi) = \frac{1}{8}v^2\left[g^2(W^+)^2+g^2(W^-)^2+(g^2+g'^2)Z_\mu^2+0\cdot A_\mu^2\right] \cdots\cdots (9)$$

As a general mass term for a massive gauge boson V has the term $\frac{1}{2}M_V^2 V_\mu^2$,

※ $\left[\frac{1}{2}(\text{mass})^2(\text{field})^2\right]$

Let`s identify Z and W mass terms in RHS of Eq. (9);

$$\begin{cases} \frac{1}{8}v^2 g^2 (W^{\pm})^2 = \frac{1}{2}m_{W^{\pm}}^2 (W^{\pm})^2 \Rightarrow m_{W^+} = m_{W^-} = \frac{1}{2}vg \\ \frac{1}{8}v^2 (g^2 + g'^2) Z_\mu^2 = \frac{1}{2}m_Z^2 (Z_\mu^2) \Rightarrow m_Z = \frac{1}{2}v\sqrt{(g^2+g'^2)} = \frac{1}{2}\frac{gv}{\cos\theta_w} \end{cases}$$

$$(\because \frac{g'}{g} = \tan\theta_w)$$

$$\Rightarrow \therefore m_Z = \frac{m_W}{\cos\theta_w}, \quad \rho \equiv \frac{m_W^2}{m_Z^2 \cos^2\theta_w} \quad [\rho\text{-(Veltman parameter)}]$$

for photon,

$$0 \cdot A_\mu^2 = \frac{1}{2}m_\gamma^2 A_\mu^2 \Rightarrow m_\gamma = 0$$

Expand around the minimum,

$$\Phi = \frac{1}{\sqrt{2}}\begin{pmatrix} 0 \\ v + h(x) \end{pmatrix} \quad \cdots\cdots (9)$$

We look for equations satisfied by $h(x)$. This choice is always possible, thanks to the local gauge invariance.

Since interactions between gauge bosons and the Higgs are

$$-\mu^2\Phi^+\Phi - \lambda(\Phi^+\Phi)^2$$

$$= -\mu^2\frac{1}{2}(v^2 + 2vh + h^2) - \frac{\lambda}{4}(v^4 + 4v^2h^2 + h^4 + 4v^3h + 4vh^3 + 2v^2h^2)$$

$$(\because \text{from Eq. (9)})$$

$$= -\frac{1}{2}\mu^2v^2 - \mu^2vh - \frac{1}{2}\mu^2h^2 - \frac{\lambda}{4}(\underset{①}{v^4} + \underset{②}{4v^2h^2} + \underset{③}{h^4} + \underset{④}{4v^3h} + \underset{⑤}{4vh^3} + \underset{⑥}{2v^2h^2})$$

$$\leftarrow ①;\ -\frac{\lambda}{4}v^4 = +\frac{1}{4}\frac{\mu^2}{v^2}v^4 = \frac{1}{4}\mu^2v^2 \quad \left(\because v^2 = -\frac{\mu^2}{\lambda}\right)$$

$$\leftarrow ②;\ -\lambda v^2h^2 = \frac{\mu^2}{v^2}v^2h^2 = \mu^2h^2$$

$$\leftarrow ③;\ -\frac{\lambda}{4}h^4$$

$$\leftarrow ④;\ -\lambda v^3 h = \frac{\mu^2}{v^2}v^3 h = v\mu^2 h$$

$$\leftarrow ⑤;\ -\lambda v h^3$$

$$\leftarrow ⑥;\ -\frac{\lambda}{4}2v^2h^2 = \frac{1}{2}\frac{\mu^2}{v^2}v^2h^2 = \frac{1}{2}\mu^2h^2$$

$$= -\frac{1}{2}\mu^2v^2 - \mu^2vh - \frac{1}{2}\mu^2h^2 + \frac{1}{4}\mu^2v^2 + \mu^2h^2 - \frac{\lambda}{4}h^4 + v\mu^2h - \lambda vh^3 + \frac{1}{2}\mu^2h^2$$

$$= -\frac{1}{4}\mu^2v^2 + \mu^2h^2 - \frac{\lambda}{4}h^4 - \lambda vh^3$$

for Higgs

$$\mu^2h^2 = -\frac{1}{2}M_h^2h^2 \Rightarrow M_h^2 = -2\mu^2 \Rightarrow \therefore\ M_h = \sqrt{2\lambda v^2}$$

↳ "negative" because μ^2 is negative.

→ Since λ is a free parameter, the mass of the Higgs boson is not predicted in the SM despite of known v ($v \approx 246$ GeV).

↳ this is the typical scale of electroweak symmetry breaking

• Higgs mechanism also endows the fermions with mass.

The full Lagrangian has terms coupling all the fermions to the Higgs field.

$$\mathcal{L} = m_f e\bar{e} - \frac{m_f}{v}e\bar{e}\,H$$

↳ Yukawa coupling (Higgs coupling to the fermion mass)

- electron: $\frac{m_e}{v} = 2\times10^{-6}$

- top quark: $\frac{m_t}{v} = 0.7$

The size of the top quark mass, m_t, seems must more natural then the mass of the lighter fermions given the value of $v = 246$ GeV.

<Higgs couplings>

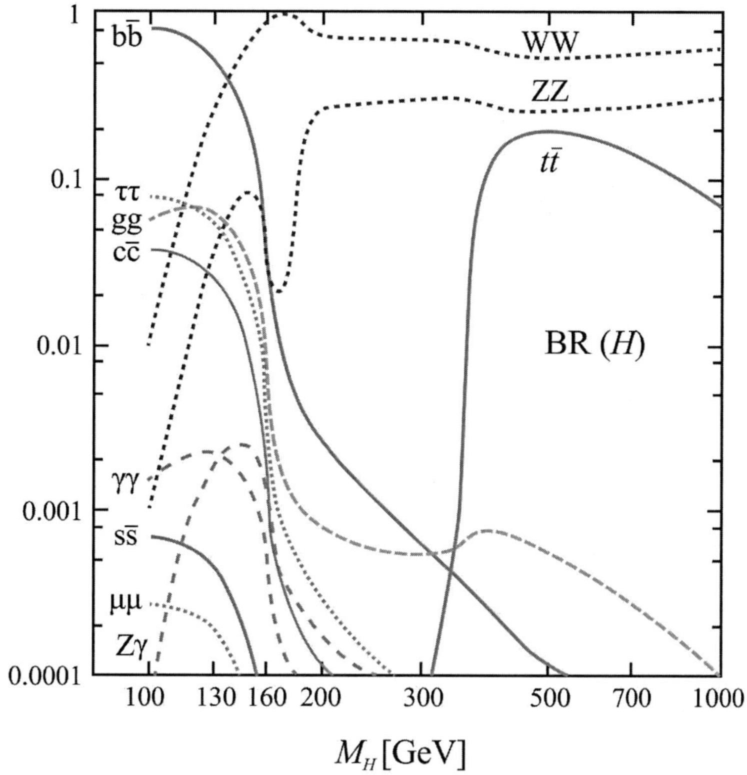

• Higgs production and detection

- The isospin doublet of scalar Higgs particles in the minimal SM yields one real particle to be found.
- Four real components of the new fields are reduced to one when three are "eaten" by the massless W and Z to produce W and Z mass.

$$\begin{bmatrix}\phi^+\\ \phi^-\end{bmatrix}=\begin{bmatrix}\frac{1}{\sqrt{2}}(\phi_1+i\phi_2)\\ \frac{1}{\sqrt{2}}(\phi_3+i\phi_4)\end{bmatrix}\qquad \begin{matrix} I_3 & Y=Q-I_3 \\ +\frac{1}{2} & +\frac{1}{2} \\ -\frac{1}{2} & +\frac{1}{2}\end{matrix}$$

- mass of the remaining physical neutral-boson is unknown
- limits on the mass can be determined
- upper limit on the Higgs mass,

  consider $\Gamma_H \sim G_F M_H^3$ (on dimensional grounds)

Since the Higgs is a manifestation of perturbation theory, Higgs self-coupling is considered not to be strong.

→ Higgs must be weakly coupling → $\Gamma_H < M_H$

$$\rightarrow M_H < G_F^{-1/2} \sim (10^5 \text{ GeV})^{1/2} \approx 300 \text{ GeV}$$

We also have a unitarity limit on WW scattering;

$$M_H < (8\pi\sqrt{2}/3)^{1/2} G_F^{-1/2} \simeq 1 \text{ TeV}$$

→ Higgs boson might be produced in e^+e^- collider.

<Higgs-strahlung process>

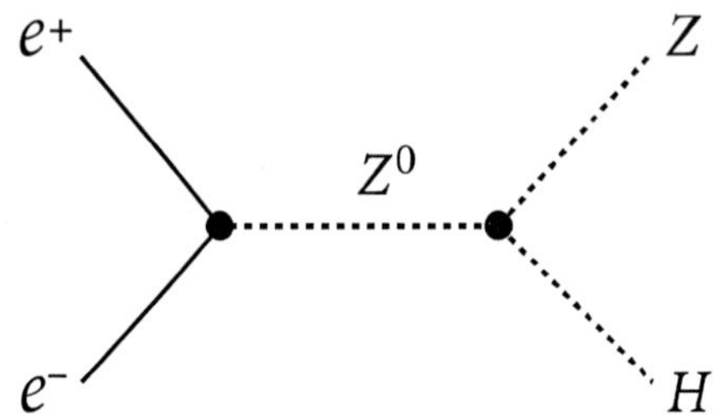

$$e^+e^- \rightarrow H^0 \ Z^0,$$

where $H^0 \rightarrow b\bar{b},\ \tau\bar{\tau},\ \cdots$, and $Z^0 \rightarrow q\bar{q},\ \ell\bar{\ell},\ \nu\bar{\nu}$

→ Higgs couples proportionally to mass, so it should decay preferentially to the heaviest possible quark or lepton.

→ In an electron-positron collider, Higgs signal would show up dramatically recoiling from a Z-decay.

※ $ZH \rightarrow \ell\ell H$ recoiling Higgs reconstruction

$p_H = p_{ZH} - p_{\ell\ell}$, ($\because$ from energy and momentum conservations)

$$\Rightarrow -m_H^2 = -m_{ZH}^2 - m_{\ell\ell}^2 - 2\vec{p}_{ZH}\cdot\vec{p}_{\ell\ell} + 2E_{ZH}E_{\ell\ell}$$

$$\Rightarrow m_H^2 = m_{ZH}^2 + m_{\ell\ell}^2 - 2E_{\ell\ell}E_{ZH} \quad (\because \ \vec{p}_{ZH} = 0)$$

$$= S + m_{\ell\ell}^2 - 2E_{\ell\ell}\sqrt{S}$$

$$(\because \ E_{ZH}^2 = \vec{p}_{ZH} + m_{ZH}^2 \Rightarrow E_{cms}^2 = m_{ZH}^2 \ \Rightarrow S = m_{ZH}^2)$$

→ With LEP now off, Higgs search proceeds at Fermilab TeVatron Collider and Large Hadron Collider (LHC) at CERN

At hadron colliders

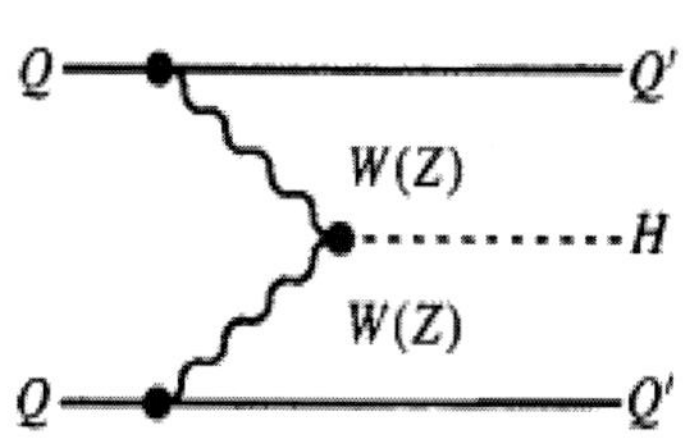

$W(Z)$ fusion diagram

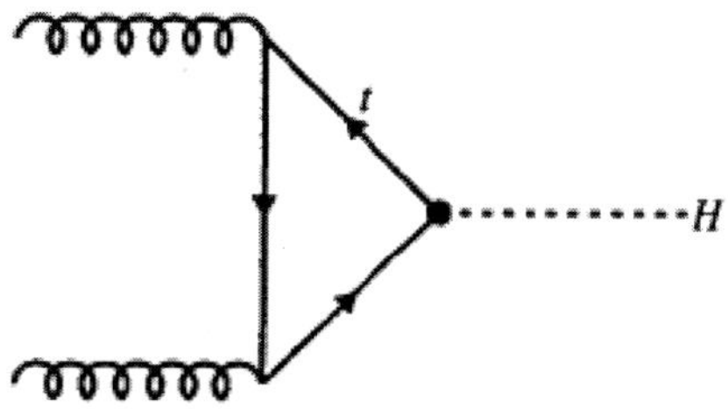

gluon fusion diagram

For $M_H = 80 \sim 150$ GeV, $H \rightarrow \gamma\gamma$ is favored.

$$= 150 \sim 1000 \text{ GeV}, \; H \rightarrow \begin{cases} \ell^+\ell^-\ell^+\ell^- \\ ZZ, \; WW \rightarrow \ell^+\ell^- + \text{two jets} \end{cases}$$

<Observation of the Higgs-like boson by CMS in $H \rightarrow \gamma\gamma$, the $\gamma\gamma$ invariant mass distribution at 125 GeV/c^2>

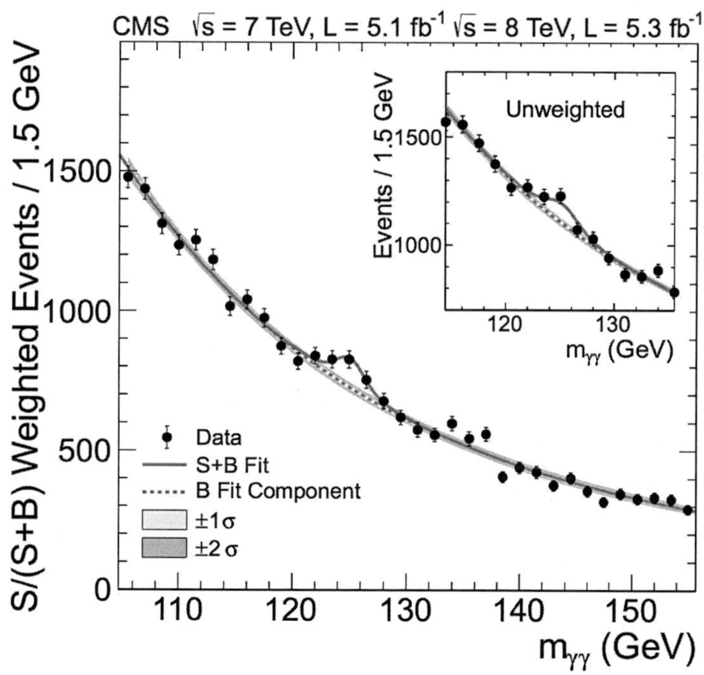

(DOI:10.1088/1742-6596/510/1/012001)

NOTE

NOTE

NOTE

NOTE

NOTE

NOTE

NOTE

NOTE

NOTE

NOTE

NOTE